Remember Me

The Civil War Through the Eyes of a Union Soldier from the 159th Regiment, New York State Volunteers

BY MICHAEL LEE KIPP

Worm Castle Publisher
8259 Leafy Way, Port Charlotte FL 33981
Copyright © 2025 by Michael Lee Kipp

All rights reserved.

No part of this publication may be reproduced, distributed, or transmitted in any form or by any means, including photocopying, recording, or other electronic or mechanical methods, without the prior written permission of the publisher, except as permitted by U.S. copyright law.

First Published 2025

Title: Remember Me: The Civil War Through the Eyes of a Union Soldier from the 159th Regiment, New York State Volunteers

Developmental Editor Barbara Reina
Book Cover by Garrett Kipp
Researcher Tamara Kipp

Identifiers: Library of Congress Control Number: 2025901444 | ISBN 979-8-9920344-1-7 (hardcover) | ISBN 979-8-9920344-0-0 (paperback) | ISBN 979-8-9920344-2-4 (ebook)

Subjects: Civil War, 1861-1865 | Civil War Correspondence | NYSV Regimental History | 159th NYSV Infantry Regiment | Army of the Golf | Battles in Louisiana | History 19th Army Corps | NYS Politics, 1861-1865 | Wartime Elections | United States Sanitary Commission, 1861-1865 | Wartime Disease & Sickness | Red River Campaign | Shenandoah Valley Campaign 1864 | Civil War Hospitals | Prisoner of War | Native Guard | The Planters Aristocracy

This book presents historical materials and letter entries from the Civil War era, reflecting the perspectives, language, and context of that time. While every effort has been made to present these materials faithfully, neither the author or publisher can guarantee the complete accuracy of the information as personal recollections may contain errors or biases. Some content may include language or views that are offensive or inappropriate by modern standards. These are included to preserve the historical integrity of the letters and do not in any way represent the views of the author or publisher. They are provided for historical and educational purposes only, and readers are encouraged to consult additional sources for a comprehensive understanding.

Author Website: michaelleekipp.com | remembermebook.com

In loving Memory of
Shirley (Mambert) Kipp and Patricia (Mambert) DeGroff,
great-great-granddaughters of John W. Mambert.

"History never repeats itself, but it often rhymes."
— Mark Twain

Contents

SECTION	PAGE
Preface	10
Acknowledgement	13
Introduction	14

Chapter One 20
- Recruit & Organize
- The Expedition
- 'The Army of the Gulf'
- Hunger Pangs & Desertion

Chapter Two 51
- Adjusting to Military Ways & Pay
- Sickness in the Ranks
- 'Infusia Diabolus'
- Worm Castles and Boiled Beef

Chapter Three 95
- Battle at Irish Bend
- A Union Victory at a Great Cost
- Foraging or Plundering?
- The Sanitary Commission

Chapter Four 132
- Port Hudson
- Battle at Brashear City
- A Confederate Prisoner
- The Planters Aristocracy

Chapter Five 175
- The Draft, Bounty & Riots
- Aristocracy-Suppressed Education
- Intoxicating Liquors 'A Crying Evil'
- The Commish

Chapter Six 215
- The Red River Campaign
- No Work for Poor Whites in the South
- Battles at Sabine Crossroads & Pleasant Hill
- The Great Dam & The Sickly Encampment

SECTION	PAGE
Chapter Seven	270

– Heading For Shenandoah valley & The Chessboard
– Battles in the Valley, Halltown, Berryville
– Battle in Winchester (Opequon) & the Battle at Fishers Hill
– Medical Facilities & Horrid Scenes from the Hospital

Chapter Eight 307
– Wartime Election
– Battle at Cedar Creek
– New York & the Presidential Election
– The Gray Ghost

Chapter Nine 360
– Provo Duty - Savannah, Ga to Morehead, Nc to Augusta, Ga
– Waltermire Takes Control
– From Elation to Sorrow
– The Native Guard

Chapter Ten 409
– General Molineux Goes Home
– 'Madison, Too Pretty to Burn'
– Mustered Out / Returning to Civilian Life
– The Death of John W. Mambert, Veteran of the Civil War

Epilogue 440

William F. Tiemann..441

The 159th NYSV Infantry Regiment 442
– Recapitulation

Roster of the 159th NYSV Infantry Regiment 448

Bibliography...494

Preface

Among the books I've read on the Civil War before writing this historical narrative, my thoughts on why the great rebellion occurred were simple: the North had a profound desire to end slavery and the South refused to comply.

However, in doing research to create this book, I reached a new conclusion. That research involved personal family history that I am now making public—the preserved letters written by my third great-grandfather while serving for the Union during the Civil War.

Stowed away for almost 160 years, still in their original envelopes, bundled and wrapped in a shopping bag, the letters were discovered among my maternal grandmother's belongings when she moved from her home. I was astounded at how well preserved they were, considering the inadequate means by which they had been kept for more than a century.

Because of my interest in Civil War history, I had an intense curiosity and needed to know what my great-grandfather wrote as a foot soldier, a man who was directly involved in the war.

In his letters, he sometimes wrote about aristocrats and the poor people of the South. Researching this further became a significant turning point for me in my understanding of history. I now realized the single biggest cause of our country's greatest tragedy was the actions of a small group of greedy individuals: the aristocrats of the South—the plantation owners. Representing a very small number as compared to the rest of the South's population, they controlled the money, slaves, poor white people, but most importantly, the politics. These individuals would make the decisions which ultimately launched the greatest rebellion dividing our country.

I had no idea there would be such a vast amount of information and details included in his writings. As I was organizing the dates and locations, I began to realize the depth of his writing and became more anxious to know his story. So, the process of deciphering them began. Anyone familiar with handwriting from that era will understand this can be a daunting process. Proving to be time-consuming, I enlisted my wife's help and had her type the letters as I dictated the words.

We were making good progress. The story of my third great-grandfather—John W. Mambert's life during his three years serving in the Union army was slowly emerging. Then, we were hit by a twenty-first century problem: my computer's hard drive crashed, and all the work was lost. The project went on hold for a few years as our family's life became hectic.

But the words in my grandfather's letters stayed with me. I often thought of what he had written. His letters inspired me to tell his story, the story of a foot soldier and his regiment. The idea for a book began to emerge.

My mother-in-law had just finished a project for a local author in our town and was willing to assist in translating the letters (once again), not that my wife would not have helped, but I believe her enthusiasm had somewhat diminished after my blunder of not saving and losing our previous work. During this time, I also began researching the 159th regiment's history.

I was advancing along nicely on both fronts (especially with my new helper) when I was offered and accepted a new position at work. The job entailed a lot of travel and my personal time slowly diminished. Over the next few years, little progress was made piecing together John and the 159th NY regiment's story. Eventually coming to a halt, it remained on hold until my wife and I retired. Once we reached this milestone in our lives, we had ample time to begin piecing together the regiment's military adventure. Through John's letters, we knew the direction the 159th NY traveled; our plans were to shadow their expeditions.

Our journey began and we headed off to every location my third great-grandfather and his regiment traveled to: from New York to Louisiana, Virginia and finally, Georgia. We visited encampments, battlefields, museums, and any facility that could possibly have a piece of their history in their archives. Then, the COVID-19 Pandemic slowed us down and many archive facilities were closed. Eventually, we were able to visit most of them and retrieve what information we could find. Some of those locations included the NYS Military Museum and Veteran Research Center in Saratoga, New York, the National Archives in Washington, DC, which housed the original Regimental Books, the US Sanitation Commission reports at the New York City Library, and the Young-Sanders Center in Franklin, Louisiana, to name a few.

The research for this historical narrative involved endless hours of scrutinizing records and documents to piece John and the regiment's story together as accurately as possible. Discovery was imperative in this process. When the smallest element was found, I had a feeling of satisfaction knowing another section of the story was slowly moving toward completion. Many discoveries helped conclude sections, but there were times when we'd hit a dead end, either nothing recorded of the event, or the records were lost. That was always a disappointment. In fact, after waiting nearly two years to get an appointment at the National Archives to review the regiment books, we found out some had been misplaced.

There were also some unexpected discoveries that were almost unbelievable. One memorable discovery was that both my maternal (John W. Mambert) and paternal (Phillip Kipp) third great-grandparents had both served in the 159th NYSV Infantry Regiment. Thinking this rather unique, I later discovered my grandchildren (Rosie & Griffin) also had their paternal (Phillip Kipp) and maternal (Lewis M. Labrie) fifth great-grandparents serve in the 159th NYSV Infantry Regiment as well.

In reading the letters, I discovered the soldiers who volunteered in the war effort lived in mostly deplorable conditions and suffered horribly through much of their service; this was not what my great-grandfather and many others were expecting. For the record, it is not my belief the Union leadership understood the extent of suffering that would occur, but because the war front expanded and moved at such a quick pace, the capability to supply a soldier's basic needs of food, clothing, shelter, and most important proper sanitary conditions, was deficient and, in some situations, impossible.

This is the soldier's story.

Acknowledgments

Throughout the journey of writing this book there are a number of people I owe thanks for their assistance and guidance. A special thank you goes to my biggest supporter who was with me assisting throughout the project, my loving partner Tammy. She was by my side wherever the research directed us, traveling thousands of miles in addition to reviewing hundreds of records at numerous archives in search of answers to many questions. She read and reread the story many times offering useful insight. I am forever grateful for her devoted encouragement while writing this historical narrative.

I would also like to thank Jim Gandy, Librarian/Archivist at The New York State Military Museum and Veteran Research Center in Saratoga, New York. Jim's assistance in providing all-important information on the 159th NYSV Infantry Regiment and Officers was invaluable as well as his advice to help locate additional information. While traveling to the many archive locations in search of answers, one of the most memorable was the Young-Sanders Center in Franklin, Louisiana where we met Roland Stansbury, the center's director. Roland shared his knowledge of the Civil War in Louisiana and provided essential information along with excellent suggestions and guidance. I am extremely grateful to him. There were certain officers within the regiment who provided a tremendous amount of recorded history, helping me narrate the story. The most important being William F. Tiemann, the regiment's historian and author of The 159th Regiment Infantry, New-York State Volunteers, in the War of the Rebellion, 1862-1865.

Introduction

A FATEFUL DAY

On April 18, 1861, the people of Columbia County, New York, awoke to news of war with the South. The headlines of the Kinderhook Herald read:

> WAR COMMENCED! Surrender of Fort Sumter...75,000 Troops called for...Fort Sumter was surrendered to the authorities of the Confederate States...In view of the present conditions of affairs the President issued a proclamation calling upon the loyal states for militia to the aggregate number of 75,000 in order to suppress after combinations against the government in the south and enforce the laws...The NY State Assembly on the 15th following the President's proclamation armed the state, a bill to enroll and equipment 30,000 volunteer militia...The bill provides for the organization of the militia into regiments of ten companies of 100 men each.[1]

On the day President Lincoln formally announced the attack on Fort Sumter, a meeting was held that evening at the courthouse in Hudson, New York, to respond to the proclamation. The crowd that gathered was so large that an adjournment was made to City Hall, which also became packed to capacity.

Spirit-stirring addresses were made by county leaders of all political parties. Colonel Cowles[2], a local prominent officer, introduced a series of resolutions adopted at the meeting which were followed by thunderous applause. The people pledged themselves without distinction of party to expend their blood and treasure, without limitation or quantity, to support

---

[1] NYS Historic Newspapers-Kinderhook Herald, April 18, 1861

[2] Colonel Daniel S. Cowels was born in Canaan, Litchfield Co., Connecticut in 1817. At age forty he became a resident of the City of Hudson, Columbia County, New York. He attended Yale, became a successful lawyer. In Hudson he held some of the highest positions in the gift of the people, and always performed his duties with the highest degree of honor and success. Colonel Cowles raised the regiment which he commanded—the 128th—by his own individual exertions. He was killed in the assault on the works at Port Hudson. The last words of Colonel Cowles were, "Tell my mother I died with my face to the enemy."

the United States government. The meeting concluded with the enrollment of all men present ready to enlist as soon as the necessary papers could be received from the Governor.[3] Tens of thousands of men flocked to enlistment and recruiting stations all over New York, ready to volunteer for the battles sure to follow. No state contributed more men to fight than New York. In the end, a total of 467,047 men served; almost nine percent of the state's total population enlisted, representing sixteen percent of the total Union forces. New York raised nearly two hundred infantry[4] regiments, twenty-six cavalry[5] regiments and more than fifty artillery[6] regiments, batteries and battalions. New York passed legislation freeing its Black citizens more than forty years prior. Many of its Black citizens enlisted in the momentous effort, adding an additional 4,125 soldiers to become the infantry regiments of the 20th, 26th and 31st.[7]

JOHN W. MAMBERT ENLISTS

John W. Mambert was a resident of the Hudson Valley region of New York State. He lived and raised his family in the town of Taghkanic in Columbia County. He was a carpenter, farmer and Justice of the Peace. He was also involved in real estate, buying, selling, and renting properties. At the time of his enlistment, he was forty-three years old and married to Maria (Sheldon) who was then forty-one. John and Maria had five children, John the oldest was twenty-one, Permilia nineteen, Martha sixteen, Jobe eleven and Walt nine.

John bade farewell to his family as he headed off to the unknown, leaving his wife and eldest son to manage the family affairs while he served his country. His life in the military officially began October 20, 1862, after he was recruited to enlist in the Union army by a local well-known politician named Homer Nelson. On November 2, 1862, the non-commissioned staff were appointed to

---

3 Columbia County NY in the Civil War (3rd Annual Report of the Bureau of Military Statistics of the State of NY, (C. Wendell), 1866. New York State Military Museum & Veterans Research Center.

4 Infantry: A branch of the military in which soldiers traveled and fought on foot.

5 Cavalry: A branch of the military mounted on horseback. Cavalry units in the Civil War could move quickly from place to place or go on scouting expeditions on horseback, but usually fought on foot. Their main job was to gather information about enemy movements.

6 Artillery: Cannon or other large caliber firearms; a branch of the army armed with cannon.

7 Military Affairs in New York 1861 - 1865, New York State Military Museum & Veterans Research Center.

the regiment and John was included as a First Principal Musician Co. G.

He served alongside friends and neighbors who lived in the surrounding villages and hamlets, people he attended church with, socialized with—some may even have appeared before him in court.

When John and the other men enlisted, most believed the conflict would be over in short order. Most men had no idea they would be gone for so long, nor imagined the rigors of travel they would endure. The thought of being killed, never seeing home again, though slightly considered, became a grim reality for John and his comrades.

Many New York men were lured into the army, envisioning themselves as the victors in glorious battles. Others were attracted by a sense of adventure. Since most of the battles would be fought in southern states, the war offered men a chance to travel and see sights they had only heard or read about. Money was another reason why men were persuaded to enlist; even though the pay of a common soldier was $13 per month as compared to the average carpenter's pay of $32.

A DIVISION WITH DEEP ROOTS

The division between North and South was already deeply rooted before Lincoln became President. It stemmed from the long-standing controversy over the South's enslavement of Black people by the North. But that wasn't the only division going on, in the two decades before the Civil War, poor southern Whites were becoming increasingly upset about their exclusion from the labor market, considering themselves an underprivileged class with few chances of upward mobility. And despite the fact that many of these people were uneducated and barely literate, or illiterate, they showed a fairly sophisticated understanding of the ways in which slavery oppressed them.[8]

Georgia Governor Joseph Brown in 1860 warned the region's poor Whites that abolishing slavery meant Blacks would become their "equals, legally and socially." Slaves were accustomed to only earning basic food and clothing in exchange for their labor, he continued, and would require little more after freedom, rendering the possibility of emancipation was terrifying

---

8 Merritt, Keri Leigh, *Masterless Men, Poor Whites and Slavery in the Antebellum South*, (Cambridge University Press 2017), Ch. 9, p. 287.

to poor Whites.[9]

Prior to the Confederacy firing on Fort Sumter, "slave owners began scaring poor whites into supporting secession[10] by warning of the dire consequences of Black emancipation. First, they predicted poor whites' wages would fall quickly and drastically – to the point of literal starvation. Second, they said that poor Whites would suddenly be the social equals of the newly freed men. Finally, masters warned that a bloody racial war between the two impoverished classes would undoubtedly follow freedom, and the poor Whites would be slaughtered by the thousands. This type of fear mongering led up to the fanatical violence of the Jim Crow era. In the years before secession, southern slaveholders literally laid the foundation for irrational, extremist, racist fear that would come to dominate the decades following the war."[11]

Alabama led the secession, with South Carolina, Georgia, Mississippi, Florida, Louisiana, and Texas withdrawing from the Union: The Confederation was taking shape. Soon after the new government was formed, the assemblage of a Confederate military force commenced, leading up to the seizure of Federal forts and arsenals within the boundaries of those seven states.

On January 17, 1861, the South Carolina Legislature passed a unanimous resolution declaring any attempt by the Federal Government to reinforce Fort Sumter would be regarded as an open act of hostility and a declaration of war. The South Carolina Legislature promised support to the Governor in all measures of defense. This included approval of swift action by their militia to fire upon the Star of the West, a federal military transport ship, if attempts were made to replenish the fort.[12]

Sumter was in critical need of provisions. President Lincoln informed southern delegates he intended to resupply the fort on April 4, 1861. South Carolinians knew any attempt to reinforce the fort meant war. Fort Sumter, located off the coast of Charleston, South Carolina, was fired upon by their state militia's artillery[13] on April 12, 1861, initiating the conflict.

---

9 Ibid., Ch.9, p. 291.

10 Secession: (pronounced *si-sesh-uhn*) Withdrawal from the Federal government of the United States. Southern states, feeling persecuted by the North, seceded by voting to separate from the Union. Southerners felt this was perfectly legal, but Unionists saw it as rebellion.

11 Ibid., Ch. 9, pp. 290-291.

12 NYS Historic Newspapers-Kinderhook Herald, Jan 17, 1861, p. 2 image 2.

13 Artillery: Cannon or other large caliber firearms; a branch of the army armed with cannon.

AFTER THE WAR

On April 9, 1865, the Civil War essentially came to an end when General Lee made the decision to surrender the Army of Northern Virginia after becoming trapped by General Ulysses S. Grant's army near the Appomattox Court House in Virginia. Confederate generals throughout the southern states followed suit with the last surrender on June 23, 1865, by General Stand Watie, commander of the First Indian Brigade[14]. The Confederate States were impoverished, the destruction upon them during four years of battles was devastating. Their infrastructure was destroyed, especially the vital transportation systems. The fields producing the majority of their economic wealth were burned and most southern cities were devastated by the Union's efforts to eliminate any manufacturing or production for the Confederate army.

President Lincoln began developing plans for reuniting the North and South after he set forth the Emancipation Proclamation, believing it would be the driving factor. He never saw the plan initiated due to his untimely death. It was now up to President Andrew Johnson, his successor, to unite the country.

Johnson's effort was weak and failed. Success with a reconstruction plan did not happen under his administration. He was constantly at odds with the Republican controlled Congress who wished to pass legislation protecting the rights of freed slaves. Johnson's inaction to assist the people of the South added to their injury. Meanwhile, freed slaves were facing persecution by groups who wanted to return to the old ways and bring back slavery.[15]

Success in unifying the country finally occurred with the election of President Ulysses S. Grant in 1868. The Fourteenth Amendment was passed, establishing birthright citizenship granting equal protection of the law to anyone born within US boundaries. He advocated for the Fifteenth Amendment, guaranteeing Black males voting rights throughout the country. Grant advocated for the establishment of the Department of Justice in 1870, tasked with investigating acts of violence against African Americans. He also supported a series of legislative acts in 1871, enhancing the federal

---

14 Brigade: A large group of soldiers usually led by a brigadier general. A brigade was made of four to six regiments. 1 company = 50 to 100 men, 10 companies = 1 regiment, about 4 regiments = 1 brigade, 2 to 5 brigades = 1 division, 2 or more divisions = 1 corps, 1 or more corps = 1 army.

15 Taylor, Richard. *Destruction & Reconstruction, Personal Experiences of the Late War*, New York 1879, p. 217

government's ability to use the military to stop acts of racial terrorism. In 1875, he signed the Civil Rights Law outlawing racial discrimination.[16]

In the end, the country paid a heavy price because of the war between the Union and Confederate states. Over 750,000 soldiers died during the conflict. New York's losses exceeded 50,000 men; equivalent to the total population of Columbia County, New York, during that time. Columbia County lost many men in the four regiments raised during the war. In the town of Taghkanic, the 159th NY regiment lost sixteen of the thirty men who enlisted. Many men returned with lifelong disabilities. Soon after being discharged, some men succumbed to diseases contracted while serving. Even with surging growth to the economy in the North during and after the war, the cost suffered by death or disability, men who never returned to their jobs or businesses, dealt a painful blow to them, their families, and communities.

JOHN'S LETTERS FROM THE BATTLEFIELDS

In John's letters, at times, he felt elated, other times anguished, but most often lonesome from his deep desire to return home, to be with his loved ones.

His thoughts and feelings, his reasons for enlisting, were constantly challenged by military life, politics and family situations. He finds military life a struggle and at times desires a release; however, unlike many others who resigned or deserted and scuttled home, he remained, though reluctant at times, staying until the conflict came to an end and he was officially mustered out.

John's letters also include information about family, friends, and other soldiers, topics of interest at the time; his relationships and emotions experienced throughout the conflict. Regrettably the letters John received from family and friends were destroyed. We can only draw conclusions on these written conversations based on John's letters home.

Remember Me is as accurate an account as could be composed of the life of a Union soldier: John W. Mambert, and his regiment while serving during the great rebellion. This is his story, constructed around the letters he wrote, while enlisted with the 159th NYSV Infantry Regiment. Thanks to his letters, it is also the regiment's story.

---

16 Article, *A Short Overview of the Reconstruction Era & Ulysses S. Grant's Presidency.* Ulysses S. Grant National Historic Site.

Chapter One

"THIS REGIMENT WILL ALL DIE OFF IN ONE YEAR'S TIME IF THEY STAY HERE"

RECRUIT & ORGANIZE

Abraham Lincoln called for an additional 300,000 troops to aid in the suppression of the rebellion on July 1, 1862. Edwin D. Morgan, Governor of New York State, was given authority by the legislators of New York to give Lieutenant-Colonel Edward L. Molineux orders to recruit and organize a regiment with headquarters in Brooklyn, New York.

In September 1862, Colonel Homer A. Nelson was given authority by the legislature in Albany, New York, to recruit and organize additional regiments of volunteers from the 11th Senatorial District of the State which included Columbia and Dutchess counties. The headquarters of Colonel Nelson's regiment was located in the rural city of Hudson, New York. Nelson, an attorney who practiced law in Poughkeepsie, New York, and served as a judge for Dutchess County, was a prominent and popular gentleman in the two counties. He was not a military man but had the spirit of patriotism which filled many men at that time.

Homer Nelson[1]

Colonel Nelson was determined to organize a regiment to aid in upholding the laws of the country violated by the Confederates. Men were enlisted in towns and villages within the two counties and sent to the recruit's depot at the fairgrounds in Hudson, transformed into a military facility.

---

1 Homer Nelson, Civil War Photo Collection United States Army Heritage and Education Center, Carlisle, PA

JOHN ENLISTS

John Mambert and the other newly enlisted men gathered at the front gate, instructed to form a line and march inside. The men were ordered to wash up before being taken to a building newly renovated with a kitchen and dining room to accommodate them. Long pine tables with benches on each side filled the greater part of the space. The men took seats and were served good bread and coffee.

Concession booths were converted into administrative offices and medical quarters where the men assembled, waiting for physicals.

John's name was called. He was led into a large room where the medical examiner and clerks were working. Ordered to remove his clothes, John's examination began. The doctor looked at his teeth, checking his mouth for any evidence of unsoundness. After being put through a variety of physical tests, John was told to dress. Next, he was weighed and measured, the color of his eyes and hair noted, including his complexion. After passing the physical examination, John was sworn into service by Lieutenant James A. Farrel[2] as a soldier for the new regiment: the 167th NYS Volunteers.

The fresh recruits were directed to the new barracks and assigned bunks. The barracks were previously constructed for the 128th NYSV Infantry Regiment, mustered[3] into service on September 4, 1862. The buildings were long and narrow, about 100 feet by 16 feet, with three tiers of bunks on each side, an alley through the middle and open at each end. The bunks were long enough for a tall man and wide enough for two men, provided they lie straight. A board was in place to keep the front man from rolling out of bed. The three buildings each accommodated 204 men.[4]

As the recruits were being transformed into soldiers, large crowds of people, including wives, children, and parents of these brave men, stood watching from outside the fence. Squads of men marched on the horse track, trying to keep in step with an officer calling out: "Left, Left, Left!" as his left foot hit the ground.[5]

---

2 James A. Farrell–Lieutenant / Adjutant–Age 37, farmer in Greenport NY. Lived and worked on the family farm with his parents.

3 Muster: To formally enroll in the army or to call roll.

4 Van Alstyne, Lawrence, Diary of An Enlisted Man, (New Haven Connecticut 1910), pp. 3-6.

5 Ibid., p. 2.

CHAPTER ONE

At the start of the war, standards for training new soldiers and officers were not "set in stone." Few veteran soldiers were available to train new recruits. Most new drill instructors trained recruits using the Hardee's Rifle and Light Infantry Tactics, a guide to the fundamentals.

Elections were held by the men within the regiment to determine officers and non-commissioned officers. This did not necessarily mean they were choosing the best leaders in battle; it was simply proper procedure. It's questionable whether this process was fully respected; some positions appeared to have been promised before the actual regiment was formed, maybe as an enticement to enlist. John himself was promised a principal musician position with pay, to join. And there was the occasional "favor" for friends or family members in the form of appointments, like the eighteen-year-old sergeant major who just happened to be related to a lieutenant colonel.

Even so, there were men in company ranks elected to officer status who received support because they took the initiative to prepare and qualify for it. Prior to enlistment, a determined individual could go to a local bookstore and purchase training manuals for different aspects of combat. This could be the reason why a nineteen-year-old country farmhand was elected sergeant. The new non-commissioned officers "elected" within the regiment were the sergeant major, commissary sergeant, quartermaster sergeant, and the color guard (one sergeant) and eight corporals. For those with ambition striving for an officer's position, there were publications available to assist, as advertised in the Poughkeepsie Daily Eagle:[6]

---

[6] *Military Book advertisement*, Adriance Memorial Library, Poughkeepsie, NY. *Poughkeepsie Daily Eagle, Oct. 25,1862*, p. 3

> **Military Books**
>
> A New Supply At
>
> ### Hickok's
>
> **New Book Store**
>
> 327 Main Street,
> A FEW DOORS WEST OF THE GREGORY HOUSE.
> U.S. ARMY REGULATIONS, latest revised edition.
> CASEY'S INFANTRY TACTICS, 3 vols.
> U.S. INFANTRY TACTICS, 1 VOL.
> HARDEES TACTICS, Watson edition. 25 cents
> Army Officers POCKET COMPANION.
> The Hospital Stewart's MANUEL.
> LENDY'S MAXIMS and Instructions on the art of War.
> FIELD MANUEL for Battalion Drill.
> MONROE'S Company Drill and Bayonet Fencing
> PATTEN'S CAVALRY TACTICS
> PATTEN'S ARTILLERY DRILL
> PATTEN'S INFANTRY DRILL, school of the battalion.
> McCLELLAN'S BAYONET EXERCISES.

On October 28, 1862, under the command of Colonel Nelson, the recruits left Camp Kelly (Camp Columbia). Now part of the Union army, the men were to continue training for a southern expedition to meet their opposing force.

The new regiment exited the gate, heading west on Columbia Street, marching down the center of the city's business district, arriving at the docks along the Hudson River. Here, the men boarded the transport steamer *Connecticut* and shortly thereafter began their trip downriver to New York City.

CHAPTER ONE

Letter #1 - October 28, 1862
To Maria Mambert

You must direct your letters to:
John W. Mambert, Fife Major -159th Regt. N.Y.S. Vol.

We have a postmaster with us. Our pay, I don't know when we will get that, but it must come sometime. Ezra Stickles wife is on the dock. I want you to look well to those two little boys Jobe and Walt, and if you can get their miniatures[7] taken, and see that they are well dressed for winter. See well to everything and I will be satisfied. When you write let me know how Permilia got home, and I want you to see well to Martha and watch over her and tell her that she must be a good girl if she ever expects anything thought of. You must see Esq. Hauver[8] and see that he has audited that house so you can pay your tax. I want you to carry out my plans as far as possible, I expect to return home again someday or other, and I hope I may find all things right. You will see that there is a heavy battle expected any day now near Fredericksburg. You will hear the worst. I wish you would send me the Star paper[9] as soon as you get it, one every week or two, I would think as much of it as a letter.

yours, John W. Mambert [10]

Colonel Nelson's regiment arrived in the city the next morning where they disembarked the steamer and marched to Park Barracks located within City Hall Park—the town square. Already at that location was the "Third Senatorial" regiment consisting of recruits from Long Island, Staten Island and Brooklyn, New York, under the command of Lieutenant-Colonel Molineux.

---

[7] Miniature: A tintype, a photograph made by creating a direct positive on a thin sheet of metal coated with a dark lacquer or enamel and used as the support for the photographic emulsion. The tintype saw the Civil War come and go, documenting the individual soldier and horrific battle scenes. It captured scenes from the Wild West, as it was easy to produce by itinerant.

[8] Esq. William H. Hauver–Age 39, farmer, and Lawyer in Taghkanic, NY. Married with two sons & two daughters.

[9] Star: *The Hudson Daily Star* newspaper established 1851-1876 in Hudson N.Y., Alex N. Webb, publisher.

[10] *Letter #1, October 28, 1862, to Maria Mambert.* John W. Mambert letters from the Civil War.

While the two regiments gathered in New York City, there were discussions in Washington, D.C., about organizing an expedition to be commanded by Major-General Nathaniel P. Banks. This expedition would consist of all troops needed for the campaign[11], including those not yet recruited. Since both regiments of Nelson and Molineux had not enlisted their full complement of men, the two regiments at Park Barracks were consolidated by authority of Albany, New York, on October 28, 1862. The new regiment officially became numbered the "159th NYSV," the letters A, C, E, G, I, given to the Hudson companies and B, D, F, H, K to those from Brooklyn.[12]

On November 1, 1862, Lieutenant R. B. Smith, U.S.A. Mustering Officer, mustered into service of the United States, the 159th NYS Volunteers "for three years, or during the war." [13]

159th Field and Line Officers		
RANK	**NAME**	**LOCATION**
Lieutenant-Colonel	Edward L. Molineux	Brooklyn
Colonel	Homer A. Nelson	Hudson
Major	Gilbert A. Draper	Brooklyn
Surgeon	Charles A. Robertson	Hudson
Assistant-Surgeon	William Y. Provost	Brooklyn
Second Asst.-Surgeon	Caleb C. Briggs	Hudson
Chaplain	Isaac L. Kipp	Hudson
Adjutant	Robert D. Lathrop	Hudson
Quartermaster	Mark D. Miller	Hudson

---

11 Campaign: A series of military operations that form a distinct phase of the War (such as the Shenandoah Valley Campaign).
12 Tiemann, William F. *The 159th Regiment Infantry, New York State Volunteers*, In the War of the Rebellion 1862 -1865, (Brooklyn New York 1891), pp. 9-10.
13 Ibid., p. 12.

159th Regiment NYS Volunteer Infantry Colors [14]

MILITARY TRAINING

The 159th NY marched to Castle Garden at the Battery in New York City, which served as the country's immigration center and recruitment office to enlist immigrants into the war effort. The men boarded a steamboat and were transported to Staten Island.[15] Upon arrival, they marched to New Dorp, a neighborhood located on the east shore, where they went into quarters at "Camp Nelson." Lieutenant-Colonel Molineux oversaw the men's training, "utmost and complete military discipline was instituted." The camp was laid out, tents were pitched, guards detailed and posted, hours for reveille, roll call, meals, drills, tattoo[16], and taps fixed. The new soldiers were drilled in company[17] formation and movements, taught the manual of arms (proper use of their rifle) and other requirements necessary on the field of battle.

Complete discipline for the regiment also included the musicians.

---

14 159th Regiment NY Volunteer Infantry | Regimental Color | Civil War, New York State Military Museum & Veterans Research Center.

15 Tiemann, William F. *The 159th Regiment Infantry, New York State Volunteers*, In the War of the Rebellion 1862-1865, (Brooklyn New York 1891), p. 10.

16 Tattoo: was a general bugle call used to notify soldiers to assemble for the final roll call of the day.

17 Company: 1 company = 50 to 100 men, 10 companies = 1 regiment, about 4 regiments = 1 brigade, 2 to 5 brigades = 1 division, 2 or more divisions = 1 corps, 1 or more corps = 1 army.

Regulations for both Federal and Volunteer regiments authorized two musicians for each company, one fifer and one drummer. Additionally, there were Regimental Principal Musicians: the Drum Major, and the Fife Major or Principal Fifer. The drum major was equal to the fife major and received rank and pay of a sergeant major.[18] Field musicians—buglers, drummers, and fifers, were considered the nerve center of the military. Army regulations allowed field musicians to enlist at age twelve, but younger boys found their way into the ranks. Few who became musicians could read music or even play an instrument. They learned. A manual was produced in 1862 by Bruce and Emmett. Notable drummer and fifer Dan Emmett was principal fifer for the 6th Infantry and George Bruce was in charge of the music school at Governor's Island. The manual focused on rudiments of music, training exercises and duty calls.[19] It was the musician's duty to transmit the various orders essential for the army's day-to-day functioning whether in camp or on the battlefield. Field music was organized into three categories: the day's regulatory calls from "Reveille" to "Taps," tactical signals and battlefield instructions.

In parades, the drum major marched in front just behind the color guard directing the entire music, while the fife major marched to the right of the first rank of fifers (who preceded the drummers). The senior principal musician was two paces in front of the field music and the other principal two paces in the rear.

The musicians played an important role during the Civil War. They had a considerable number of responsibilities, many of which involved more than just playing music. In addition to playing for dress parades, guard mounting, morning colors, regimental and divisional reviews and funerals, musicians were stretcher bearers on battlefields. During battle, musicians served as medical orderlies, helping surgeons with amputations and other functions in medical field services behind the lines. They loaded the wounded onto ambulances and helped bury the dead. Musicians were called for unpleasant duties related to disciplinary actions. Two to four fifers and drummers were detailed for these jobs. Unruly women were drummed out of camp in the morning to the unaccompanied drum beat of 'fatigue.' Men dishonorably discharged, their heads shaved, wearing signs around their neck, were drummed out of camp

---

18 Poulin, David, *FIELD MUSIC OF THE CIVIL WAR Including extracts from, The 1861 Revised Regulations, Enactments of Congress, And Customs of Service*. Organization and Duties, p. 13.

19 Ibid., p. 24.

CHAPTER ONE

with a fife and drum tune called 'The Rogue's March.' Minor offenders were paraded around camp to the same music with signs hung about their necks indicating their specific offense. In camp, musicians were quartered on the end of company streets so the officers and noncommissioned officers could easily find them.[20]

The 159th NY trained for three weeks at Camp Nelson, becoming further proficient and familiar with the military and its daily routines. Regrettably during their time at camp, the regiment lost a considerable number of men by desertion. Still close to home and realizing what they enlisted for, eagerness to participate in the glory and excitement of battle faded just as they did. Though guards were given strict instructions to pass no one, sometimes the guards themselves deserted.[21] In total, approximately one tenth of the regiment was gone before they even departed for their destination—about 117 men disappeared; Thirty percent from Columbia / Dutchess counties and the remaining from Brooklyn, probably higher because of the close proximity to their families.

While the remaining men learned to be proficient soldiers, Major-General Nathaniel P. Banks was directed to New York and Boston with orders to collect and organize a force of 30,000 new recruits drawn from New York and New England. As former governor of Massachusetts, he was politically connected to the governors of those surrounding states and the recruitment effort was successful. The 159th NY was attached to the new army known as "Banks's Expedition."

The regiment broke camp at New Dorp on November 24, 1862, marched to the wharf and was taken by steamboat back to the city. From there, the regiment loaded onto the United States steam transport *Northern Light* at pier 3 in the North River.

The following day, Colonel Nelson resigned his military commission, winning election by his district to serve as a Democrat in the thirty-eighth Congress. He remained in office from March 4, 1863, until March 3, 1865.[22] Lieutenant Colonel Molineux was appointed in his stead and took command

---

20 Poulin, David, *FIELD MUSIC OF THE CIVIL WAR Including extracts from, The 1861 Revised Regulations, Enactments of Congress, And Customs of Service*. Disciplinary Actions, p. 22.
21 Tiemann, William F. *The 159th Regiment Infantry, New York State Volunteers,* In the War of the Rebellion 1862 -1865, (Brooklyn New York 1891), p. 14.
22 Homer A. Nelson, Biographical Directory of the United States Congress.

of the 159th NY regiment—which he essentially had from their beginning.

The new commander, Edward Leslie Molineux, had a strong interest in State Militias. He studied military matters for many years and corresponded with governors of different states and heads of educational institutions concerning training the youth of our land in thorough practical military instruction. He was the Lieutenant-Colonel of the "Brooklyn 23rd" State Militia and served as a volunteer in the 2nd Company "New York 7th" State Militia when that regiment was sent to Washington, D.C. during the outbreak of the Civil War. By his studies and experience, he was extremely qualified to organize and command the 159th NYS Volunteer regiment.[23]

THE EXPEDITION

On December 2, 1862, Company K of the 161st NYS Volunteer regiment was loaded onto the *Northern Light* joining the men of the 159th NY. Two days later, the ship cast off and the transport entered the channel, setting sail for Ship Island, Mississippi, 1,967 miles from New York City. "We saw New-York fade away in the gray distance, as the staunch and gallant North Star, Capt. LEFEVRE, plowed its way majestically through the waters, freighted with its souls, toward whose fate and destination the whole country is doubtless, at this moment, looking with such absorbing interest "Where are we going?" was the universal question on board, and one which seemed as far from solution there as it had been for the two last months on land, while the expedition was preparing."[24]

Eleven days later, they arrived at their destination joining a large fleet of transport ships loaded with Union troops waiting for orders. Ship island was long and narrow, flat, and sandy with a partially constructed fort at one end. Located in the Gulf of Mexico, it is seventy-two miles west of New Orleans and in a direct line with the mouth of the Mississippi River.

On December 14, 1862, the transport took on board a pilot navigating them up the Mississippi River headed for New Orleans. The route sailed past several rebel gunboats which had been destroyed: John and his fellow comrades saw for the first time the unpleasant consequences of war. As they sailed by the

---

23 Tiemann, William F. *The 159th Regiment Infantry, New York State Volunteers*, In the War of the Rebellion 1862-1865, (Brooklyn New York 1891), p. 7.
24 *On Board Steamer North Star, Gulf of Mexico*, December 29, 1862, Page 1, New York Times.

plantations, the slaves waved their hats and aprons while the planters stood by in resentful silence, scowling, giving disapproving looks to the Union invaders. That evening, the regiment sighted the lights of New Orleans and set anchor.[25]

During the new recruits' sea voyage along the eastern coast, a fast-approaching craft put the men on high alert. Chaplain Kipp wrote about it in a letter to his hometown newspaper the Fishkill Journal: "A few of the line officers and myself were on the deck, enjoying a social chat casting my eyes over the water, in the very point where sky and water seemed to meet, I discerned a sail. Merely calling the attention of my companions to it, we resumed our conversation. Full half an hour afterwards the mate came hurrying towards us and raised the stars and stripes. Looking towards the vessel, we saw that she was gaining on us; and soon a flash, a cloud of smoke, and the dull report signaled us to come to. Being rather slow in minding, not many minutes elapsed before signal No. 2 reiterated the command, the shot this time falling much nearer to us. We slacked our speed, but not until the third shot struck the waves within half a mile of the steamer, the engines stopped, and all eyes were strained towards the fast approaching craft, whose nationality was a subject of universal anxiety on board our ship. Opera and field glasses and telescopes turned inquiring eyes to that one object. After a while I heard our Captain say, she floats the stars and stripes, now; but how long she may keep them up, we cannot tell! Ou she came, her sails all set, flying before the wind. Soon we could distinguish our flag. Her character was unmistakable—a regular war dog, guns all out—and altogether an ugly customer to have any disagreement with. Yet it was a splendid sight: and despite the anxiety, which every moment increased, to know whether we were to go on our course or be led back like sheep to some rebel port, I could not but admire her bearing. The vessel drew nearer, and almost within hailing distance, rounded to, took in sail, and then down came the United States Flag. I can tell you there was perturbation on board the *Northern Light*. We watched them as they lowered a small boat, and each one breathed more freely when the lieutenant, from the vessel, answered our inquiries as to his mission—"United States Ohio." We gave him some papers and committed our letters to his charge. And so ended our adventure." [26]

On December 16, 1862, the regiment's steamer, along with five other transports accompanied by six gunboats and ironclads, sailed up the Mississippi 130 miles

---

25 Tiemann, William F. *The 159th Regiment Infantry, New York State Volunteers*, In the War of the Rebellion 1862 -1865, (Brooklyn New York 1891), pp. 15-17.
26 Blodgett Memorial Library, Fishkill NY–*Fishkill Journal, Rev. Isaac L. Kipp, Jan. 1, 1863*, p. 2.

to Baton Rouge, arriving the next day. Finally, after traveling for nearly fifteen days and 2,292 miles, the men reached their destination. During the voyage, more than 1,200 men, including Major-General N.P. Banks and his staff, Lieutenant-Colonel R.B. Irwin Assistant Adjutant-General, and Lieutenant-Colonel William S. Abert Assistant Inspector-General, were closely packed on board the *Northern Light*, a two hundred fifty-four-foot-long wooden side-wheel steamer with three decks, designed to accommodate 800 passengers maximum, 900 if you packed steerage.[27] Sleeping accommodations were poor; the men slept on deck or wherever they could find room to lie down. Inefficient arrangements for cooking meals made it almost impossible for the men to get something to eat. Fortunately, the weather was good. The plain food and fresh sea air made "almost" all on board feel well and healthy. The troops were vaccinated just after leaving New York and as they arrived in Baton Rouge, few if any were seriously ill.

A reporter following the 161st NYS Volunteers shared the steamer with the 159th NY regiment. He sent the following telegraph from New Orleans once they arrived:

> On board Northern Light, off New Orleans, Dec 15, 1862 according to me promise, I sit down this evening to inform you of the events transpiring in this quarter as far as they have come under my observation. The company to which I am attached is company K of the 161st NYV. Company K was raised from an old regiment and was transferred to the 161st. We form part of Banks expedition. We embarked on board the Northern Light, at NY City, Dec 2d, and set sail Dec.4th; arrived here Dec. 14th in the evening, being about 12 ½ days on the passage. When we came on board the ship at N.Y., we found the 159th N.Y. in possession of all available sleeping bunks, staterooms &c., and, it being late in the evening, the 161st had to sleep on the open deck, which was rather uncomfortable, the weather being cold and freezing. However, further along in the same correspondence he wrote: The weather was extremely warm off of Key West, and we of the 161st could now enjoy on deck the beautiful soft moonlight evenings and pity the poor fellows of the 159th confines between the closed decks with the air hot and suffocating.[28]

---

27 Description of ship *Northern Lights*, The Maritime Heritage Project, Ship Passengers: 1846–1899.
28 New York State Military Museum, 161st Regiment Newspaper Clippings, *Army correspondence. From New Orleans. On board Northern Light, off New Orleans*, Dec. 15, 1862. Editors Telegraph.

'THE ARMY OF THE GULF'

Major General Benjamin Butler marched into Louisiana on May 1, 1862. It was only a year prior on January 26, 1861, during the Secession Convention held at the State House in Baton Rouge, that Louisiana became the sixth state to secede from the Nation. President Lincoln's election in 1860 created a deep fear of economic ruin for the state with the threat of slavery possibly coming to an end under his administration. Many Unionists from Louisiana did not favor the drastic measure of secession. Compromises were discussed to prevent it, but in the end, secessionists captured fifty-three percent of the vote. Immediately after the convention, Alexander Mouton, chairman of convention, proclaimed the connection between Louisiana and the United States to be dissolved. In a symbolic demonstration of this change in status, the delegates lowered the American flag in the chamber and replaced it with a flag depicting a pelican feeding her young, the image on the state seal. Louisiana held its independent status through March 21, 1861, when it conveyed its allegiance to the Confederate States of America.

General Butler took back control of the most important cities in Louisiana including the naval bases in the southern state—this was the start of the Union strategy. From the beginning of the war, the plan was to block southern seaports and take control of major rivers like the Mississippi. Union planners believed "the army of the gulf" would starve the South into submission. By controlling the river, the Union could block the export of cotton, eliminating a valuable source of revenue for the Confederacy.[29]

BANKS'S UNEXPECTED ASSIGNMENT

General Butler served the administration well, but the time for his reign over Louisiana came to an end. President Lincoln decided a change was needed and offered General Banks command of the "Army of the Gulf." The assignment was wholly unexpected by Banks who had not solicited for it.

The reason Lincoln wanted him in Louisiana was due to the success Banks had when he was given command of the "Defenses of Washington."

---

29 Confederacy: Also called the South or the Confederate States of America, the Confederacy incorporated the states that seceded from the United States of America to form their own nation. Confederate states were Alabama, Arkansas, Florida, Georgia, Louisiana, Mississippi, North Carolina, South Carolina, Tennessee, Texas, and Virginia.

Banks restored order within the city in less than forty-eight hours after being given the assignment. He cleared from the streets of Washington 30,000 wounded men and convalescents who knew not where to go, and stragglers with no intentions of going where they were needed. Banks ended the chaos. Because of his accomplishment, less than two months later in the closing days of October 1862, President Lincoln sent for Banks and said, "You have let me sleep in peace for the first time since I came here. I want you to go to Louisiana and do the same thing there."[30]

Banks arrived in New Orleans, Louisiana, with his newly formed large force of raw recruits. He was ready to replace Butler as the new commander of the Department of the Gulf. Butler disliked Banks, but he did welcome him to New Orleans, fully briefing him on the civil and military affairs that were important. Banks learned he had to contend not only with Southern opposition to the Union occupation in New Orleans, but also with politically hostile Radical Republicans in the city and in Washington D.C., —critics of his moderate approach to administration.

When General Nathaniel Banks[31] assumed command of the Union forces in south Louisiana, his first orders were to send some 8,000 to 10,000 Union

---

30 Irwin, Richard B. *History of the 19th Army Corps*. G. P. Putnam's Sons New York, (The Knickerbocker Press 1892), ch. 5, par. 6.

31 Nathaniel Banks-a self-educated man. Nat, as he was known, started his education in a one room schoolhouse in Waltham, Massachusetts. Because he came from a large family it was expected of him to drop out of school, get a job at the mill where his father was employed and help support the family. Banks did not wish to end his education, but understanding the family needs he did and began working at the age of fourteen as a bobbin boy at the mill. Because of his desire to continue learning he became involved in the local debating societies where he developed an interest in politics. His oratorical skills were noted by individuals within the Democratic Party which he became part of. Eventually he ran for political office and became a state legislator in 1848. In 1852 he won a Democratic seat in the U.S. House of Representatives. Years later Banks would be forced to leave the Democratic party due to his strong anti-slavery views and other disagreements on key issues. Due to his abolitionist views, he became a member of the newly emerging Republican Party. Eventually he became Speaker of the United States House of Representatives 34th Congress after two months of voting and 133 ballots; it was later stated that Banks was one of the best Speakers the House ever had. Political connections earned him his military position as Major-General.

CHAPTER ONE

troops under the command of General Cuvier Grover[32] (one of the more senior generals in the army) to reoccupy Baton Rouge.

The enemy, learning of the Union approach in large force, decided to evacuate while the *USS Monitor*[33] gave them a parting salute with her guns. The *Northern Light* pulled alongside the Natchez, a river steamboat used as a landing stage where the men could disembark onto the banks of Baton Rouge, Louisiana. The Stars and Stripes were hoisted into the breeze from the Capitol building that day.[34] Once ashore, the troops marched one mile and camped on the United States Arsenal grounds, previously occupied by the US regular army until surrendered to Louisiana State Militia forces January 10, 1861. (Later that year, the arsenal was transferred to CSA (Confederate States of America) control until Union forces took back the barracks and arsenal for the duration of the war.) As additional troops continued landing, they were assigned various positions best adapted to cover and guard the approaches to the city. Company, regimental, and brigade drills[35] became the daily directive for the men.

For the next three months, John and his comrades continued training to become proficient using a musket, practicing all the essential drills of a disciplined, effective army. Few senior officers in the volunteer service were better adapted than Colonel Molineux for this task. He was quick, energetic, and ambitious to do his own duty and keep every man in his command busy— the true success of a disciplinarian.[36] The work of making thorough soldiers began in New York and continued in Baton Rouge. Soon, their newly learned talents would be put to use. Only twenty-two miles away in Port Hudson, it

---

32 Cuvier Grover– born in Bethel, Maine, was the younger brother of Governor and Senator La Fayette Grover of Oregon. A graduate of the United States Military Academy, he was stationed in the western frontier in Utah, but was immediately recalled and assigned to a command in the Army of the Potomac to help the defense preparations in Washington, D.C. at the outbreak of the Civil War. In the latter part of 1862, he was transferred to the Department of the Gulf. During the battle of Cedar Creek in late fall of 1864 he was injured, losing his arm. Harper's Weekly November 12, 1864.

33 *U.S.S.* Monitor: The first ironclad warship in the United States Navy, commanded by Admiral John L. Worden. The vessel had a large, round gun turret on top of a flat raft-like bottom, which caused some spectators to describe it as a "cheese box on a raft."

34 Duffy, Edward B., *History of the 159th Regiment, N.Y.S.V.* (New York 1890), p. 5.

35 Drill: To practice marching, military formation and the steps in firing and handling one's weapon.

36 Ibid., p. 6.

was reported 15,000 Rebel troops were garrisoned.[37]

John was impressed with the cannon engagement of the little gunboats at Baton Rouge, it was his first experience witnessing the effectiveness of the specially designed vessels.

Letter #2 - December 20, 1862

Baton Rouge
To Mrs. John W. Mambert:

I am as well as I can expect after my voyage. We left New Orleans on the morning of the 16th; they said we had to capture Baton Rouge further up the river. The next morning at 9 o'clock we arrived in front of Baton Rouge and the little mud turtle fired the first gun. I was down in the cabin and went up on deck, the Captain turned around and we were off with our ship 1/4 of a mile and then I went in the cabin and put on my sword and went back to the upper deck where I could have a fair view but just then they fired again from off a gunboat by our bridge and it passed us close and right through the city. The other little mud turtle fired away, the way the smoke and ball and shell flew it did me good and the rebels left for the woods on a keen run. We fired 17 or 18 times, that is in throwing shell and every time the cannon went off about 2 seconds you could hear the shells crack. It did me better than anything I had seen since I left Camp Columbia. Then it was not long before the rebs came running down from the city with a white flag and gave up. The Capital of the State is a very beautiful building so our folks went up to the top and put up the stars and stripes amidst the cheer of our fleet. And down came the Secession flag, we came unexpectedly on them. They never fired a shot but the Sunday morning before they shot 22 balls through one of our vessels killing one mate. One of our shells killed one man, took his southern head off. A lot of towns in the country are filled with aristocracy. We don't hear any news and the negroes are coming in by 100s, their masters have left them, and they are left to starve. I have seen no grain of any kind from the mouth of the river up to here except a few ears

---

[37] Tiemann, William F. *The 159th Regiment Infantry, New York State Volunteers,* In the War of the Rebellion 1862 -1865, (Brooklyn New York 1891), pp. 18-19.

of corn, some molasses cane or sugar cane, oranges and lemons, some sweet potatoes. You must direct your letters to John W. Mambert, 159th Regt. NYS Vols. Banks Division, Baton Rouge, La., in care of Capt. Sliter[38], Company F. You may send me a few postage stamps. We have 13- or 14,000 men soldiers and heavy artillery. We have a splendid place for our camp. It is very warm here in the daytime and cool nights. The soldiers fetch beef cattle and butcher them, but our regiment is scanter for food and suffer considerable. Tell our folks not to volunteer under no consideration whatever, not on account of being killed, but on account of suffering after being drafted. You see it will be some time before a letter can go and find its way here, if you look on the map you will see where I am. You can tell those who talk about the south they know nothing about it, this country in the south is mean and all destroyed and no one to blame but themselves. We don't know anything up there about the destruction of war here. I must say that you must look after everything til I get back, I think we may be home in April or May. Send my paper. I don't have too much to eat about these days, I have lost 25 lbs. since I went aboard the ship. We were 19 days on the ship, and I was sick all the time, they said they never saw a man fail as fast as I did.

J.W. Mambert [39]

Mud Turtles [40]

---

38 William H. Sliter, Captain, 159th Co. F–Discharged July 30, 1863. Age 24, farm laborer in Chatham, NY. Lived with mother and sister.

39 *Letter #2, December 20, 1862, Baton Rouge, To Mrs. John W. Mambert.* John W. Mambert letters from the Civil War.

40 *Mud Turtles, The Descent of the Mississippi*, published in Frank Leslie's Illustrated, December 1861.

IRONCLADS

John wrote about the "little mud turtles" that were strategic in the Union's plan to reoccupy Baton Rouge. This was a recommendation from Rear Admiral Farragut to General Banks upon his arrival in New Orleans on December 16, 1861. Three days later, Farragut ordered his transport to proceed directly to Baton Rouge with two gunboats (ironclads) to cover the landing.[41]

The need for ironclad gunboats came shortly after President Lincoln issued a proclamation declaring a blockade of Southern ports on April 19, 1861. It would extend from South Carolina to Texas covering 3,000 miles of coastline.[42] In May, General Winfield Scott proposed "a powerful movement down the Mississippi to the ocean." The idea was to connect with the blockade and surround the Confederacy. Known as the Anaconda Plan, the strategy was simple and believed to be the means toward a swift victory. The Union Navy would assist in the effort to split the Confederacy along the line of the Mississippi River. Control of the river and its tributaries would allow the Navy to support the army with amphibious gunfire and transport.[43]

With the Union Navy advancing on the Mississippi River, the Confederacy began building new defenses while strengthening old ones. The Union promptly realized taking control of the waterways would require a differently designed gunship. On August 3, 1861, Congress authorized the Secretary of the Navy to appoint a board to review plans and specifications for the construction of iron or steel-clad steamships or steam batteries, appropriating $1,500,000.[44]

Four days later, the War Department contracted with J.B. Eads of St Louis to construct seven shallow-draft iron-clad gunboats. Eads, a respected shipwright, was well positioned for the task, having owned multiple sawmills and foundries. He partnered with Samuel M. Pook who worked with the Navy on warship design. The ironclads known as the City Class were named after cities along the Ohio and Mississippi rivers, Cairo, Carondelet, Cincinnati, Louisville, Mound City, Pittsburgh and St. Louis.

On September 16, 1861, the Ironclad Board reported:

"for river and harbor service we consider ironclad vessels of light draught,

---

41 Civil War naval chronology, 1861-1865. Sec. 1. p. 114.
42 Ibid., Sec. 1. p. 17.
43 Ibid., Sec. 1. p. 20.
44 Ibid., Sec. 2. pp. 29-30.

or floating batteries thus shielded, as very important."

They recommended construction of three additional ironclads, the Monitor, Galena and New Ironsides.[45]

The seven city class gunboats were put into commission on January 16, 1862. Manned by 1,000 sailors, they became known as Pook's turtles or mud turtles. These ships, and those that followed, revolutionized naval warfare.

THE CLOSE OF 1862

On December 30, 1862, the 159th NY regiment was attached to the 31st Massachusetts, 25th Connecticut and 26th Maine regiments organized into the 3rd Brigade. The other regiments were also assigned to brigades and as a whole constituted a division commanded by General Grover. On January 5, 1863, by order of the War Department, troops in the "Department of the Gulf" constituted the 19th Army Corps[46] to be commanded by Major-General Nathaniel P. Banks.

By year's end 1862, the 159th NY mustered in only two months prior with the strength of 1,027 soldiers and officers, was down to 877 men. Nine men were returned home under orders of a Writ of Habeas Corpus,[47] five succumbed to sickness, one to alcoholism, fifteen were discharged, ten having disabilities. But the biggest and most immediate loss to the regiment was the 121 men who deserted from their commitment to the Union army.

A CHANGE IN NY POLITICS

The end of 1862 also brought about change in New York's political atmosphere. The electors chose a Democrat for Governor, voting-in Horatio Seymore, replacing Edwin Morgan, a Republican. Morgan chose not to seek reelection, instead running a successful campaign for senator.

Seymore won by a narrow three percent against Republican candidate James Wadsworth. Propelling him to the highest position in New York

---

45 Ibid., Sec. 2. p. 34.

46 Corps: (pronounced *Kohr* or *korz*) A very large group of soldiers led by (Union) a major general or a (Confederate) a lieutenant general and designated by Roman numerals (such as XIX Corps).

47 Writ of Habeas corpus: the constitutionally authorized means by which a court may immediately assume jurisdiction and inquire into the legality of an individual's detention.

was Seymore's platform that the U.S. government should have supported and passed the Crittenden Compromise. He believed this would have permanently protected slavery in the United States Constitution, making it unconstitutional for any future congresses to end slavery. This, he felt, would have diverted the country from war. Though Seymore promoted himself as a loyal Union man during the election, he was sympathetic to the copperheads[48] who wanted to oppose the war and damage the Union effort by opposing the draft and encouraging soldiers to desert.[49]

MOLINEUX'S 'SPLENDID, CUSHIONED CHAIR'

On January 11, 1863, it was reported the enemy was approaching in force from Vicksburg, and all troops were drawn in to protect the city. Gunboats also assisted in their defense. The troops stood under guard from early daybreak to 6:30 p.m. for a week until the report of attack was proved false. On January 14, 1863, the brigade was reorganized with the 31st Massachusetts transferred out and replaced by the 13th Connecticut; Colonel Henry W. Birge assumed command.

Birge, from Norwich, Connecticut, was a merchant before the war. He was Major in the 4th Connecticut Infantry and then the 1st Heavy Artillery. In May 1861, he received his commission as Colonel of the 13th Connecticut Regiment. During this preparatory training, he displayed a decided taste and aptness for military pursuits as a rigid disciplinarian, quick observer, well drilled, dignified, courteous, brave, fond of making a good show and

---

48 Copperheads were citizens in the North who opposed the war policy and advocated restoration of the Union through a negotiated settlement with the South. The word Copperhead in reference to the snake that sneaks and strikes without warning. Nearly all Copperheads were Democrats, but most Northern Democrats were not Copperheads, their strength was mainly in the Midwest where many families had Southern roots. In addition, groups opposed to conscription and emancipation—e.g., the Irish population in New York City, who feared that freed Southern Blacks would come north and take jobs away.
The Copperheads' interest was to maintain the existence of the Democratic Party and defeat Republican opponents for public office. By the end of the war, the terms Democrat and Copperhead had become virtually synonymous throughout much of the North. As a result, even though the Copperheads failed to exercise any significant influence on the conduct or outcome of the war and even though most Northern Democrats supported Lincoln and the war effort, the Democratic Party carried the stigma of disloyalty for decades after Appomattox.

49 Brummer, Sidney David, A.M. *Political History of New York State During the Period of the Civil War.* (New York 1911), p. 255.

CHAPTER ONE

possessing a remarkable quality: the gift of silence.[50]

On January 22, 1863, the 159th NY moved out on the Port Hudson Road about one mile from the battleground, their brigade being the advance of the division as it had been from the beginning. The new location became their camp. It was virtually impregnable, surrounded by thick forest and deep bayous: no opposing force could advance on them.

Colonel Molineux designed the 159th NY's encampment near Baton Rouge. The following two pages were his personal illustrations for the camp. The diagrams show the location of his tent and dining tent a far distance away from the soldiers' dog tents. The first sketch shows his comfortable amenities including a "splendid, cushioned chair" and brick fireplace for cool nights. The second sketch provides a great visual of the camp layout. Note the position between soldiers, noncommissioned staff and officers, the musician's tents, hospital tent and stables.

---

50 Sprague, Homer B. (Homer Baxter), b.1829., *History of the 13th Infantry Regiment of Connecticut Volunteers, during the Great Rebellion,* (Hartford Conn. 1867), p. 11.

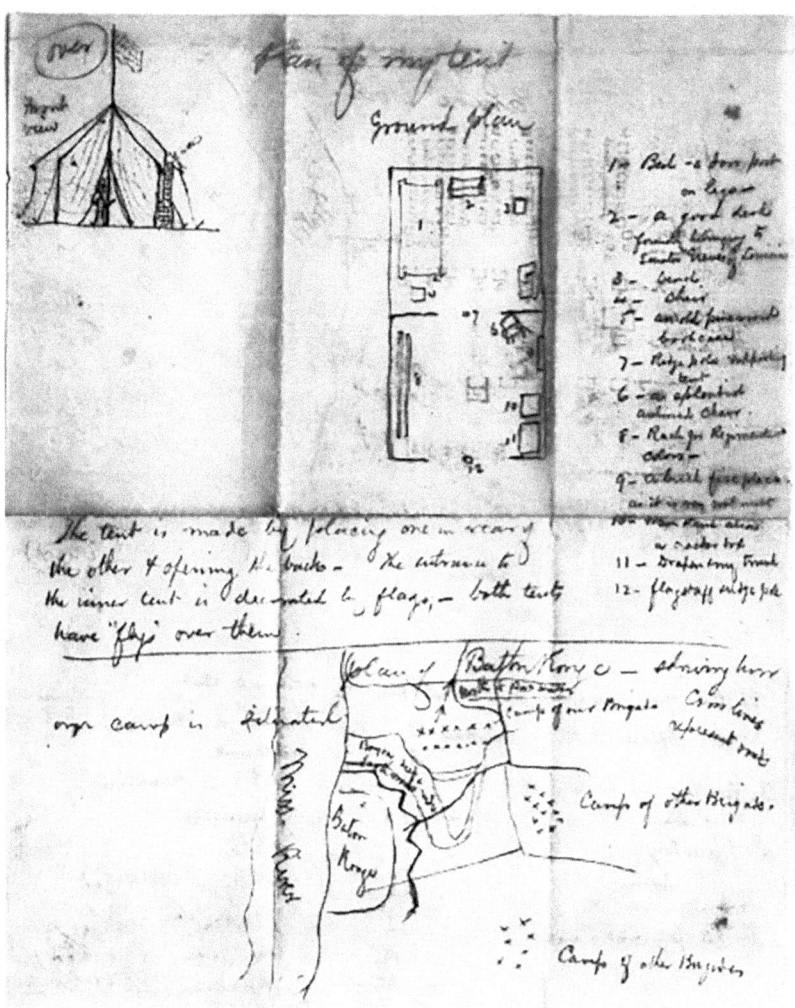

Sketch of 159th Camp in Baton Rouge by Colonel Molineux

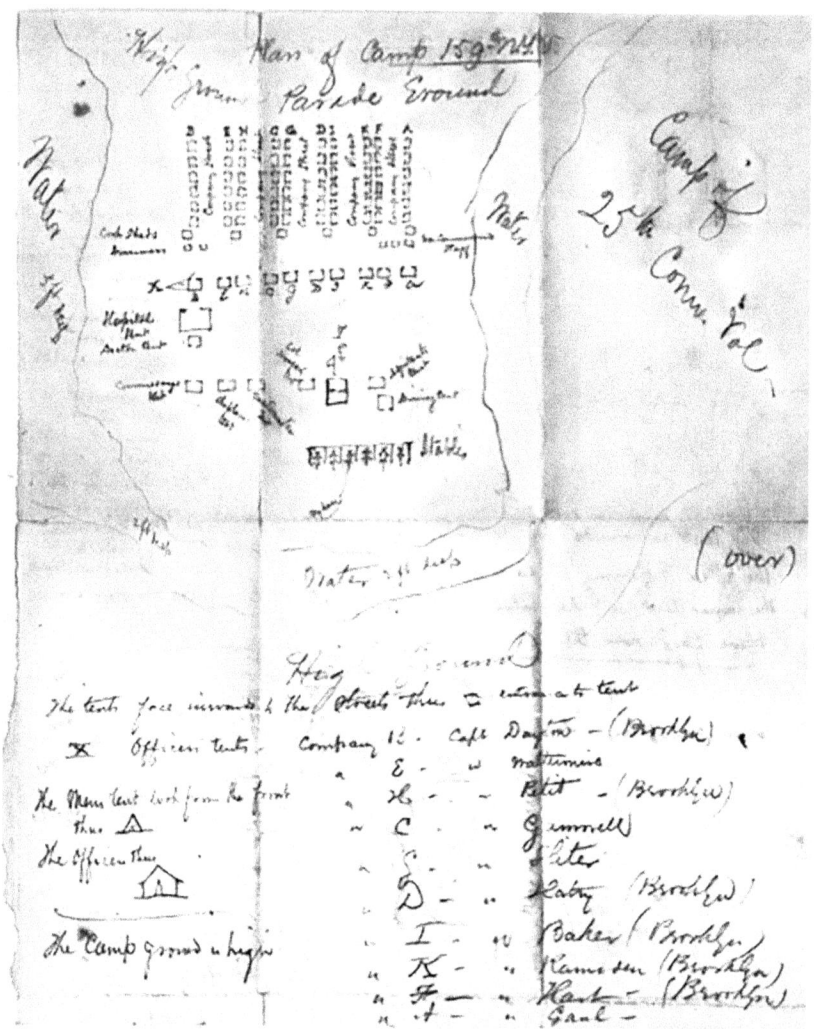

Sketch of 159th Camp in Baton Rouge by Colonel Molineux [51]

HUNGER PANGS AND DESERTION

While the regiment was encamped near Baton Rouge, affairs within their brigade appeared mismanaged, making it impossible for the men to get shoes, socks or other clothing. For two weeks, the men were short on rations

---

[51] *Molineux Camp Sketches*, James I Robertson Jr Civil War Sesquicentennial Legacy Collection at The Library of Virginia – Edward L. Molineux collection.

and for three days they had neither bread nor flour, surviving on morsels. Initially, the men received adequate rations of coffee, salt pork, salt beef, beans, fresh beef and flour or soft bread. Now, John and his comrades were experiencing what it was like to go hungry out of necessity while marching through enemy territory, especially after other regiments had "foraged" previously, leaving no provisions to help support the stores.

On January 25, 1863, while Second Lieutenant Tiemann was on picket, the officer of the day met him at 3:00 p.m. with information stating thirty of the regiment's men had schemed a plot to desert. He was given instructions to be doubly careful of who passed through the lines. Shortly thereafter, a company from the 13th Connecticut was sent to assist, and guards were posted all along the road with orders to halt all soldiers they saw. Nothing was observed and at daybreak, the company was marched back to its quarters; two men had escaped, both from Claverack, N.Y.[52]

On January 26, 1863, the troops were reviewed by Major-General C. C. Auger, commander of the 1st Division. After a general inspection, Colonel Molineux complimented the regiment for having the cleanest camp in the division, adding that they should take great pride in their accomplishment.[53] Apparently, the "officer of the day" was off by a few days on his mass exodus prediction, as more men from the regiment hightailed it back to New York.

Captain Edward Wardle[54], Second-Lieutenant Jacob Finger[55] and musician Charles Clark, all from Co. I. were discharged. Later in March, Colonel Molineux mentioned Mr. Finger in a letter home: "I learn you have been called upon by (former Lieut.) Finger who gives you a precious account of the regiment. I have never met a more unprincipled rascal than that man, a robber of poor men and a disgrace to humanity. I had him dismissed the service. If he enters the store again, I should order him out."[56] Second-Lieutenant Jacob

---

52 W.F. Tiemann Collection at The N.Y.S. Military Museum, *Letter Jan 26, 1863, Baton Rouge.*
53 Tiemann, William F. *The 159th Regiment Infantry, New York State Volunteers*, In the War of the Rebellion 1862 -1865, (Brooklyn New York 1891), pp. 20-22.
54 Edward Wardle, Captain 159th Co. I – Discharged by order of General Banks January 16, 1863. Age 26 a tobacconist from Hudson NY., no family.
55 Jacob Finger, 2nd Lieutenant 159th Co. I – Discharged by order of General Banks January 26, 1863. Age 33 a carpenter from Smokey Hollow, Claverack, NY. Married with two children ages 9 & 7.
56 James I Robertson Jr Civil War Sesquicentennial Legacy Collection at The Library of Virginia – Edward L. Molineux collection, *Letter written March 25, 1863.*

Finger was discharged by order of General Banks.

John, along with the other men of the 159th NY, were suffering the pains of hunger, weakness and frustration from the regiment's shortage of provisions. John writes in his letter home that the situation was too much for some of the men to bear, causing them to flee for home.

Letter #3 - January 28, 1863

Louisiana

To John Mambert and Marie Mambert (don't give it to him, see him yourself and tell him all about it) In this is a letter to John M. Welch[57] if you can't do anything with N.T. Post[58] about me on reasonable terms then go to Welch and see what you can do with him. Be careful to see what is the least he will take for what's in this letter to you. Better not give him the letter, I have authorized you as my attorney to get me out or be released and he may charge just what he chooses. I therefore want you to agree on a price and if he cannot get me out then he must have nothing. I feel as though I would pay $100 if I was back home, for I have been abused like a dog since I have been here. I have suffered a great deal, the provision is crackers and coffee twice a day and cold beef for dinner, I can't get fat on that for I am weak and lean. I did not go out on Saturday, only yesterday and today, I don't feel able to do my duty. I want you to send me some money, at least $2.50, which I leant out at Camp New Dorp. I feel as if I could not live in this country on such living as we get. This regiment will all die off in one year's time if they stay here, several of our boys ran away last night and the night before. If you send a box, send some writing paper and stamps and envelopes. If you could send some apples, they sell them for 5 cents apiece; plugs of tobacco 1.00, writing paper 3 small sheets for 5 cents. This is the dreariest country you ever saw, there is nothing here. You must write about what the prospects are about the war's ending.

---

57 John M. Welch – Age 37, attorney who practiced and lived in Hudson NY. Married with one child.

58 Nathan. S. Post 1st Lieutenant, Co. D – Discharged in Baton Rouge by orders of General Banks. Age 37, attorney who lived in Livingston, NY.

Andrew Brush[59] is as fat as a pig. I have not received a letter from home yet. If you can't send a box, then send me some money til they pay me. You must not allow any of you to come here for you can't stand it, don't attempt it unless I send for you. Tell everyone you see not to enlist and if they get drafted tell them to leave for Canada, don't suffer yourself to be drafted. Tell Harm not to enlist, tell Rams and Ole and the rest of them. Today it is cold here, I have 10 cents left, I don't know when I will get more. How happy I shall be when I meet you all again. My love to you all and to Marie Mambert and the rest of you all. Trust in God for my deliverance.

J.W. Mambert [60]

Surprisingly, food and supplies were actually plentiful, but the administration's ineffectiveness in logistics caused delays in delivering necessities to the soldiers. There were other reasons the men went hungry, the most obvious that they were responsible for cooking their own meals. Most, if not all of them, had little if any experience with cooking, relying on their mother, wife, or someone else to provide their meals. The men put no effort into foraging for food to supplement their lack of supplies. In the beginning, regiments were given strict orders not to take food from civilian supplies. As the war pressed on, foraging became the normal routine.

BATON ROUGE

The port city of Baton Rouge, Louisiana, was touched by the ravages of war during the early stages of the Federal Government's takeover of the southern waterways. After Lake Pontchartrain and the surrounding rivers were occupied by the Union Navy's light-draught gunboats, New Orleans became nearly invincible to rebel forces. It was at this time General Butler and Admiral David G. Farragut

---

59 Andrew Brush - Private, 159th Co. E – Discharged New Orleans La. Jan. 9, 1864. Age 55 (at enlistment stated 44), farmer in Churchtown NY. Married with six children ranging from age 7 to 23 years old.

60 *Letter #3, January 28, 1863, Louisiana, To John Mambert and Marie Mambert.* John W. Mambert letters from the Civil War.

decided to take the main fleet to Baton Rouge, seize the city and occupy it.[61]

On May 7, 1862, a section of Farragut's gunboat fleet sailed up the Mississippi River with two regiments. That evening, Union Naval Commander James S. Palmer reached Baton Rouge aboard the gunboat *USS Iroquois*.[62] Palmer ordered the city to surrender, but the mayor refused to comply. Realizing there was no threat from within the city he weighed anchor, landed a force and took possession of the arsenal, barracks and other public property of the United States.[63] The next day, Brigadier-General Williams and his troops arrived and in a show of force: They disembarked from the steamers, marched to the arsenal, then promptly returned to their transports. Determining there was no need to leave a force, they left for Vicksburg the next morning.

On May 28, 1862, Admiral Farragut's fleet (including the *Hartford* and several other gunboats) arrived in Baton Rouge on their return from Vicksburg. Farragut learned from the officer in charge that a party of brazen guerrillas rode down to the water's edge the day before and fired into one of his boats, wounding an officer and two men. The admiral, irritated by the attack, ordered his gunboats to open their batteries on the enemy, doing extensive damage to the buildings near the river, including to the capital. It was decided the city could not be left unoccupied.

General Williams and his troops returned to Baton Rouge the following day, taking over the ordnance depot, occupying the capitol, and pitching tents on its grounds. Farragut left two gunboats, *Kineo* and *Kennebec*, to support the troops holding the arsenal.[64]

Confederate General Breckinridge was given orders to attack the Union soldiers positioned at Baton Rouge with a force of 6,000 men along with the gunboat *Arkansas*. Because of epidemic disease, Breckinridge's force was reduced to 2,500 men by the time they reached Baton Rouge. The *Arkansas*, after arriving within a short distance, became unmanageable from an

---

61 Butler, B. F. (Benjamin Franklin). (1892). *Autobiography and personal reminiscences of Major-General Benj. F. Butler: Butler's book: a review of his legal, political, and military career.* Boston [Mass.]: A.M. Thayer & Co. pp. 454-455.

62 O.R., Series I, Vol. 15. *Butler to Stanton May 8, 1862.*

63 Civil War Naval Chronology, 1861-1865. Part I – 1861. p. 61.

64 O.R., Series I, Vol. 15. *Reports of Brig. Gen. Thomas Williams, Army, commanding Expeditionary Corps, operations May 26- Aug 2, instructions from Major-General Butler. Headquarters Second Brigade, Baton Rouge Arsenal, 29, 1862.*

unrepairable engine and was destroyed by the captain. With a reduced force and no gunboat,[65] Breckinridge attacked Baton Rouge.

To oppose the enemy, General Williams had a force of about 3,500 men (half on the sick list), one cavalry regiment, four gunboats and three batteries. Williams formed his lines at daybreak August 5, 1862, nearly a mile beyond town, hoping to protect the citizens and the city.

The enemy made their attack in two divisions on the right and left. Massing their forces, extending their lines and flanking them, they tried to force William's men back. But their strategy backfired when a mistake by one of the rebel regiments forced their own men back instead. The Confederate force made three attempts during six hours of continuous fighting to gain on William's men, but failed. After heavy losses, Breckinridge decided to retreat from the field.

Ninety Union soldiers were killed and 250 wounded. Among the dead was Brigadier-General Williams, killed while leading his men on the battlefield. His death was considered a great loss on a day of victory by Union forces at Baton Rouge.[66]

With the loss of so many men on both sides, the Union decided to abandon Baton Rouge. This allowed Confederate forces to advance until John's regiment (part of General Grover's division) arrived on December 17, 1862.

Encampment at Baton Rouge, La 1862 [67]

---

65 O.R., Series I, Vol. 15. *Major-General Earl Van Dorn, Headquarters District of Mississippi, Jackson, Miss., September 9, 1862.*

66 O.R., Series I, Vol. 15. *Maj. Gen. Benjamin F. Butler, U. S. Army, commanding the Department of the Gulf of engagement at Baton Rouge, La., with orders and resulting correspondence. Headquarters Department of the Gulf, New Orleans, La., August 10, 1862.*

67 *Encampment at Baton Rouge, La 1862*, E. L. Molineux Photo Collection, NYS Military Museum.

While some enlisted men from Louisiana joined the Union cause, it was John's observation that the common people of this southern state did not want to be involved in the conflict. John's education on the aristocracy of the South was about to begin.

Letter #4

Baton Rouge Louisiana

in care of Capt. Sliter Company F,159 Redg. NYS vols

I don't see how the people live, they raise nothing but sugar cane and no crops to eat. Flour is $25 per barrel and can't be had at that. Baton Rouge is situated on a rise of ground; it once was a beautiful place, but now it is all destroyed. Some places there is nothing but chimneys standing, but there are yet some nice buildings, a good many of them brick with a cannonball hole in one side and out of the other side. There are entrenchments all around the city. The city looks desolate, most all house entry doors wide open and no one there or at home, and some full of furniture. Some of the Union men say they would give everything they had if the Union could be preserved. There are 2 Union regiments from this state with us, this shows that if they had their own way there would be no war. The rebs draft from 14 to 60 in age, they have ravaged the country, there is nothing left but negroes to starve to death. We don't hear any news, and you don't see as much as an almanac. Please just send me my almanac and put a wrapper around the almanac as a newspaper and direct your letters to Baton Rouge Louisiana in care of Capt. Sliter Company,159 Regt. NYS vols. It is warm here with hopping flies, crickets, birds, and everything, like in September there. Look on the map and you will see where I am. Write to me and let me know how you all do. I am doing well, send my paper.

John W. Mambert, Major Fifer [68]

---

[68] *Letter #4, Baton Rouge, Louisiana, in care of Capt. Sliter.* John W. Mambert letters from the Civil War.

The two Louisiana (Union) regiments John refers to were organized while General Butler was in charge. The 1st Louisiana infantry regiment formed in New Orleans on July 30, 1862. They remained until January 1863, when they were sent to Baton Rouge in support of the 19th Army Corps. They continued serving through the Siege at Port Hudson and the Red River Campaign. Additionally, the 2nd Louisiana infantry regiment became part of the 19th Army Corps and remained with them throughout the time they were positioned in Louisiana. They became the 2nd Louisiana U.S. Mounted Infantry in September 1863, eventually transitioning to the Cavalry division during the Red River Campaign. In the fall of 1864, when two Divisions of the 19th Corps departed Louisiana and went north to the Shenandoah Valley, Virginia, both Louisiana regiments remained in Baton Rouge and New Orleans for the remainder of the war. In total, there were twelve Federal regiments raised in Louisiana consisting of men loyal to the Union in addition to the black regiments that were created.

Though John is far from home and unable to mentor or care for his children, especially Jobe and Walt, he appears to have adopted some young men in the regiment. Befriending them, he provides the attention only a father can give.

Letter #5

Baton Rouge Louisiana

I have been looking for a letter for some time now, and some papers, for a letter is as good as a good dose of medicine. You know how anxious you are to receive a letter from me and hear how I get along, just so I long to hear from you. Tell Permilia that a yankee from Long Island, who is an old bachelor of 28, saw her miniature and said he was a joining to have her if he got back, he asked consent, I said yes. He is a farmer, and he wants me to come down and buy a farm on Long Island, he said I could fish too. Clams, oysters, flounder, and all kinds of shellfish and raise sweet potatoes and he said fruit grows there plenty. At all events he said I should come down if I returned and visit him and he would help look for a farm. He has a small farm of 70 acres, he said even a good drought man can live half on fish and hunting. He said as soon as he came home, he was coming up to see us, he calls me Old Daddy. Then there is a boy in Company G, he calls me Father, he looks like Jobe, red hair, he was so sick, and he called me Father, he was 17 years old, I felt sorry for him he seemed almost like my own son, he is better now, he calls me father. Permilia's man name is Amos Benjamin Laws[69], tell Martha that she must be a good girl, and when I come back, I will take you all down to Long Island. Tell Jobe that he must go to school and take good care of the sheep, cows, and chickens. Tell Walt he must help Jobe and he must take good care of Mig[70] and go to school. And John must take good care of you all, and see that all the things are well and punctually done, no more at present.

Yours, may God bless and protect you.

From John W. Mambert
to his wife and family

Direct your letters as before to the
Banks Expedition instead of Division.[71]

---

69 Amos Benjamin Laws, Corporal 159th Co. F–Enlisted in Brookhaven NY., transferred to the Navy when they were looking for recruits in May 1864. Age 29, a boatman from Brookhaven NY. Lived with his parents and eight siblings.

70 Mig: The Mambert family dog.

71 *Letter #5, Baton Rouge, Louisiana.* John W. Mambert letters from the Civil War.

Chapter Two

"OUR REGIMENT IS ABOUT HALF SICK AND UNFIT FOR DUTY"

ADJUSTING TO MILITARY WAYS

The enlisting men were no different than men who joined the Continental Army ninety years earlier—average age of fifteen to forty years, coming from all walks of life, farm laborer to professional. For the average person prior to the breakout of the Civil War, understanding what it meant to be a soldier was not fully understood. The newly enlisted men from the Hudson Valley learned quickly that life as soldiers was not easy. Taking orders from men with rank who controlled their every move from morning to night was a situation the men were having trouble enduring.

John, similar to many of the soldiers, was much older than the "wet behind the ears" commanders, in many cases, old enough to be their father. Foot soldiers holding positions of authority in civilian life were suddenly the ones taking orders. It was especially difficult to salute an officer who may have been your hometown postmaster only a few weeks prior.

Soldiers were also at war with sicknesses like measles and dropsy (edema), cold weather and a lack of food. In his next letter, John advises others not to enlist, stating that it's not "the fear of fighting" but instead "the treatment a soldier gets" that is the most unbearable.

Letter #6 - January 29, 1863

Baton Rouge, Louisiana

Since writing yesterday and this morning, I received two letters from you. I was glad to hear from you; it makes me feel better to hear that you are all well and enjoying yourselves. The papers for me and Boswell are in with the letter "B" or else in with the DEEDS in the bureau rear drawer. The suits not paid are Blap Proper and Williams, Chargely is paid. You see Ryphenburgh and he will know how much they have paid and find out where E. Miller or L. Hawver have left the Executors. The Fanning Mill belongs to me and the grindstone too, and whenever you think fit you can fetch it. The place of Boswell[1] you must see that he pays the rent and I think you better keep the place till I come back unless you can get $300 for it cash down. Today I felt a lot better. At night it has been quite cold, it froze ice. It is rumored here that we shall move up to PORT HUDSON, if we do then the fur will fly in a heavy battle. I enclose in this letter to you a letter to John M. Welch, I don't want you to give it to him but to go and see him and tell him all about it. I have written to give him the power of attorney to get me released if it cost $100 but try and see how much deposit he will do it for. If he doesn't get me released, he must have nothing. I first spoke to N. L. Post and sent a letter with him to you. I told him to try and get me out, but he might charge me more than I should wish to pay, and therefore I want you to see how much he would ask. If he cannot get me out, then he shall have nothing also. First try Post and then see Welch. When I hear from you, I feel better, I have not received the almanac, send another until I do get it. I asked for a box, if I thought I would get paid I would not ask for the box. I have been taking medicine for some time now, our regiment is about half sick and unfit for duty. Pierce Allen has got the dropsies, he is puffed up like a puff ball. Andrew Brush is as fat as a pig, but I think it is the dropsies[2]. You must tell Welch that he can make a Writ of N. L. Post in my case as it regards deceiving me about the $41 per month pay. On this I could be released alone. I suppose that Post would hate to swear to that on his own account, for he has no right to deceive anybody, in such cases attend to that matter. Don't advise anybody to enlist, they need not fear the fighting, that is least of all, it's the treatment the soldier gets. It is impossible for a man to stand it unless he has a gizzard like a crocodile and can sleep right out in the open air in the rain and cold and heat and live on dry crackers and coffee. The

---

1 Benjamin Boswell–Age 66, a painter from Smokey Hollow, Claverack, NY.

2 Dropsy–Edema, swelling caused by fluid in the body tissues. It usually occurs in the feet, ankles and legs, but it can involve the entire body. Causes can include salt, sunburn and standing or walking a lot in warm weather.

volunteers are deceived by the kind of danger of fighting and not of their treatment. Captain Wardle, Captain Jacob Finger, Lieutenant and several others have been discharged. When they get back you and Grandpap Mambert go and see Captain Wardle in Hudson and talk with him. He promised me to get me out when he gets back. Our Coronel is the meanest man ever God let live. Young Cameron[3] is an object of pity, he has been sick for some time with the measles. Steve Wheeler[4] and Houghtail[5] and quite a number are dead, and plenty of others will die. If a man gets sick that is the end of him, and the doctors would not discharge a man till he is dead. I want no one to read my letters public until you and John first read them in secret for, I don't want everybody to know my business. Let John read them first to you and you be the judge. Write a letter every week the mail leaves New York 10/17/27 of each month.

John W. Mambert [6]

MILITARY PAY

John described Colonel Molineux as "the meanest man ever" likely because John's opinions were not appreciated or desired by the commander. Upon enlistment, John held a public position of authority making it somewhat difficult for him to adjust to the "taking orders" life of a foot soldier, and likely the reason he, and the colonel, did not have a harmonious relationship. John and the colonel were likely having intense conversations over his disputed military pay.

The military pay scale did not indicate principal musicians were paid $41 per month, so either Mr. Post exaggerated the rate while he was enticing John to enlist, or John misunderstood him. This was unlikely due to what was stated in the letter. Whatever the reason, John was an established businessperson and

---

3 Robert V. Cameron, Sargent 159th Co. C–Discharged for disability on August 23, 1863. Age 19, a farm laborer who lived and worked for parents in East Taghkanic, NY.

4 Stephan Wheeler, Private 159th Co. C–Died in Baton Rouge La. hospital February 1863. Age 21, left behind his parents in Gallatin, NY.

5 William Houghtailing, Private 159th Co. C–Died of disease in Baton Rouge La. in January 1863. Age 20, left behind his parents in Taghkanic, NY.

6 *Letter #6, January 29, 1863, Baton Rouge, Louisiana.* John W. Mambert letters from the Civil War.

a Justice of the Peace in his community. His monthly income was likely over $40 per month, based on the average pay for his employment during the 1860s. The fact that John took leave from his business and family for less than half his normal income was unlikely, and it appears he would remain relentless in his efforts to get what he believed was due to him.

Military pay was "in theory" received by the soldiers every two months. In practice, soldiers were not so lucky, as the paymaster unsuccessfully tried keeping up with troops moving swiftly over long distances. When the paymaster arrived with current and back pay it was a cheerful day in camp.

The following table compares income per month of each rank. [7]

ARMY PAY PER MONTH Artillery and Infantry	
POSITION	RATE UNION
Private	$12
Corporal	$14
Sergeant	$17
Principal Musician	$21
Musician	$12
Wagoneer	$14
First Sergeant	$20
Quartermaster Sergeant	$21
Sergeant Major	$21
Second Lieutenant	$45
First Lieutenant	$50
Captain	$60
Major	$70
Lieutenant Colonel	$80
Colonel	$95

---

7 United States hotel guide and railway companion for 1867: being a directory to the best hotel in nearly all the principal cities and towns throughout the United States. With supplement containing useful and interesting information, p. 28.

SICKNESS IN THE RANKS

John stated young Cameron had measles—one of the many diseases flourishing in camp. Most enlisted soldiers were Caucasian males who grew up in relative isolation and didn't experience common childhood diseases. Now, many were falling prey to them. General Robert E. Lee, experiencing the hindrance of sickness in the ranks, wrote to his wife in August 1861: "The soldiers everywhere are sick. The measles are prevalent throughout the whole army...." By September 1, battle readiness is still hobbled by infected troops. Those on the sick-list would form an army."

When soldiers got sick, they did not take care of themselves, causing many to die from complications. This was due to ignorance of the causes and treatments of these diseases by both patients and physicians. Regiments paid dearly with high numbers of sick men unable to perform their duties.

Throughout John's letters, he mentions men who are sick, dying, or dead from one of the illnesses plaguing the regiment; This letter is no different. John states the regiment is sickly, death is inevitable. This time, he gives his family back home a word picture of Baton Rouge and its surroundings, describing the devastation the war had caused on the city.

Letter #7 - February 6, 1863

I want you first to read my letters in secret or let John read them to you for I don't want everyone to know my business.

Baton Rouge
159th Regiment N.Y.S.VOLS.

To Mrs. John W. Mambert and Family,

I am quite well since I wrote my last letter. I am gaining weight quite fast; my appetite is good I could eat the scales of a crocodile. I am hungry all the time. I have been taking fever agone medicine although I have no fever, I think now that I shall get along quite good. I received your letter and paper and almanac dated 11th of January and I can tell you that I was glad to hear from home, and to hear that you were all well and enjoying yourselves. I don't want you to feel uneasy about me, your question if we had been in a hard fight, our division was not in the fight. The fight was at Vicksburg up the river but between us and Vicksburg is a place called Port Hudson and that place we expect to take, then move on to Vicksburg and attack them from both sides. Port Hudson is a strong fortified place and no doubt we will have a dusty time there. We are expected to move any day and it is only 14 miles from here. We are near Baton Rouge about half a mile from the city on the east side of the city our pickets[8] are two miles out. There are a great many people who returned to Baton Rouge which had left when we first came here. This country is nice, and I think that when this war ends carpenter work will be from two to three dollars per day here. The city is nearly half burnt down. The houses are pierced with cannon ball holes. The brick buildings you can see where a large cannon ball went one side in and the other out and go around and you can see the street where we went in another and out on the other side. Go into the woods where the battle was fought, and you will see all the trees cut down by cannon balls and musket balls. Trees as large as a barrel, a sharp cannon ball shot in one side and out the other. This country is all laid to waste. The army burns all fences. Clean and nice large houses which were deserted torn them down and burned them up for cooking. Yesterday there was a large house about 3/4 miles from here the 124th said it was all in flames. It is so frequent an occurrence that I do not stop to notice it. I think that slavery has been a great curse to this country. I feel as if the constitution could be preserved as it was it would be better. The regiment is weakly and sick we buried a man yesterday from Smoky Hollow by the name of Alexander Kells[9]. Pierce Allen is in the hospital with the dropsies. I expect to be paid off in a few days, if that is so then I will send money home with the American Express to the express office in Hudson and you can get it then. Nathan T. Post has resigned, and I sent a letter with him to you, but I don't want him to do anything for me unless

---

8 Picket: Soldiers posted on guard ahead of a main force. Pickets included about 40 or 50 men each. Several pickets would form a rough line in front of the main army's camp. In case of enemy attack, the pickets usually would have time to warn the rest of the force.

9 Alexander Kells, Private 159th Co. E–Died of disease on February 2, 1863 in Baton Rouge La. Age 28 from Smokey Hollow, Claverack, NY. Married.

he does it for a stated sum. If we are nine-month men then I don't care anything about it and I don't want Post to bother himself about me. Capt. Edward Wardle has resigned, he will be home in a short time, and he said that I was discharged. I want you to go to see him when he comes home, and he can tell you all about the treatment we all receive. I also wrote a letter to John M. Welch and wrapped it in the other letter, you need not give it to him, but you can inform him of the contents. I want you to write to let me know, how you know we are only nine-month men; I hope that is so. I also gave Andrew Brush who sent to his wife a small note of $3 which I bought his boots for. I was to pay for it here, but he is out of money, and he thought if you could pay Mag Brush she might send it here, but you need not pay it unless you can conveniently; if I get paid then I will pay him here. They say they are paying the other regiments now. We have as near as I can learn about 35,000 troops from here to New Orleans, this is by a mere guess. The Rebels are quite peaceable, once in a while they shoot at us, but we don't care, we always shoot back again. I want you to write back every week on Friday night and send me all the news you can and be careful how you do your business and carry out my plans as near as possible. Let Wes Bortle[10] pay you as much on the principle as he can and pay that on Mags note and the interest, see that the Rans note is paid and see that A. Decker[11] pays that $20 note and interest and pay that to Chip Potts[12].

Yours,
John W. Mambert

Direct your letters as before.[13]

---

10 Sylvester "Wes" Bortle–Age 36 farmer in East Taghkanic NY. Married with one child.
11 Peter A. Decker–Age 34 farmer in Taghkanic NY. Married with two children.
12 John "Chip" Potts–Age 60 farmer in Livingston NY. Married with four children ranging in age 10 to 30 years old.
13 *Letter #7, February 6, 1863, read my letters in secret.* John W. Mambert letters from the Civil War.

UNMASKING BOGUS DISEASES

To develop a strong army required men fit for military duty mentally, morally and physically. Recruiting officers were responsible for confirming that men who voluntarily enlisted qualified, so that new recruits would not become a burden on the military.[14]

Early in the war, a thorough examination of recruits was not always a priority, resulting in many soldiers unfit for duty, becoming liabilities to their units. So, to verify the overall health of the men, recruits were given an extensive examination. Sober and clean, they were required to strip to the skin in front of the enrollment board. John experienced this at Camp Columbia when he enlisted.[15]

In the beginning of the war, men were filled with enthusiasm and eagerness to support the cause. As the war progressed and men were less eager to be soldiers, surgeons were faced with the task of unmasking bogus diseases and conditions. Although any sickness among soldiers was frightening, some men considered feigning illness an opportunity to be discharged. Surgeons had to be watchful of malingerers in their care. Constant surveillance was the best way to uncover a patient in the hospital faking an illness. Doctor Robert Bartholow wrote <u>A Manual of Instruction for Enlisting and Discharging Soldiers.</u> In it he stated:

"to promote the continued efficiency of the army these principles must be constantly applied rendering it necessary on the one hand to use the utmost skill and ingenuity in detecting feigned factitious aggravated and exaggerated diseases and on the other to discharge those disqualified by infirmities which escaped the observation of the examining surgeon."[16]

Doctor Bartholow wrote of the many tricks malingerers used to avoid work or get discharged from duty. Here are a few of the more popular feigned illnesses.

Deafness as a disability was almost always faked. It was so popular, surgeons had to develop a few tricks to expose the sham. Talking very loudly on a topic of interest to the patient, then unexpectedly changing to almost a whisper caught many a dodger in his deception. Other telling indicators were the patient's reaction to the sound of dropping coins on the floor behind

---

14 A Manual of Instruction for Enlisting and discharging Soldiers with Special Reference to the Medical Examination of Recruits and the Detection of Disqualifying and Feigned Diseases, by Roberts Bartholow A.M., M.D. pp. 9-10

15 Ibid., pp. 167-168.

16 Ibid., pp. 9-10

him, or a loud noise waking him from sleep.[17]

Chronic rheumatism was frequently faked with 8,283 discharges in the first two years of the war—791 cases were discovered as false. Convalescent hospitals were filling up with malingerers looking to spend time in bed while recuperating from their "suffering" having no evidence of the disease. Diagnosing these malingerers was done by secretly observing the faker. For instance, a soldier who walked with great difficulty and needed a cane miraculously ran to catch his boat, not limping at all, no longer needing his cane. Something needed to change, so a General Order was issued by the War Department in 1862 prohibiting discharges for rheumatism.

The most shameless of the malingerers were those masquerading as imbeciles. A sudden loss of memory and the slacker is incapable of remembering his military duties, becoming listless and immobile. By observation at unexpected times, the imposters revealed evidence of intelligence and coherence rather quickly.[18]

Some feigned diseases were grotesque. One soldier was "feigned sick with piles" enjoying two months of hospital rest, until his deception was discovered: He placed the heart of a turkey into his rectum to resemble hemorrhoids.

Surgeons were constantly challenged by men wishing to avoid duties or trying to get discharged from service. Medical officers enlisted nurses and trusted patients to spy out supposed slackers. It was their duty to report to the surgeon any glimmers of unusual healthy activity exhibited by the patient contrary to his illness. Restless, sick, wounded patients were also willing to expose the slacker, often jealous of his success in evading duty.[19] Once a malingerer was exposed, the surgeon informed him that his imposture had been discovered, depriving the malingerer of hope and certainly contributing to his cure.[20]

'INFUSIA DIABOLUS'

Colonel Molineux wrote in his diary about a surgeon from the Connecticut regiment (believed to be Benjamin Comings of New Britain, Connecticut) who did not allow himself to be played by these "laxy bummers." An unsympathetic man, the doctor had a reputation as an advocate for abstinence

---

17 Ibid., pp. 107-109.
18 Ibid., pp. 116-126.
19 Ibid., pp. 101-102.
20 Ibid., p. 100.

from alcohol and studied how it affected the brain.²¹ The doctor had, as part of his medical reserves, an elixir he named, "Infusia Diabolus." This remarkable medicine soon became popular with the medical professionals of his brigade. Molineux attended the Connecticut regiment's sick call and was amazed with the small number of convalescents. He was curious about the remarkable state of good health, contrasting as it did with the 159th NY.

As individuals were summoned during sick call, some of the cases were clever. The surgeon decided whether the soldier required medical attention and relief from duty. One soldier in particular, Patrick Donovan, was a regular during sick calls. His history of avoiding work or battles started at Bull Run: He ran at the first shot fired; his company didn't see him for two days. When he found his way back to the regiment, he was full of brilliant stories of close escapes and great deeds at duty's call. His comrades did not believe the tales and laughed at his cowardice. One day, the sergeant in charge told the doctor that Pat was up to his old tricks—playing sick, avoiding work, but regular at mealtimes.

The doctor asked, "Well, Pat, is it the old trouble?"

"No doctor, dear me, the indigestion is all right, but it's the rheumatism doubling me up," Pat responded. "I cannot stand, and it's worse when I bend over the handle of a shovel. All I want, doctor, is a little lotion and to lie down on my back for a day or two."

The doctor told the hospital steward, "Give Donovan a dose of Infusia."

"For the love of saints, doctor, don't make me take that broth of hell! Never mind me rheumatism serg., I'll go with the boys and do me best," said Pat, walking away, miraculously cured.

The doctor gave Molineux the recipe, which included asafetida, ipecac, and castor oil, offering him a dose. Molineaux refused and chose to first try it out on someone under his command. The dose to be given was enough to make a patient want to die. Once a man received the medicine, he never applied for a second dose of the "heavenly inspired elixir." It was immediately adopted into the volunteer service. Rumor has it, a load of whiskey once passed through a cavalry camp without losing a drop because it was labeled: "Infusia Diabolus."²²

---

21 Sprague, Homer B. (Homer Baxter), b.1829., *History of the 13th Infantry Regiment of Connecticut Volunteers, during the Great Rebellion,* (Hartford Conn. 1867), p. 15.

22 James I Robertson Jr Civil War Sesquicentennial Legacy Collection at The Library of Virginia, *Edward L. Molineux collection, Infusia Diabolus, War stories.*

NINE-MONTH MEN

The Militia Act of 1795 allowed the president to call individual states for troops, but restricted federal service to only three months. In June 1862, New York Governor Edwin Morgan along with other governors of the Union urged President Lincoln to call upon them for additional troops to speedily crush the rebellion. Following their advice, (knowing full well the Confederacy would not be defeated in ninety days) the president approved the Militia Act on July 17, 1862. The bill allowed the federal government to require a draft for states failing to recruit its militia quota. On August 4, 1862, Lincoln called for a draft of 300,000 of these men for nine months of service.

New York's responsibility for new soldiers was 59,705 men. Governor Morgan made his plea for volunteers stating:

"he believed the insurrection is in its death throes; that a mighty blow will end its monstrous existence." He went on to say: "A languishing war entails vast losses of life, of property, the ruin of business pursuits, and invites the interference of foreign powers."

New York State deemed it necessary to resort to a draft, but delays ensued. Eventually, the draft was suspended, resulting in impressive numbers of enlistments. New York obtained more men completing their quota, a significant number enlisted as three-year volunteers (three-year men counted for four nine-month men). So, New York was viewed as sending soldiers of a higher value off to support the Union cause. For this reason, New York provided only one regiment of volunteers for nine months, the 177th Regiment of Volunteer Infantry. They served in the 19th Army Corp and were engaged at the Port Hudson siege. Before being mustered out of service, they lost 161 men. [23] Based on this information and the fact John enlisted into a three-year regiment, his hopes of being released as a nine-month man were soon dashed. However, the combined Northern states did raise a good number of nine-month regiments, sending twenty-two with General Banks on his expedition to Louisiana.

During Banks's failed June 14 multi-pronged advance on Port Hudson, the soldiers assigned to capture the citadel, spike its great guns and heave them into the river, were a select group detailed as stormers. The nine-month

---

23 New York State Military Museum and Veterans Research Center, *Nine Month Men, Military Affairs In New York 1861—1865.*

troops led the advance on the works and left many corpses lying on the field. Their bodies remained on the field for more than forty days until the garrison's surrender.[24] So, it was no surprise when General Banks appealed for 1,000 volunteers for "one last storming operation" on Port Hudson. Three-year regiments provided the needed men, but from the nine-month regiments, only four volunteered.

THE QUARTERMASTER

The men of the 159th NY settled-in, naming their new tent city, "Camp Grover." The brigade was busy practicing drills which proved to be tiresome work. Lacking ambition and energy, their skills improved at a slow pace. While performing drills, the men appeared to go through the movements as though a weight was dragging behind them. Exhausted, unable to get shoes or stockings, having no extra shirts or drawers, and two weeks short on rations, the men were miserable. For three days, they had neither bread nor flour, which almost caused a mutiny in camp.[25] If the men had not been malnourished, officers may have gotten better results from the men. The paymaster managed to pay some of the regiment, hoping the men would take more interest in what they were ordered to do.

One month after John mentioned the problem of "soldier abuse," it continued on. The men were sick, hungry, and uncomfortable, not the conditions expected when they enlisted only four months earlier. Colonel Molineux was an empathetic man whose frustration matched that of his soldiers. He desired the situation to be corrected. His conversations with Quartermaster Mark Wilber were equally frustrating, since Wilber had the power to squelch the outcry happening within the colonel's command.

The quartermaster's position held many responsibilities; the most important was supplying soldiers with essential needs. Although the supply issue causing the regiment to suffer was the fault of the government, it was Wilber's responsibility to resolve it. His challenge: distance. With every mile the military marched, it became increasingly difficult to find space in whatever transportation mode

---

24 Bacon, Edward [1830-1901], *Among the Cotton Thieves*. Detroit 1867. p. 152.
25 W.F. Tiemann Collection, *Letter Feb 20, 1863, Camp Grover, Baton Rouge*. The N.Y.S. Military Museum and Veteran Research Center.

was needed to ship the items.²⁶ Utilizing the railroad, rugged topography forced box cars to lighten their loads. This placed a high premium on available space. Essentials like food could be left on the dock in favor of more whiskey. Whatever the reason, basic necessities were not getting to the 159th NY, and the colonel was slowly losing control of his command. The men were at their breaking point.

Wilber was finally able to successfully resolve the supply chain issue (but not before creating headaches for Molineux, and hunger pains for the men.) In fact, as the regiment was about to head out on its next excursion, Wilber was given praise for resolving more than just the regiment's concerns. A correspondent from the New York Tribune paid the following tribute to his friend, Mark D. Wilber, "The Quartermaster:"

"It will be seen that he found no difficulty in solving the great transportation problem, which has puzzled the brains of so many of our military men."

"The Quartermaster, Lient. M. D. Wilber deserves heaps of compliments for his alacrity in providing forage for the teamsters and teams too, for the forage. Of course he does not conciliate Secesh much by confiscating horses and mules but he has helped along the expedition amazingly by his facility for providing transportation."²⁷

*Quartermaster's Store House 159th NYSV, titled, "U.S. Hotel D' Grub."*²⁸

---

26 Ibid., p. 28.
27 *M.D. Wilber, Quartermaster 159th NYSV. Daily Eagle on April 10, 1863*, p. 3. Adriance Memorial Library, Poughkeepsie, NY.
28 E. L. Molineux Photo Collection, *Quartermaster's Store House*, NYS Military Museum & Veterans Research Ctr.

Regimental quartermasters received appointments from state governors. Mark D. Wilber, from Poughkeepsie, was assigned to the 159th NYS Volunteers. Wilber remained in position until December 5, 1863, when he resigned and was replaced by John H. Charlotte,[29] a native of Hudson, New York.

The president appoints quartermasters who assist brigades, divisions, corps, or army staff. The job of quartermaster was complex. Men in those positions were a select group of officers providing fundamental supplies and support for the army. They were responsible for procuring and managing military transportation during the war, organizing, and managing dozens of depots across the country. These depots stored and distributed the materials, including animals needed by the military forces in the region.

WASHINGTON'S BIRTHDAY

The birth of the "Father of this Country" was celebrated February 22, 1863, and the men were given two days off. Sunday morning, the usual regimental inspection occurred. At noon, there was a salute of thirty-two guns fired by two land batteries, and a gun boat. In the afternoon, brigade regiments marched to headquarters where an address was delivered by Chaplain Kipp.[30] The men were pleased with the whole affair, especially after learning they would have the next day off as well.

The following day, the men planned to have a grand time as the 25th Connecticut challenged the 159th NY to a game of baseball. All preparations were made for the match and two teams were selected. Officers of the 25th came to check if the 159th were ready, but shortly afterward they withdrew their challenge. The men of the 159th were bound to have fun with or without the 25th and decided to make two teams among themselves. Major Draper took one side and Colonel Molineux the other. After an exciting four-hour match, Molineux's team came off victorious. The rest of the day passed quietly and peacefully.[31]

---

29 John H. Charlotte, quartermaster, 159th Regiment, Age 19, bookkeeper in Hudson NY. Lived with his parents and three siblings.

30 Isaac L. Kipp, Chaplain/Captain 159th regiment, Age 27, Pastor in Fishkill, NY. Enlisted in NY City just out of Theology school. Married, after the war and became Pastor at the Reformed Church in Schodack, NY.

31 W.F. Tiemann Collection, *Letter February 23, 1863, Baton Rouge*. The N.Y.S. Military Museum and Veteran Research Center.

PRINCIPAL MUSICIAN AND A TENT OF HIS OWN

After four months serving the Union, John received his promised position as principal musician. In March, he writes home, feeling good, hopeful that the war will end within the year. At the time of his writing, John comments, "Sickness is the most of our danger."

LETTER #8 - MARCH 2, 1863

Headquarters - 3rd Brigade.

Baton Rouge

To Mrs. Mambert

I feel very good today, better than I have since I have been in Louisiana. I feel first rate. There has been much discussion about my position, the officers of the Regiment were very much dissatisfied with the conduct of certain officers, that is with the Colonel and Drum Major. They finally all met and consulted on the matter and all the officers were in my favor except the Colonel, he found out that if he did not give me my position, he could not hold me, so he said he would see the Pay Master. The Pay Master told the Colonel he was entitled to two Principles musicians, so he concluded to give me my position. He would have to discharge me if he did not for I had council with the Quarter Master Wilber[32] and he told them so. I now have my position and get $21 per month; I don't know if I will get it from the 20th of September or from the 1st of January. So, you see I will be all right, the Pay Master came and paid four Companies, A, B, C, and D, then he had no more money and went off, that is the last we have seen of him. I don't expect he will pay us until at least the middle of March and when he pays me, I will send it home to Churchtown, to Alberts. You will be careful who you send the letters, I was taken from my Company and put on the field with the noncommissioned staff. Now I don't have to carry a knapp sack, only a blanket. So, you see my wages will be $252 per year and if you include saving and take good care together with interest you will make over $400 per year. So,

---

32 Mark D. Wilber, Quartermaster 159th Regiment, age 33 an attorney in Poughkeepsie NY.

you see we ought to make something over every year. Pretty nice
NY work which I have to do about an hour a day. I sleep very good
and comfortably; I see you every once in a while, in a dream and
we have a good talk together and that makes me feel very good.
I go to bed about half past 8 and get up about half past 5 o clock
am. The weather here is very nice and warm. This morning, the
weather is the nicest I ever saw. The peach trees are all in bloom
and it is just like the last of May or June. We go swimming once a
week, it is full of snakes here and large ones too. I sleep with Eli[33]
and Tommy Miller[34], I have a tent of my own, William Moon[35] was
in my tent and he got ugly and saucy and so I told him to take up
his bed and walk, he did not want to do that. I went to his Captain
and told him he must find a place for Moon; he told me to send
him on and so I told him to be off and so he went. The captain said
he had just the place for such saucy[36] fellows and he put him in
with just such a group of saucy fellows. I think we will be home,
if we live, in less than a year. The yellow fever is in New Orleans,
our regiment is very sickly, there died one a few days ago from
Hillsdale by the name of Shearwood[37]. They are dying off. Jones
Coons[38] is miserable but I think he is a little better. Bob Gardner[39] is
getting better. Ezra Stickles[40], Ephran Stickles[41] and several others

---

33 Eli Miller, Musician 159th regiment Co. G, age 45, discharged with disability in Baton Rouge, 1863. Germantown, NY.

34 Thomas B. Miller, Musician 159th Regiment Co. A, age 15 ½, promoted to 2nd Principal Musician January 28, 1864. Farm laborer in Livingston, NY. Lived with his parents and four siblings.

35 William Moon, Private 159th regiment Co. I, age 45, died of disease August 21, 1864. Enlisted in Milan, Dutchess County, NY.

36 Saucy: Impertinent boldness or sassiness; disobeying the rules of decorum; treating superiors with contempt.

37 John H. Sherwood, Private 159th Regiment Co. E, age 19, died of disease in Baton Rouge, La. February 25, 1863. Enlisted in Hillsdale, NY.

38 Jonas Coons, Private 128th regiment Co. D, age 26, absent without leave at muster out. Farm laborer in Livingston NY. Lived with his parents and two siblings.

39 Robert R. Gardner, Sergeant 159th regiment Co. A, age 23, lived in Hudson, NY with the Ed Roreabeck family.

40 Ezra Stickles, Private 159th regiment Co. C, age 30, deserted in Baton Rouge March 1863. Farm laborer in Taghkanic, NY. Married with a seven-year-old son.

41 Ephraim Stickles, Private 159th Co. C, age 49, deserted in Baton Rouge March 1863. Farmer in Gallatin NY. Married with four children, ages range from 2 to 11.

have run away. Tommy Miller is as fat as a pig and Eli I think will be discharged, he has not played fife since we first came. Old Andrew Brush will be discharged too. We think this war will soon end. Tell Steve Miller[42] I send my best respects to him and Samuel Miller[43]. Tell Steve he ought to be here when the war ends, he can get $3 per day and work all the year round for it is summer here all the time. Tell Harm he must not allow himself to be drafted for he won't stand it here. Tell Grandpap Shelden I send my love to him and tell Granny Shelden[44] I send my love to her. Tell her that she must pray for me that the blessed lord will guide and protect me and bring me back home again to greet her and my Blessed family, my faith in Christ is strong and I feel his protection. Tell Albert Shelden[45] and wife I send my love to them tell them to remember me in prayers. My prayers are with you all that we may meet again upon earth. Sickness is the most of our danger. Tell Ves Bortle I send my love to him, and I thank him for his kind letter, and I hope he will write again. Write to Sarah and let her know that I am well, and to Mary and to Rams and Steve. You don't know what peaceful hours you all enjoy if you have seen what I have seen, you wouldn't get a better experience in the best school. Tell Mart Best[46] I wish he was here with me; our drum corps is nothing nor never will be. Last night and the night before are the first I slept in a tent in a week, I slept on the ground on my rubber blanket and overcoat and another coat over me, but I have not caught a bit of cold. The mosquitoes and wood ticks are the worst. I don't know how soon we will be off again; they are packing up again and I must finish

---

42 Steve Miller, age 34, farm laborer in Taghkanic, NY. Married with six children ranging in ages from 1 to10.

43 Samuel Miller, Private 159th regiment Co. I, age 34, deserted Park Barracks NY on November 2, 1862. A miller who lived in Livingston, NY. Married with five children ranging in age from 1 to 14.

44 Granny Sheldon, Mother to Job Sheldon, his daughter Mariah married J.W.M. no history on Job's mother.

45 Albert Sheldon, age 37, auctioneer with business located in the hamlet of Churchtown N.Y. where he also resided.

46 Martin Best, Private 159th regiment Co. G, age 42, discharged at Staten Island Nov 18, 1862 by Writ Habeas Corpus. Farmer in Taghkanic NY. Married with four children ranging in age from 3 to 12.

and pack too. Continue to write and pray for my safety. I have great faith in prayer. Our regiment was praised after all for being the bravest regiment in Baton Rouge. Maria you must write, perhaps it is well that I have come here, I might have stayed home and been accidentally killed while building. God moves in mysterious ways, his wonders to perform, and perhaps it is all for the best only he can tempt and hope for the better, trust in God. It rains here now, it is showery. Write, they think we will go to Texas soon. Our regiment is mildly healthy.

To my dear beloved wife and family. John, take good care of things and the little boys.

John W. Mambert

Major Fifer-159 R.[47]

Colonel Molineux believed he had proper authority to fill positions as he wished, and under normal circumstances, he was correct. State of New York's general order No. 113, paragraph XXXI stated, "...field and company officers of regiments, now or hereafter in service without the State, will recommend, through the Regimental Commander, candidates for vacancies." However, this regiment was unique because it was a consolidated regiment. Under the same general order, paragraph XXVI it also stated, "...if delay shall occur, in the filling up of a company, or regiment, to the minimum standard, the Governor may consolidate the incomplete organizations, assigning officers according to rank, or, where rank is of the same date, it will be determined by lot. Officers rendered redundant by consolidation, and who cannot be assigned to any vacancy, will be immediately mustered out of the service." The paymaster understood this to mean if they did not assign John the second principal musician position, they would need to release him from service.[48] Presumably, Colonel Molineux decided to give John the position instead of losing another soldier in the ranks. Besides, John enlisted believing that would be his rank—but at a higher than stated pay.

---

47 *Letter #8, March 2, 1863, Headquarters, 3rd Brigade, Baton Rouge, To Mrs. Mambert.* John W. Mambert letters from the Civil War.

48 *G.O., No. 113, General Headquarters State of N.Y. Adjutant General Office, Albany Nov. 26, 1861.* NYS Military Museum and Veteran Research Center.

THE MOST PECULIAR YELLOW FEVER INCIDENT

Fear of contracting a serious disease was present in the minds of soldiers, and now there was a breakout of yellow fever. Several epidemics of yellow fever during the Civil War caused substantial numbers of soldiers to suffer.

In his letter, John stated the disease was in New Orleans. Ironically the most peculiar yellow fever incident during the war was an epidemic that did not happen—in New Orleans.

New Orleans had commercial ties to Latin America and the Caribbean where yellow fever was endemic. Because of this, New Orleans was known as the yellow fever center of the South.

Shortly after General Butler took possession of New Orleans, he learned the rebels were hoping the yellow fever would decimate the Northern troops. Like attractive, tasty morsels to infected mosquitos, soldiers with fresh lips and clear complexions were usually the first victims of the fever. Butler also noted church leaders praying: "that the pestilence might come as a divine interposition on behalf of the brethren." And there was word (though at first not believed) that General Lee relied on the depletion of Northern troops by yellow fever for the defense of Louisiana and the recapture of New Orleans. This was later confirmed by correspondence from M. Lovell, the commanding Confederate major general to Governor Thomas Overton Moore. Lovell wrote of his plans to:

"organize a central force of 5,000 men which in connection with corps of Partisan Rangers might succeed in confining the enemy to New Orleans and thus subject him to the diseases."

The letter was submitted to General Lee and agreed to by Governor Moore; there is nothing in the war correspondence showing Lee objected to it.

Instances were reported of efforts to dampen the morale of homesick officers and soldiers with fearful stories detailing the scourge of yellow fever. In one account, a soldier from Maine, longing for the pure air and bright landscape of home, was pacing his weary beat. Naturally, he listened to the conversations going on around him. Two newsboys stood near him. They had found a listening ear in this naive soldier to spread their worrisome gossip:

"Jack, have you heard the news?"

"No Tom, what is it?"

"Got the yellow fever prime down in Frenchtown, two Yanks dead already. It will sweep them all off."

Panic seized many officers. Morale was low. Every day, officers were requesting a leave of absence to go home under every excuse, pretense and reason for furlough. General Butler knew that no surgeon in his army had ever seen a case of yellow fever or knew how to win this battle. In August 1853, yellow fever killed 30,000 people in Louisiana. One in every four died, leaving the dead laying in heaps because of the inability of the living to inter them.

So, Butler asked an old New Orleans physician if there was any way to keep the fever out of the city. He said there was none. Once the fever got into the city (occupied by armed forces for many months) he saw no reason why the disease would not spread with irresistible fury since so many unacclimated persons would be confined there, most likely dealing with unsanitary conditions. But a quarantine could help.

The doctor informed Butler of a book on the peculiarities of the disease, The Description of the Rise, Progress and Decline of the Disease in 1853, by Professor Everett. In his book, Everett recounted that the fever seemed to originate around the French market. Butler searched for and found a map showing hot spots in the city where incidents of yellow fever were high. Examining the French Market and a number of other localities, he detected why it raged in those spots: They were astonishingly filthy with rotting matter. In the French Market, the stall women would allow all the unwanted remains from gutting birds, animals and fish to fall to the floor...leaving it there. The fact that the disease flourished so much in the vicinity of decaying and putrid animal matter led to the conclusion that this prolific cause of the fever must have something to do with vomit.

Butler turned his attention to the filth and concluded it could be disposed of. New Orleans had no sewers, but it did have drains above ground that were easily accessible. Open drains sent sewage from homes and the city flowing to the lake. The commercial water works pumped water from the Mississippi River into the lake through these open drains. Realizing these ditches and drains had not been cleaned for many years, Butler spearheaded a massive cleanup campaign in the city. Two thousand men worked for thirty days to clean the drains and waste from the public streets, squares, and vacant land. He also initiated a system of refuse collection.

Butler established a quarantine station seventy miles below New

Orleans. Vessels were required to stop below Fort St. Philip, the quarantine establishment, to be inspected by the health officer who would report the condition of the vessel, passengers, crew and cargo directly to Butler. The officer at Fort St. Philip was instructed to allow no vessel to go up without his personal order, which was not given unless the quarantine physician reported a clean bill of health. If any vessel attempted to evade quarantine regulations and pass without being examined, the vessel was to be stopped. If necessary, cannon shots would "send the message" to execute the health order. Any vessel found with sickness on board of any malarial kind or with ship fever was required to stay down forty days and not come up again until reinspected. Also, no vessel coming from an infected port (a port where yellow fever was prevailing) would be allowed to come up under forty days.

The quarantine officer was to make true reports of the condition of the vessels after a full and intelligent examination. And as the health and lives of so many depended on truthful reports, he was notified that any neglect of his duty would incur the heaviest punishment known.

New Orleans was never so clean; Butler's attempt to stop the spread of the disease worked surprisingly well. Due to his rigid quarantine restrictions, he managed to keep the infected mosquitoes from reaching the city. The feared epidemic did not materialize, but one year later, after Banks replaced Butler, the fever was back in the Crescent City.[49]

CONTROL OF THE RIVER

On March 7, 1863, orders were given for the regiment to begin preparations for a move. Prior to receiving the orders, the Confederate forces captured, and now possessed the *Queen of the West*, a sidewheel steamer, originally attached to the Union ram fleet in the upper Mississippi River. In addition, the Indianola, a Union ironclad gunboat, was attacked off Palmyra Island by the *Queen of the West* and sunk, completing removal of the Union navy. Thus, control of the long reach of the Mississippi River between Port Hudson and Vicksburg passed into the hands of the enemy.

When Union naval commander Admiral Farragut learned of the situation,

---

49 Butler, B. F. (Benjamin Franklin). (1892). *Autobiography and personal reminiscences of Major-General Benj. F. Butler: Butler's book: a review of his legal, political, and military career.* Boston [Mass.]: A.M. Thayer & Co. pp. 396-402.

he was furious. Perched high on the bluffs of Port Hudson, rebel guns stood between him and the Confederate flotilla. Farragut knew he needed to look into the barrels of Confederate cannons to pass these batteries. The admiral devised a plan to move part of his fleet above Port Hudson. After reviewing the plan with General Banks who promptly agreed with the Admiral's strategy, the commanders discussed the details. While Farragut was reviewing all needed arrangements and getting his fleet in order for the undertaking, Banks concentrated all his disposable force in Baton Rouge. On March 7, 1863, Banks left General T. W. Sherman to cover New Orleans, and General Weitzel to keep a strong hold on Lafourche Parish. This left him with a marching column composed of Augur's, Emory's, and Grover's divisions—15,000 men strong.[50]

It was necessary for the Confederate force, with their artillery positioned high on the bluff overlooking the Mississippi River at Port Hudson, to be silenced. The Union needed to secure the river and the 110-mile stretch of territory between Port Hudson and Vicksburg if they were going to succeed in dividing the Confederacy. Additionally, the Red River serving as the Confederacies supply line emptied into the Mississippi River halfway between the two enemy strongholds: losing this ability would devastate the enemy.

The Confederates knew command of the Mississippi River was in jeopardy, and additional defenses were needed at Port Hudson. So, they assigned a new commander for the garrison on December 28, 1862, Major General Franklin Gardner.[51] During the interval that elapsed since its first occupation a formidable series of earthworks had been built, commanding not only the river but all the inland approaches that were deemed practicable. With Gardner in command his first plan for land defense was mainly against an attack expected to come from the direction of Baton Rouge. Accordingly, four miles below Port Hudson a system of earthworks was constructed, the lines were over four miles long, running in a semicircular sweep from the river below Port Hudson to the impassable swamp above. Following this line after leaving the river on the south, the bluff is broken into irregular ridges and deep ravines with narrow plateaus and difficult gullies; beyond these for hundreds of yards their course lay through fields and over hilly ground to a ravine at the bottom where Sandy Creek runs,

---

50 Irwin (n1), ch. 11, par. 23.

51 Major General Franklin Gardner was born in New York City and graduated from West Point. He married the daughter of former Louisiana governor Alexander Mouton, who compelled him to join the CSA when the war began.

and the earthworks ended.

By the end of March, the garrison consisted of 1,366 officers, 14,921 men of all arms present for duty, making a total of 16,287. The main body was organized into five brigades. The fortifications on the riverfront were mounted with twenty-two heavy guns, ranging from 10-inch columbiads down to 24-pounder siege guns, manned by three battalions of heavy artillery. Thirteen light batteries with seventy-eight pieces were available for the defense of the surrounding lines.[52]

Lieutenant General Pemberton, commander of the Confederate fortress in Vicksburg began to feel the weight of Grant's pressure in May and called on General Gardner for reinforcements. Gardner responded, in fact the preparations and orders had been given for the evacuation of Port Hudson; but now the Confederate chiefs that were to seal the doom of Vicksburg began to rethink Port Hudson. Gardner, who was the last to move out of the garrison, had barely arrived at Clinton when he was met by an order from Pemberton to return to Port Hudson. Gardner, with only a few thousand men, was ordered to hold the fortress to the end.[53]

General Banks's plans were to advance on Port Hudson, but rumors and information from spies convinced him otherwise. In May Banks reported to Halleck,

"The works are defended' by a garrison much larger than generally represented. There appears to be no want of ammunition or provisions on the part of the enemy. The fortifications are very strong, and surrounded by a most intricate tract of country, diversified by ravines, woods, plains, and cliffs, which it is almost impossible to comprehend without careful and extended reconnaissance."[54]

Banks decided to delay his attack for the disparity between the relative strength of Banks's army and that of the garrison was too well known to justify the thought of an actual attack upon the works.[55]

The 159th NY, along with additional troops for the expedition were given their orders. Tents were struck and rations issued. Unfortunately, it

---

52 Irwin, Richard B. *History of the 19th Army Corps*. G. P. Putnam's Sons New York, (The Knickerbocker Press 1892), ch.8, par.13.

53 Ibid., ch.16, par.2.

54 OR. *Banks to Hollock*. Ser. I, Vol. XXVI, p.44

55 Ibid., ch.8, par.12.

would be six months before the troops were under their tents again. The troops gradually began concentrating in Baton Rouge, making it evident an important move was about to commence.[56]

John writes home, keeping the family informed of the regiment's health. Several men will be returning home due to disabilities. Sickness is taking its toll on the regiment's overall strength. There are many ill and John predicts some will not survive.

Letter #9 - March 11, 1863

Baton Rouge

I thought I would write to let you know the sad news that Pierce Allen[57] is dead. He is buried today and McCaly or Mc Kally from Punderson & Tiffneys Drug Store and also a young man by the name of Terry[58] was buried yesterday. We are under marching orders our tents are all down and moved and my trunk with my well and pen is gone north with my baggage, and I don't know if I will see it again or not. My ink, paper and pen are in it, so I write with a pencil just to let you know we slept all night on the ground. I don't know how much longer. We have three-day rations, Indian rubber blanket and one other blanket, an overcoat. I suppose our destination is Port Hudson, I suppose before this reaches you the battle will be fought, may it be for better or worse. Eli Miller is in the hospital; I don't think he will ever come home. I am getting well, I feel good. The boats are going up the river. I have ordered my trunk to come home by express. In case I don't see it again go to the express office or to Ed Roreabecks.[59] We have several regiments on the way and are building bridges which the Rebels have burnt

---

56 Tiemann, William F. *The 159th Regiment Infantry, New York State Volunteers*, In the War of the Rebellion 1862 -1865, (Brooklyn New York 1891), p. 23.

57 William Pierce Allen, Private 159th regiment Co. C, age 35, died March 10, 1863 in Baton Rouge LA Hospital. A woodworker in Taghkanic, NY. Left behind a wife and three children ranging in age from 3 to 10.

58 Austin H. Terry, Corporal 128th regiment Co. F, Infantry, age 23, died of disease on March 10, 1863. Fishkill, NY.

59 Edward A. Rorabach, age 35, owner of a boarding house on Diamond St. in Hudson, NY, run by him and his wife. (John sent letters to this address).

and last night there was one man killed in working at the bridge of the 25th Connecticut Volunteers. Our Brigade is Regiment 159th NY, 13 & 25th Conn. and 26th Maine. General Birge who commands our regiment is sickly. We expect to start every moment. Andrew Brush is coming home and several others. Last night it rained most all night, but I and Tommy fixed our boards and slept on one rubber blanket and one over, we slept dry with our overcoats for blankets.

The night before it rained and we slept under our tents, they were not down, I slept dry and good, we don't expect to sleep in tents again very soon. You must write and direct your letters as before in the manner as stated in my last letter. I will write as soon as the battle is fought if the blessed Lord spares my life on whom all my faith and dependency rests. I hope you will trust in him as I do, and he will do all things well. Pray to him that he will guide and protect me, that I may return to greet you again before death shall separate us. We have a large army, if I have to guess I should say upwards of 40,000 men. It is a very nice day here today, a little cooler now for some time back. We have had no pay yet as I stated in my other letter. I sent a map in the other letter; you can see where the road to Port Hudson is. Our men go to cut off the railroad as you see they may have taken it by this time. May God guide and protect you. To you all from your Husband, Father and friend.

John W. Mambert, Major Fifer

N.B. May says Vicksburg is taken and I think it is, you will hear.

Principal Drummer, George D. Dayton
Principal Fife, John W. Mambert
Drum Seargent, Harry Durham [60]

TROOP MOVEMENT

On the morning of March 10, 1863, troops were prepared and ready to move. Repairs delayed Admiral Farragut's fleet and the last vessels reached

---

60 *Letter #9, March 11, 1863, Baton Rouge.* John W. Mambert letters from the Civil War.

Baton Rouge two days later. On March 13, 1863, a separate force consisting of the 159th NY, three companies of the 26th Maine, the Louisiana (loyal) Cavalry, and one section of Nim's Massachusetts battery[61] commanded by Colonel Molineux deviated from the main body of the army as a provisional brigade, constituting the right wing. Orders were specific: Open and keep clear the Clinton Road, preventing a flank[62] movement by the enemy. To do this, Molineux rapidly pushed his brigade forward to prevent the destruction of bridges and rebuild those already destroyed. The brigade pushed on until reaching Redwood Bridge, then crossing a deep bayou east of Port Hudson. The bridge was a strategic point of great importance. The orders were to "hold the bridge and if that was impossible, destroy it, and fall back to a secure position in line with the main army."

ADVANCING INTO ENEMY TERRITORY

The 159th NY, along with the advancing troops from Baton Rouge, began marching at 4 p.m. As they filed past waiting regiments, cheer after cheer greeted them along the entire line expressing good luck for their welfare. Within a few minutes, they were on the road described as the most beautiful road they'd ever seen. It ran a straight line for thirty-three miles, flat as a pancake. Thick woods bordered each side, and an occasional plantation made the scene more beautiful.

John and his comrades were in enemy territory.[63] The Cavalry was in advance, followed by two companies of infantry to skirmish.[64] The remaining followed as reserves.[65] The companies deployed on both flanks, marching through impenetrable dense brush. At dusk, several shots were heard in the front. They were fired at a Negro attempting to signal the enemy's advancing troops

---

61 Battery: Usually, six cannons that worked as a group all similar in size, it took 100 men to work a battery in battle.

62 Flank: Used as a noun, a "flank" is the end (or side) of a military position, also called a "wing." An unprotected flank is "in the air," while a protected flank is a "refused flank." Used as a verb, "to flank" is to move around and gain the side of an enemy position, avoiding a frontal assault.

63 W.F. Tiemann Collection at The N.Y.S. Military Museum, *Letter Mar 23, 1863, Baton Rouge*.

64 Skirmishes: Their purpose is to harass the enemy, engaging them in only light or sporadic combat in order to delay their movement, disrupt their attack, or weaken their morale.

65 Reserve(s): Part(s) of the army which were withheld from fighting during a particular battle but ready and available to fight if necessary.

with a torch—his light, however, was hastily extinguished by a few precise shots.

After making bridge repairs and clearing the road, the advance column halted late in the evening at Cypress Bayou Bridge—which was burned. Troops bivouacked for the evening. At 2 p.m. the signal officers[66] opened communication from Springfield Landing, signaling Admiral Farragut's fleet was anchored near the head of Prophet Island, just south of Port Hudson's garrison. Before sunrise, orders to fall in were given, followed by the sound of a hurried march as soldiers crossed the bridge reconstructed during the night by the 26th Maine regiment. Almost two miles beyond, the brigade came to a rebel outpost. Troops began a continuous drive on them while Nim's artillery opened fire on their camp. The rebels ran, leaving weapons and unfinished breakfast behind. For security, Nim's artillery fired two more shells into the woods, then sent troops off at a quick pace. They barely moved before halting again, their scouts spotted a strange looking object in the road ahead. Indistinguishable, Nim's men fired a shot and sent soldiers forward to see what the object was. It turned out to be a huge iron boiler, mounted on stocks, resembling a huge cannon from a distance. The men destroyed the counterfeit. Continuing six miles they arrived at their destination: the Redwood Bridge.[67] The bridge was eighteen miles from Baton Rouge on a direct line east from Port Hudson. Coming to a halt, Molineux allowed the men to rest, eat, and most importantly make coffee.

Confederate forces burned the bridge, moved beyond it and made a stand against Union troops. In a line of battle, the Confederates opened fire upon the approaching Union detachment. Although the enemy force was greater, the rebels were forced to retreat in disorder after a short but sharp contest.[68] Colonel Molineux reported his men collected thirty horses and ten bales of cotton from the enemy, who shot and wounded some of his soldiers. He was instructed by the General to return the horses to the wives of the men who

---

66 The Signal Corps - A signal system using flags and torches was invented by a U.S. Army surgeon, Major Albert Myer, in the 1850's and adopted by the U.S. Army in 1860. The new signal technology permitted rapid communication across the battlefield and farther using flags or torches to talk to each other. The flagmen appeared to be merely waving a flag back and forth, hence the name "wig wag." Myer became the U.S. Army's first chief signal officer. National Park Service.

67 Ibid., *Letter March 23, 1863, Baton Rouge.*

68 From the collection of E. L. Molineux, *unknown newspaper copy.* NYS Military Museum & Veteran Research Center.

shot his soldiers. Molineux reluctantly returned the horses, knowing they would be in use by the enemy again.[69]

Because the bridge was destroyed, the detachment fell back six miles to Cypress Bayou Bridge, as instructed. Molineux received information: a force of 1,200 Confederate cavalry was somewhere between Clinton and Baton Rouge. Strong detachments became necessary to observe all approaches and hold the road and bridges in the rear. This was important to secure a safe withdrawal of his army when the demonstration was complete.

ARTILLERY NEAR THE RIVER

General Banks needed to concentrate his artillery near the river at Port Hudson in a way that would distract or divide the attention of the enemy's lower batteries.[70] Around 10 p.m., Farragut's fleet weighed anchor and began their move upriver. The flagship *Hartford*, a sloop-of-war steamer, took the lead, with the *Albatross*, a screw steamer, lashed to her port side. Next, the *Richmond*, another sloop-of-war steamer, then the *Genesee* and *Monongahela*, both steamers and the *Kineo*, an Unadilla-class gunboat designed for close inshore operations, and lastly the side-wheeler *Mississippi*.

One hour later, a rocket shot from the bluff. Instantly the Confederate batteries opened fire and were joined by long lines of sharpshooters. To avoid the shoal jetting out widely from the western bank, and escape the enemy's fire of musketry and artillery, the Union ships hugged close the eastern bluff—so close, the yards brushed the leaves from the overhanging trees, and the voices of men on shore could be distinctly heard by those on board.

Watchfires were lit by the Confederates to show the ship's range. This did more harm than good, since the smoke from the fires, and the guns ashore, rendered the Union fleet invisible—unfortunately, the ship's pilots were unable to clearly navigate.

The *Hartford*, meeting the swift eddy at the bend where the current creates a right angle, narrowly escaped being driven ashore. The *Richmond* was

---

69 From the collection of E. L. Molineux, *extracts from his diary, horse return*. NYS Military Museum & Veteran Research Center.

70 Irwin, Richard B. *History of the 19th Army Corps*. G. P. Putnam's Sons New York, (The Knickerbocker Press 1892), ch. 8, par. 4.

disabled by a shot through her engine room. The *Kineo* lost the use of her rudder, and the *Monongahela* ran aground. The *Kineo*, drifting clear, grounded but was soon afloat again. With her assistance, the *Monongahela* swung free after nearly a half hour of imminent peril. Then the *Kineo*, cast loose by her consort, drifted helplessly downstream while the *Monongahela* passed until a heated bearing brought her engines to a stop and she too drifted with the current. The last of the fleet, the *Mississippi*, was initially invisible in the smoke and therefore safe from the Confederate guns. And yet, unable to see friend, foe or landmark, she was carried by the current. As the smoke cleared, she lay alone and helpless under concentrated aim of Confederate batteries. She was abandoned and set on fire by her captain. Morning came and the *Hartford* and *Albatross* rode safely above Port Hudson, while the *Richmond, Monongahela, Genesee* and *Kineo*, battered and injured, lay at anchor near Prophet Island. The *Mississippi* perished in a blaze of glory.[71]

The passage of Farragut's Union Fleet past the fortifications of Port Hudson Louisiana, March 14, 1863 [72]

---

71 Irwin, Richard B. *History of the 19th Army Corps*. G. P. Putnam's Sons New York, (The Knickerbocker Press 1892), ch. 8, par. 10.

72 *Farragut's Union Fleet past Port Hudson,* Harper's Weekly Newspaper. Drawing by Hamilton, April 18, 1863

The Union troop movement was an attempt to attract the garrison's attention and enable Admiral Farragut's fleet to proceed upriver past the cannons at Port Hudson. This bold move by the Admiral took away the Confederates' control of the Mississippi River above Port Hudson and hampered the use of the Red River for transporting enemy supplies. The successful defense offered by General Banks's army took pressure off the navy. Now, the enemy was led to believe that Union movement around Port Hudson was with the intent to attack and reduce the garrison.[73]

After the engagement, the Union army was isolated in a bad location, an easy target for the enemy. Nothing would be gained by staying there.

On the morning of March 15, 1863, word came from the active and adventurous signal officers: "all was well with the remaining fleet." Banks took up the line of march heading south along the Mississippi River. Later that day, the troops went into bivouac in great discomfort on the soggy borders of Monte Sano Bayou just north of Baton Rouge. Rain began to fall in torrents, saturating the ground, leaving the men ankle deep in mud and water, seeking shelter on knolls and under trees. Finally, the rain stopped just after daybreak and the brigade was moved to a dryer location.

The new encampment was a more secure and comfortable location. There was no lack of food: chickens, ducks and turkeys, fresh beef and mutton were plentiful. The men were able to have a well-deserved meal, feasting on the locally foraged provision.

COLONEL MOLINEUX...CAPTURED?

On March 16, 1863, the camp was startled by a badly wounded Negro reporting Colonel Molineux had been captured. Before the search and rescue mission got underway, Colonel Molineux came riding in, greeted by cheers from his men.

Molineux explained what happened: While riding up the road with the Captain of the Cavalry followed by an orderly and Negro servant, they heard several shots behind them. Immediately, they hid in the woods, watching as ten of the enemy Cavalry rushed past at full gallop. The orderly was not

---

73 Duffy, Edward B., *History of the 159th Regiment, N.Y.S.V.* (New York 1890), p. 7.

located and presumed captured. The servant was shot three times.[74]

Colonel Molineux's provisional brigade rejoined the main body of the army on March 20, 1863, and proceeded to march to their old camp in Baton Rouge.[75]

DISGRUNTLED

Colonel Molineux was a studied and trained military man whose participation in the war was for one reason only—victory to the Union. He had no interest in collecting cotton or other commodities for profit and took issue with the distraction from what he knew was important. He expressed his frustrations in a letter home.

"In my letters you must have misunderstood me, I never will shrink from my duty. I am disappointed in two things: first, high officers use us for their own gain in cotton, this is to an extent that it is sickening. Secondly, my officers will not, as a rule, look after their men and do not care for their responsibility to God and their country for their "children" placed in their care. I have not a single officer who fully supports me."[76]

Molineux, obviously troubled over the Redwood Bridge incident, had a strong opinion about his commander's directives. He expressed his feelings to loved ones back home:

"we are about to embark on board transports, i.e., our Division (Grover's 4th) to go to Algiers or Donaldsonville from whence we go to Brasher City, LA., away back in the wilderness due west of New Orleans toward Texas. From there it is supposed we march overland Northwest to the Red River, a three week tramp. Alas, the day we are not to open the noble Mississippi and no further attack is to be made on Port Hudson. We retire without striking a blow at it—another masterly retreat of the self-made, highly praised "Banks." You had better obtain one of Lloyd's maps of Southern states to see our movements. There will be a little fighting, but plenty of Sugar and Cotton stealing for high officials."[77]

Colonel Molineux also struggled with fostering his company and field

---

74 Tiemann, William F. *The 159th Regiment Infantry, New York State Volunteers*, In the War of the Rebellion 1862 -1865, (Brooklyn New York 1891), p. 24.

75 Ibid., pp. 24-25.

76 James I Robertson Jr Civil War Sesquicentennial Legacy Collection at The Library of Virginia – Edward L. Molineux collection, *Letter written Mar. 20, 1863*.

77 Ibid., *Letter written Mar. 27, 1863*.

officers into competent leaders capable of executing the job for which they were assigned. The officers' attitudes, visible to the soldiers for whom they were responsible, affected the performance of those men. Molineux's annoyance with his officers was ongoing and obvious to his company in the field and his family back home.

"The greatest is the ignorance of men and imbecility and jealousy of officers and truly one cannot make himself beloved in a consolidated regiment and must fall back upon fear and respect. Of my officers I have only eleven doing duty cheerfully, the rest are sick or pretend to be and are discontented, homesick and disagreeable--all times grumbling at government and colonel and wanting to have their resignations accepted."[78]

To his foot soldiers, Colonel Molineux was an outstanding commander who demonstrated what he demanded of his officers from the beginning. Though John and Molineux had their moments, in general, the colonel showed himself to be a kind and empathetic leader.

On occasion, Molineux wrote home about the needs of his soldiers' families as shown in this excerpt from his letter dated March 27, 1863:

"a private who is well and employed in the ambulance Corps, told me his wife was distressed and without friends. The poor fellow has not been paid yet. Perhaps you can help her a little. Ann Kelly, 71 Hudson Avenue, Brooklyn."

He proved himself to be a compassionate leader who gained the respect of his soldiers; an example his officers needed to follow.

CAMP FEVER—TYPHOID

Sickness was on the rise. Men in the 159th NY were succumbing to Camp Fever, better known as Typhoid, an intestinal infection caused by contaminated food or water.

Typhoid fever often came with abdominal pain, severe diarrhea, fever—an overall ill feeling. Some soldiers developed "rose spots," small red spots, a rash on the abdomen and chest. Bloody stools, chills, agitation, confusion, delirium, seeing or hearing things not there, difficulty paying attention, nosebleeds, and severe fatigue were among the symptoms.

---

[78] James I Robertson Jr Civil War Sesquicentennial Legacy Collection at The Library of Virginia – *"cheerful duty"*, Edward L. Molineux.

The situation for John's regiment being "nine-month men," looked rather dark." Many of the men were sick and dying. John filled in the details in his next letter home.

Letter #10 - March 27, 1863

Baton Rouge

I received your letters, as I mentioned in a letter before this, and one from Sarah. I mentioned in my last letter that we were packing up, but I expect we will start tomorrow I think for Chattanooga Bay to reinforce Weitzel[79], that is below New Orleans toward Texas. When I get to the place of destination I will write and let you know where I am. My health is middling good, I was obliged to take some pill last night for the bloat again. I feel as though my clothes were so tight around my stomach that my stomach was bruised. It is the poison boils which create this and bloats me, and the pills I think remove it. Our regiment continues to die off, I was to a funeral yesterday, there were five buried, a fellow from Fishkill, and I think Parmers boy who used to work for Mutt Houghtailing[80]. Eli Miller has gone home, got his discharge, Tommy is sick, he has not played for some time, I play all alone on our expeditions. The other week the boys lost all their drums, so we have no music, I go out and play without the drums sometimes. Things have changed and the Drum Corps

---

79 Major Godfrey Weitzel: Born in Cincinnati, Ohio, November 1, 1835. A career U.S. army officer and civil engineer, Godfrey Weitzel graduated from West Point in 1855 with a commission in the Corps of Engineers. After the Civil War started, Weitzel served under various engineering commands tasked with improving the defenses of Cincinnati and Washington. He was then appointed chief engineer of the Department of the Gulf during Gen. Benjamin Butler's expedition that captured New Orleans. In August 1862, he earned a promotion to brigadier general of volunteers, and the following year commanded a division of troops under Major General Nathaniel P. Banks during the assault and siege of Port Hudson, Louisiana. Weitzel assumed command of XVIII Corps and led his troops in the capture of Fort Harrison, a key component of the defenses of Richmond. After his promotion to full major general of volunteers in November 1864, Weitzel took command of XXV Corps, composed of U.S. Colored Troops, and led them in a land assault on Fort Fisher, North Carolina. In February 1865, Gen. Grant directed Weitzel to take command of all Union forces north of the Appomattox River. After the fall of Petersburg, Weitzel's troops entered the Confederate capital of Richmond on April 3, 1865. National Park Service, Richmond National Battlefield Park.

80 Peter (Mutt) Houghtailing: Age 46 a farmer in Taghkanic, NY. Married with three children ranging in age 13-23.

is played out and I stand firm in the eyes of the regiment, they have obtained confidence in me, and the rest are played out. You said we were nine-month men, I hope it is so, but it looks rather dark unless Governor Seymour[81] call the regiment back, it looks dark. You said N. L. Post could get me out and that he would send the papers by the 1st of April, they have not come yet. If I positively knew we were nine-month men, I would stay my time, but I fear not so, if he can get me out then you see what it will cost. Be sure to make a bargain if he gets me out, if not I don't expect to pay him anything. I want you to continue to write and direct your letters as before until I write to you to direct them elsewhere. The time we fired at the rebels which I mentioned in my other letter, killed three and wounded one, no wonder they were in such a hurry. Calvin Finkle[82] is sick in the hospital, very sick and deaf, caught cold in the swamps. Henry Finkle is also sick and, in the hospital, but he is on the gain. Andrew Brush has gone to New Orleans to the hospital, I think he will be discharged and soon be home. My medicine is the best thing I ever had; I think it has saved my life. We expect to get paid off every day, but it doesn't come. When it does come it will be good for sore eyes.[83]

The fever had a high mortality rate, about thirty-six percent. Opiates, turpentine, quinine, capsicum, and ammonia were among the common medicines used in treatment. Quinine was a popular drug, considered a miracle medicine for many illnesses. Most likely, it was the pill John received for the bloat.

At the 159th NY encampment, conditions were perfect for escalating the disease. Bad water and unsanitary conditions fostered the spread of sickness throughout the camp. Under these conditions, the sickness thrived. The men couldn't grasp the concept: Sanitary conditions and clean habits contribute to good physical health.

Daniel M. Holt, surgeon of the 121st NY regiment, recorded the devastation of the fever firsthand:

"last night a man died of typhoid fever, and quite a number look as if they

---

81 Governor Horatio Seymour: Democratic Governor of NYS from January 1, 1863–December 31, 1864.

82 Calvin Finkle, Corporal 159th regiment Co. C: Age 34, died at Irish Bend on April 14, 1863. Farm laborer in Taghkanic NY. Left behind his father and four siblings.

83 *Letter #10, March 27, 1863, Baton Rouge.* John W. Mambert letters from the Civil War.

would soon follow. Poor fellows, when I see the way in which they lie and the lack of all earthly comforts…I wonder how any get well" He later wrote in his journal noting, "Very little can be done for a man while he lies upon the ground with typhoid fever…When I order one to the hospital, it seems almost equivalent to ordering his grave dug."[84]

The assistant surgeon Briggs and John were acquaintances before the war. Briggs's private practice was a few miles from John's home in the bordering village of Claverack. Because John was a musician, he was likely recruited by Briggs to care for the sick in the hospital. John must have been afraid of contracting typhoid among all the ill.

MORE LOSSES

The 159th NY was ready to move at a moment's notice. On March 28, 1863, the time came. The regiment embarked on the transport Laurel Hill, a screw steamship, and was taken to Donaldsonville, fifty miles south of Baton Rouge, on the west bank of the Mississippi River at the junction of the Bayou La Fourche.

The steamer *Diana* was sent to scout the enemy's position and strengthen the Navy on the lower Teche. Unfortunately, as it continued down the Atchafalaya River, instead of returning by Grand Lake as intended, it ran into the arms of the enemy. She was easy prey—*Diana* fell into Confederate hands. Several men were killed. Nearly 150 were taken prisoner and later paroled.[85]

This gave the Confederates three rather formidable boats in the Atchafalaya and the Teche.[86]

On March 31, 1863, the regiment was detailed to guard the division supplies and property onboard the transport Empire Parish as it sailed up the bayou arriving at Thibodaux midafternoon. The regiment disembarked and

---

84 Surgeon Daniel M. Holt as quoted in, Coryell, Janet L., James M. Greiner, James R. Smither, and Janet L. Coryell. A Surgeon's Civil War the Letters and Diary of Daniel M. Holt, M.D. Ashland, (The Kent State University Press, 2012), p. 33.

85 Parole: A pledge by a prisoner of war or a defeated soldier not to bear arms. When prisoners were returned to their own side during the War (in exchange for men their side had captured) the parole was no longer in effect, and they were allowed to pick up their weapons and fight. When the South lost the War and the Confederate armies gave their parole, they promised never to bear weapons against the Union again.

86 Irwin, Richard B. *History of the 19th Army Corps.* G. P. Putnam's Sons New York, (The Knickerbocker Press 1892), ch. 9, par. 9.

proceeded to march to the back of the levee where they set up camp.[87] Second Lieutenant Tiemann described the regiment's location in a letter to his father:

Thibodaux is a small place but very pretty. The inhabitants are mostly creoles and here I saw some of the most beautiful women that I ever saw. They are all secesh but say they would sooner have our army in occupation than their own. If you pass a lady in the street, she immediately drops her veil or brings their parasols down so as to cover their faces. They all dress elegantly and in extreme taste. Everything is cheaper here than at Baton Rouge. The slaves are paid for their work and consequently there are not so many who run away. The house we stayed in belonged to Brig. General Bush of the rebel army. [88]

UNIQUE LETTER FROM JOHN

The following letter from John is unique. Written from the field to a non-family member—Attorney William Hauver, John's letter is a response to Hauver regarding holding his office as Justice of the Peace in the town of Taghkanic. John was elected to a four-year term in 1853, reelected in 1857 and 1861.

For an unknown reason, John's letter to Hauver remained in the Mambert family letter collection. (Early on, John wrote a letter to Attorney Nathaniel Post, but directed the family to hold onto it and let Post read it. That letter was not included in the collection.) The question remains: Did Hauver ever read John's letter?

Letter #11 - April 2, 1863
Thibodaux, LA

To Wm H. Hawver, Esq.
Dear Sir,

I received your kind letter and was glad to hear from you. I am writing on my knee sitting on the top of a building. I can't write very good under such circumstances; I've got but a few moments to write. I am well except the diarrhea which is a prevalent disease here. You said you wished me good luck, thank you. I am not homesick, but I think it

---

87 Tiemann, William F. *The 159th Regiment Infantry, New York State Volunteers*, In the War of the Rebellion 1862 -1865, (Brooklyn New York 1891), p. 26.
88 W.F. Tiemann Collection at The N.Y.S. Military Museum, *letter describing Thibodaux to his father.*

an imposition upon me that all my friends who started with me have not had courage enough to stay. I feel just as I did when I was with you. My heart is filled with the same spirit that animated the hearts of our fathers of 1776 I have felt no disposition to yield, but one thing has sickened me and that is the treatment of the poor soldiers, they are shamefully abused and if you want to hear go to Capt. Wardle in Hudson who is home, I suppose this had caused my heart to yearn over them. As for battles I have no fear or dread for that enlivens my spirit. We went on a march the 159th up East of Port Hudson while the gunboats shelled the port and burnt a bridge, we drove the rebels away by firing canons, we came on them unexpected they were just eating breakfast by the side of the road, we killed 3 and wounded 13 by the bursting of the shell they left in the woods on a double quick, we picked up a sword and 1 carbine and all their breakfast that was spart, there was a negro said they had formed a line of battle across the road and was told not to go out after we fired, he said they has clear the road now I guess. I can't write much, I would like to write but postage stamps can't be got here unless you send them from there, if you had $25 in your pocket you could not get one, so if you write to me, send a stamp and then I will answer it. We are on the march now to Berwick Bay to reinforce Weitzel. My position I have got at least, and I draw that pay, the officers found I was a joining to leave and so they put me down at principal musician's pay. Our regiment is very unhealthy, diarrhea and fever. I am doing very well; you must not believe all you hear about me being so tired of being here. I heard about the affidavit which if it is not right you can change it about my office, I shall hold it and I expect to be home again, the Lord being my helper. I could write for a week, but it is dark, and I write by moonlight. Tomorrow morning, we move again, I can't write anymore tonight, excuse mistakes and give my respects to all except cowards.

Truly yours, John W. Mambert
Fife Major 159 Regt. N.Y.S. Vol.

Direct your letters 159 Regt. NYS Banks Ex., LA, BR or elsewhere.[89]

---

[89] *Letter #11, April 2, 1863, Thibodaux, LA, To Wm H. Hawver, Esq.* John W. Mambert letters from the Civil War.

It appears, from what is written, John's ability to retain his position as town justice may have been in jeopardy. No further records indicated he was a justice for the town after his 1861 win.

John wrote to Hauver, "you must not believe all you hear about me being so tired of being here," since John's family members were already talking with several people, including local officials, about getting him released from military service. It's probable that Hauver knew of John's desire to be released from the army due to the injustices he perceived were committed against him.

THE REGIMENT IS ON THE MOVE

The regiment remained in Thibodaux for three days until April 3, 1863, when they marched to Terrebonne, a small station on the New Orleans and Opelousas Railroad. The following day, the regiment traveled by rail cars arriving at Bayou Boeuf. It was nothing more than a railroad station with a few plantations nearby: most of them empty with no sign of life. Bayou Boeuf is seventy-three miles west of New Orleans, and seven miles east of Brashear City. The regiment was to join the rest of the division there. The enemy had been bold since the troops left Baton Rouge, three pickets were shot, and it was feared they would be attacking their current location in force. Weitzel's division was in advance at Brashear City. From there, they could see the enemy on the other side of the bay in Berwick performing drills and watering their horses.[90]

Transport ships and gunboats came through the gulf and up the Atchafalaya River arriving at Brashear City. During the first week in April, General Banks massed a force of 18,000 men at the location. On April 6, 1863, orders were issued to store soldiers' unnecessary baggage in a local sugar house. The men were allowed to carry a small roll of blankets and only necessary utensils.[91] Regrettably, on June 23, 1863, Union Lieutenant Colonel Duganne was forced to make a bitter decision which proved devastating for the 159th NY.

---

90 W.F. Tiemann Collection at The N.Y.S. Military Museum, *Letter describing enemy actions, April 7, 1863*.

91 Tiemann, William F. *The 159th Regiment Infantry, New York State Volunteers*, In the War of the Rebellion 1862 -1865, (Brooklyn New York 1891), p. 26-27.

DUGANNE'S BITTER DECISION

Duganne and his small force were garrisoned at the sugar house stores. They were about to be attacked by a rebel cavalry out of Texas who previously captured Brashear City, along with a million plus dollars of Union stores.

The large sugar house was filled with army supplies, officers' trunks, extra baggage, arms, and military appurtenances of all kinds with an estimated value of at least a half-million dollars. Its worth may have been much more for the trunks, boxes, desks containing money, watches, jewelry, and other valuables left behind by the regiments; John's included. Duganne had no intentions of letting the Confederate force receive another prize, adding it to the pillage they were rejoicing over in Brashear City. So, with a sigh over the necessary sacrifice, he ordered the stores and personal belongings to the flame of a federal bonfire: the 159th NY lost everything.[92]

Although losing treasured keepsakes, John still retained his most valuable possessions: faith and family, common themes in his letters. News from home was precious to John. Getting updates from the family was important, including if they heard any word of the war ending.

Letter #12 - April 6, 1863

Bayo Beauff

To Mrs. John W. Mambert Family, very weak this morning,

I think that Brandy and Gin and other medicine has been the preservation of my health, by the help of God in whom I trust. Last night I was paid off, $43.75, this was at $13 per month from the time I was enlisted to the 1st of January. But this is not all due to me from that time, the next time I get paid I will get $21 per month from the 1st of November which definitely correct on the other pay roll. So, I am back $24 now and from the 1st of January I will continue to draw $21 per month. I will send in this letter $30 and the rest I will keep, I borrowed $4 or nearly that. I am under the

---

92 Duganne, A. J. H. *Camps and Prisons, Twenty Months in the Department of the Gulf.* (New York J.P. Robens, Publisher 1865), p157-158.

obligation to let you know that Henry Finkle[93] and Calvin Finkle are both dead. Both died in one day, I borrowed $1 from him, and I want you to pay that dollar to his poor Mother without fail. There have been several more dead whose names I don't remember now. Gladia Rockefeller[94] is as good as dead, when I see so many men fall of disease it makes me shudder, but I trust in God, and I know he is able to deliver me from all harm. I want you to make the best use of this money, you may think fit and proper. I am sorry I could not send more but I can't, do the best you can with it. When we will be paid again, I don't know. We are all comfortable situated except those sick, they die off pretty fast. One thing I must say is that in all my laying out I have not taken any cold and my lungs I think are stronger than they ever have been, I don't cough a bit and I feel good thus far. I have not received a letter in some time, send me letters for I love to hear from home, it is better than all the medicine. Let me know if we are positively 9 month-men if you can. It is nice here and warm, poor water, no society, and a slave county and in the midst of the enemy. I can't write much for I have no time, let me know what N.L. Post is doing and whether they are going to draft. If they do draft don't suffer yourself to be sent into harm, for they can't make it here, I never expected to come here when I enlisted, if I had, I would not have done it... Let me know what the prospect is about the war ending. I have just come in now from guard mounting, let me know how you get along with totaling up the business and whether that note for Rons is paid. In short give me all the news you can, John Allan[95] is well and Mart, tell Albert Sheldon that the men whom he had reference to they are in the 125th Langdon's Company[96]. Our eating, it is almost impossible

---

93 William Henry Finkle, Private 159th Co. C: Age 23, died in New Orleans LA. Farm laborer in Taghkanic, NY. Left behind his parents and five siblings.

94 Mortimer (Gladia) Rockefeller, Private 159th Co. C: Age 18, wounded at Irish Bend La. April 14, 1863. Farm laborer in Taghkanic, NY. Lived with John and Matilda Davison the farm owners.

95 John E. Allen, Musician 128th regiment Co. K: Age 41, died of disease at Barracks Hospital La. on August 4, 1863. Nursery laborer in Claverack, NY. Born in England he lived and worked for his cousin Eldridge Studley and family.

96 125th NYSV: Recruited from Rensselaer County, mustered in on August 27, 1862, they fought in many of the northern battles including Wilderness and Gettysburg.

to live on the food which the Army gets to eat, at least here in this regiment. It will do well enough so long as the stomach is right, but it cannot keep right, and if you are sick it is all day. It takes all of my attention to keep and preserve my health, by the help of the Lord and I hope and trust that he will deliver me from this place to my native land to greet you again. And that he will preserve you all that I may see you all again. Trust in God.

From your Husband and Father
John W. Mambert, Major Fifer, 159th N.Y.S.V.[97]

WORM CASTLES AND BOILED BEEF

Often, John complained about military food—either a lack of it, or that it was just bad. The men received their food in two ways: marching rations and camp rations. Marching rations were issued when troops were traveling from place to place and consisted of roughly three items: salted meat, hardtack, and coffee.

Hardtack was a flour biscuit renamed "tooth-dullers," "worm castles," and "sheet iron crackers." John Billings, who wrote <u>Hard Tack and Coffee</u> described them as,

"a plain flour and water biscuit. Two which I have in my possession as mementos measure three and one eighth by two and seven eighths inches and are nearly half an inch thick."[98]

Hardtack could be eaten or prepared in a number of ways, toasted over a fire, crumbled into soups, or crumbled and fried with their pork and bacon fat. In his book, Billings wrote,

"a dish akin to this one which was said to make the hair curl and certainly was indigestible enough to satisfy the cravings of the most ambitious dyspeptic was prepared by soaking hardtack in cold water then frying them brown in pork fat salting to taste. Another name for this dish was skillygalee."

Hard Tack often came infested with maggots and weevils, the second being more abundant than the maggots. Weevils, little slim brown bugs an

---

97 *Letter #12, April 6, 1863, Bayo Beauff, To Mrs. John W. Mambert Family.* John W. Mambert letters from the Civil War.
98 Billings, John D., Hardtack and Coffee. Published 1888. p.113.

eighth of an inch in length, great at boring small holes, completely riddled the hardtack. The biscuits were "alive" and well covered with the webs these creatures left to insure condemnation. Under these circumstances, hardtack was best eaten in the dark. [99]

Camp rations were issued when men were in range of resupply. It was similar to marching rations, but with molasses or vinegar, a necessary ingredient to hide the taste of rancid salted beef. Food rations were meant to last three days while on active campaign and were supposed to be based on the general staples of meat and bread. A single ration of what a soldier was entitled to have in one day consisted of twelve ounces of pork or bacon, or one pound-four ounces of salt or fresh beef, one pound-six ounces of soft bread, or flour, or one pound of hard bread, or one pound-four ounces of cornmeal. With every hundred rations there should be distributed: one peck of beans or peas, ten pounds of rice or hominy, ten pounds of green coffee or eight pounds of roasted, and ground, or one pound-eight ounces of tea, fifteen pounds of sugar, one pound-four ounces of candles, four pounds of soap, two quarts of salt.

Meat usually came in the form of salted pork, or on rare occasions, fresh beef—usually when foraged. Rations of pork or beef were boiled, broiled, or fried over open campfires. There were large quantities of stale beef or salt horse served, and rusty unwholesome pork. [100]

Other food items included rice, beans, dried fruit, and potatoes. Baked beans were a northern favorite, when time could be taken to prepare them, and a cooking pot with a lid could be obtained.

When the army was settled for a long stop, company cooks did the cooking. But there was no uniformity about it. Each regiment's rations were cooked by persons specially selected by their officer, grouped into a "mess" to combine and share rations. While on active campaign, rations were usually prepared by each man to the individual's taste. It was considered important for the men to cook the meat ration as soon as it was issued so that it could be eaten cold if activity prevented cook fires. A common campaign dinner was salted pork sliced over hardtack with coffee boiled in tin cups. [101]

Worry over the health effects of bad food was second only to the fear

---

99 Ibid., pp.115, 117.
100 Ibid., pp.110, 111.
101 Ibid., p.131.

of dying from disease. The anxiety men felt daily was justified, records regarding regiment health reveal between mid-March and the first week of April 1863, the 159th NY lost sixteen men who succumbed to one of the many prevailing illnesses.

THE IMPORTANCE OF MAIL

John was no different from any soldier in the army, they all looked forward to getting mail. It was a treasured link between camps or battlefields and "back home." John had no problem expressing his anxiety to his family, especially when the wait between letters from home was longer than he had expected. Spending the majority of their time sitting in camp, waiting and thinking, no doubt John, along with the others, felt lonesome and neglected when the wait was long to receive word from family or friends.

Partial Letter #13 - Sent in April 1863

I will send it home when I get it and when I want another box, I will let you know. When you send letters send some postage stamps for I can't get them here if I had $100 in my pocket, I could not get a postage stamp for we can't get them here, they have all got to come from home. We can send letters without stamps, but you have to pay 5 cents. When you take them out of the office, and they are not so much noticed as letters with stamps on. A few lines to my dear Maria. I am very happy to think that you can write good, and I hope you will continue to learn and write as often as possible and that Jobe writes, he does well. Tell Walt to write too. I want you to write to me about your health whether you are healthy as when I was home, and I want you to put your hopes in God and pray to him that I may be protected and permitted to return home again to meet you, the time has not seemed long to me but I suppose it has seemed long to you, but I hope and trust in God that I shall soon be home again, and if I do I shall never venture upon so hazard a trip again. I should like to see you all. Give my love to all and take good care of yourself and I will of myself and I hope and trust we shall soon meet again. Write to me. From your kind Husband John W. Mambert [102]

---

102 *Partial Letter #13, Sent in April 1863.* John W. Mambert letters from the Civil War.

One soldier described to his brother the scene of joy or disappointment when mail arrived at camp:

"when the mail comes and the boys all huddle around the Captain's tent to get their letter it is a pleasure to watch the crowd all look pleased and as each ones name is called they sing out and step forward, then you ought to see the Big grin they will get as they pass through the crowd to their tents. The crowd gradually grows smaller until perhaps 20 or 30 only will be left when the Captain stops calling and the mail is all given out, there are a lot of different looking countenances about as long as a 18 inch boot leg, and you will hear expressions like these, what none for me, dam strange, sure I haven't got one Captain, well I written 3 or 4 now and damned if I write any more until I get one. If some of them folks at home could see how disappointed these poor fellows are, they could sit up nights to write."[103]

Recognizing the mail's importance for morale, the Union assigned personnel to collect, distribute and deliver soldiers' mail. Wagons and tents served as traveling post offices. Some soldiers wrote home weekly while others appeared to spend all their free time writing. A letter from home might be tucked into a pocket close at hand where it could be read again in moments of loneliness. Many soldiers carried prepared letters in their pockets to be forwarded to loved ones if they were killed in action.

The U.S. Post Office introduced several improvements during the Civil War making it easier to send and receive mail. Occasionally, soldiers had trouble acquiring postage stamps or keeping the gummed bits of little paper from congealing into sodden lumps. John mentioned many times in his letters the ongoing issue with stamps.

In July 1861, a change was made allowing soldiers to mail letters without stamps. Simply by writing "Soldier's Letter" on the envelope, postage would be collected from the recipient.[104] John mentioned this service, but continually requested stamps, maybe because he did not trust the letter would reach its intended destination without one.

---

103 Morgan City Archives, 501 Federal Ave, Morgan City, La. *Letter from Sam white to brother Solo, May 15, 1863 Brashear City.*
104 United States Postal Service, *Mail service and the Civil War*, USPS.com

Chapter Three

"THERE WAS A LARGE MAJORITY FOR THE UNION, BUT THEIR HEAD MEN MADE FALSE RETURNS AND THE MAJORITY COULD NOT HELP THEMSELVES, HE SAID THE SOUTH WAS GONE AND RUINED"

– CONFEDERATE PRISONER

BATTLE AT IRISH BEND

The idea was to begin preparations for an advance into western Louisiana and defend Union occupancy in Lafourche Parish. It was General Banks's next military move, an idea proposed by General Weitzel to General Butler. Based on this idea, Butler created Admiral Farragut's Naval gunboat flotilla. With the newly created flotilla, Weitzel approached Banks with the next idea:

"With such a naval force in that bay, in co-operation with a suitable land force, the only true campaign in this section could be made. Look at the map. Berwick Bay leads into the Grand Lake, Grand Lake into the Atchafalaya, the Atchafalaya into Red River. Boats drawing not more than four feet and in the force I mention [10 or 12], with a proper land force, could clear out the Atchafalaya, Red River, and Black River. All communications from Vicksburg and Port Hudson cross the line indicated by me."

Banks agreed with Weitzel's position, but not entirely. Short on time, Banks was looking for a faster route. Weitzel felt the time was right and urged his plan again on February 15, 1863, but this time to his commander General Christopher Auger:

"Let two of the brigades be moved to and landed at Indian Bend, while the two others are crossed and attack in front. If Mouton escapes (which I think, if properly conducted, will be doubtful) we form a junction at Indian Bend. We proceed to attack and with much superior force, because I do not believe Mouton and Sibley united will exceed 6,000 men. We can defeat them, pursue our success to Alexandria and of course Butte a la Rose."

CHAPTER THREE

The plans were set in motion...[1]

Banks instructed General Grover to secure a landing site on the shores of Grand Lake and march his force to the city of Franklin. Depending on the circumstances, he was to cut off Taylor's retreat, or attack the Confederate's force. The 159th NY departed for Brashear City on the morning of April 8, 1863, reaching their destination in the afternoon. They encamped for two days waiting for additional divisions of the 19th Corps—Generals William Emory, and Godfrey Weitzel. Once arrived, General Banks had 16,000 men assembled. The troops were instructed to move across Berwick Bay via transport ships.

The flotilla for the task was commanded by Lieutenant-Commander Cooke. It comprised the flagships *Estrella* (a paddle steamship) the *Arizona*, *Clifton*, and *Calhoun* all sidewheel gunboats and *Laurel Hill*. It ferried Emory and Weitzel's divisions of 12,000 men. Next, it delivered Grover's force of 4,000 men to their destination on the shores of Grand Lake. The fleet was at the ready with guns to protect each division's landing and movements. The 159th NY marched onboard *Laurel Hill* while the remaining troops waited for additional transport. The boat was so crowded with men they could hardly move, but they were in good spirits as they sailed off on April 11, 1863.

The next day, while the flotilla was sailing up the Atchafalaya River looking for a landing area to intercept a Confederate retreat, *Arizona* became grounded. Four hours were lost attempting to get her afloat. Rather than delay the Confederate pursuit any longer, Cooke decided to take the remaining flotilla full steam ahead, leaving *Arizona* to take care of herself.

Union gunboats and transports were anchored below Miller's Point at Oak Lawn. Grover, under orders, was told he might find a good shell road leading straight to the Teche. The road would allow troops to cross the bayou, and head toward the middle of the arc in the river where Irish Bend (also referred to as Indian Bend) was located. Grover ordered Colonel Fiske, with two companies of the 1st Louisiana (loyal), ashore to survey the area. At midnight, Fiske returned to the gunboat and reported the road was under water and quite impracticable for travel.[2]

---

1 Irwin, Richard B. *History of the 19th Army Corps*. G. P. Putnam's Sons New York, (The Knickerbocker Press 1892), ch. 9, par. 1-6.

2 Ibid., ch.11, par. 5.

That evening, Banks's force arrived outside Fort Bisland's defenses. Forming a line of battle, they patiently waited for sunrise. The whole line advanced on the fort within musketry range. Combat began after 11:00 a.m. April 13th. General Banks had three brigades deployed—Weitzel on the left, Colonel Halbert E. Paine with the 4th Wisconsin on the right and Colonel Timothy Ingraham commanding the 38th Massachusetts in support.

Opposing the Union forces south of the Teche was the "Arizona Brigade" commanded by General Henry Hopkins Sibley.

North of Bayou Teche, the Union brigade of Colonel Oliver P. Gooding, with the 31st Massachusetts, faced off against Mouton's Confederate brigade.

Now in enemy hands, the ex-Union gunboat *Diana* was shelling Union troops. The Confederate gunboat was positioned in front of the Confederate earthworks,[3] slowing Emory and Weitzel's men as they approached. *Diana*'s gun was swiftly silenced: a 30-pound parrott gun from the 1st Indiana fired a shot taking out the boat's engine room. *Diana*, taken out of the conflict, slowly moved up the bayou.[4]

It was midday. General Taylor received information that Union gunboats and transports were heading up Grand Lake with a division landing behind his army looking to cut off his retreat. Taylor immediately sent Colonel Vincent with the 2nd Louisiana cavalry and a section of battery to Verdun landing four miles behind Camp Bisland. Vincent's orders were to observe and oppose the Union movement.[5] With the gunboat *Diana* removed, the whole Union line closed upon the Confederate works. The engagement became fierce with non-stop artillery and musketry.

At daybreak above Fort Bisland, Fiske, and his men, back aboard the fleet, sailed six miles up the lake anchoring at Magee's Point. General William Dwight sent two staff officers and a small detachment from the 6th NY, to examine the plantation road leading to the Tech. The road was found practical for all arms and the debarkation began at 8 a.m.[6]

---

3 Earthwork: A field fortification (such as a trench or a mound) made of earth. Earthworks were used to protect troops during battles or sieges, to protect artillery batteries, and to slow an advancing enemy.
4 Ibid., ch.10, par. 4.
5 Ibid., ch.11, par. 9.
6 Irwin, Richard B. *History of the 19th Army Corps*. G. P. Putnam's Sons New York, (The Knickerbocker Press 1892), ch.11, par. 4.

Meanwhile, Taylor sent the 4th Texas Cavalry to join Vincent in delaying the Union division's progress. Vincent found the location where Grover's army was landing, positioning his men with two batteries.[7] They fought unsuccessfully in destroying the transports. Lieutenant-Commander Cooke's guns quickly drove Vincent's men off. Two companies of the 1st Louisiana Infantry (loyal) landed and were deployed as skirmishers. While advancing across a field, they were under heavy fire by enemy cannons and skirmishers.

Using flat boats, the 159th NY was second to land. They were under scattering fire as they disembarked, rapidly forming a line of battle. The 13th Connecticut joined the 159th NY in support of the skirmishing 1st Louisiana. Grover sent a battery and a troop of cavalry for additional support. Through a cold rain, the regiments marched toward the woods skirting the field while supporting the line. Forcing the rebels into retreat, the men were kept in front all day.[8]

Because the transports could only get within 100 yards of the beach, the remaining troops' disembarkation proceeded slowly. Foot soldiers were able to jump overboard and scramble ashore, as did their horses. A bridge of flat boats had to be made for the guns and caissons of the artillery regiments. It was late afternoon before the whole division assembled on shore at McWilliams's plantation. At the upper reach of Irish Bend, the plantation was situated where the road split in two directions. Irish Bend was the way to Franklin, and to Franklin, Grover was ordered to go.

Dwight's troops seized a bridge crossing the Teche approaching Madame Porter's plantation from the north. His men arrived in time to extinguish a fire Vincent's men lit to destroy the bridge. Their cavalry traveled down the left bank of the Teche and seized another bridge, likely on the shell road where Grover was originally instructed to disembark. The remaining troops were waiting for their haversacks to be filled with hard bread and coffee. They were without wagons except for a few loaded with reserve ammunition.

All the delays were having an effect on Grover's calculations, making him unsure of General Banks's movements or his situation. At 6 p.m., while the sun was

---

7 Battery: The basic unit of soldiers in an artillery regiment; similar to a company in an infantry regiment. Batteries include six cannons (with the horses, ammunition, and equipment needed to move and fire them).

8 Tiemann, William F. *The 159th Regiment Infantry, New York State Volunteers*, In the War of the Rebellion 1862 -1865, (Brooklyn New York 1891), p. 27.

low among the treetops, Grover made the decision to take up a line of march with Birge and Colonel Kimball and head for the Teche. Crossing the upper bridge, the division went into camp on the plantation of Madame Porter.[9]

Banks was growing impatient and feared the campaign was failing—he did not know Grover's location. He was concerned everything gained that day would be lost should the Confederates move out of the Fort under the cover of night. Unexpectedly, there was a burst over the lines of Bisland. It was from a shell shot from the gunboat *Clifton*'s bow gun. With it came news: Grover was on land and marching toward his location. Banks, with a sense of relief, recalled orders for an assault and drew his line back out of fire so the men could rest.[10] General Taylor began evacuating supplies, men and weapons from Fort Bisland leaving a small force to slow Banks's movement.

Grover's decision to halt at the first bridge left the lower bridge wide open for Taylor and his men to escape. Taylor later wrote in his memoir:

"not a picket nor scout was found, and Yokely Causeway and Bridge were safe. From the farther edge of the wood, in open fields, Federal camp fires were visible. It was a wonderful chance. Grover stopped just short of the prize, thirty minutes would have given him the woods and bridge closing the trap on his force."[11]

Early on the morning of April 14, 1863, Banks and his troops found the fort abandoned, while at the same time, Grover's movement toward Irish Bend resumed. The 159th NY took the advance position in the brigade column, while the 25th Connecticut deployed as skirmishers a short distance in front. At Irish Bend, where the bayou makes a turn forming a horseshoe, the regiment marched eastward toward Franklin believing the Confederate force was still at Fort Bisland—but Grover's army soon discovered the true position of Taylor's force.

General Taylor formed his line of battle waiting for Grover and his troops. He positioned Major Clack's Confederate guard, two pieces of Cornay's battery, Colonel Reilly's regiment, followed by Colonel Vincent's regiment with a second section of Cornay's guns south of the bayou road.

The Confederate's task was simple but desperate: hold off Union troops

---

9 Irwin, Richard B. *History of the 19th Army Corps*. G. P. Putnam's Sons New York, (The Knickerbocker Press 1892), ch. 11, par. 13.

10 Ibid., ch.10, par.10,11.

11 Taylor, Richard. *Destruction & Reconstruction, Personal Experiences of the Late War*, (New York 1879), p. 115.

until they had safe passage behind the barrier of heavy vegetation growth. Then fall to the rear of the Confederate's main column and attempt their escape from Grover's army.

Shortly after Grover's troops began their march, Birge discovered the enemy's position viewing his skirmisher's engagement. He ordered the reserved battalion of the 25th Connecticut to adjust front forward and move across the field against the Confederates' left. Colonel Bissell led his men of the 25th Connecticut within 100 yards of the woods where they lay down under the partial cover of a ditch. They began firing with the 26th Maine supporting Bissell's left. Suddenly the spray of fire from the Confederate line became strong, lasting only until Lieutenant Bradley took two cannons, and at a gallop moved opposite their infantry, opening fire at 500 yards range. For a while, this engaged all the attention of the Confederate cannoneers.

The 159th NY continued marching along the levee road. Located off the road was a large sugarhouse, "McKerall's" used as a hospital during and after the engagement. The field was crisscrossed with deep ditches and planted with sugarcane reaching a foot in height. Rebels were posted in dense woods, protected by a strong fence bordering the bayou, giving them and their gunboat *Diana* good cover.[12]

Grover, riding with Birge, sent in the 159th NY positioning them left of the 26th Maine, giving orders to seize the woods: the 13th Connecticut reformed their line on their extreme left.

MOLINEUX PAINFULLY WOUNDED

Molineux promptly deployed his troops and gallantly led the men forward at a double-quick to short point-blank range of the enemy's musketry. Noticing the men were exhausted by the rapid advance over the rough and heavy ground, and suffering severely from bullets of the enemy, he ordered them to "halt, lie down," and "throw off their blankets and overcoats." He then ordered "commence firing" and the 159th NY opened a vigorous, well directed fire at the enemy's line. In preparing for the counterattack, Confederate forces reduced their fire.

Molineux, sensing his opportunity, began the command, "Forward!" when

---

12 Tiemann, William F. *The 159th Regiment Infantry, New York State Volunteers*, In the War of the Rebellion 1862 -1865, (Brooklyn New York 1891), p. 29.

a bullet struck him in the mouth. He fell, painfully wounded. The command of the regiment transferred to Captain Dayton.[13]

The Battle of Irish Bend Louisiana sketched by William Hall of the 22nd Maine.[14]

Colonel Gray, with his regiment the 28th Louisiana, arrived at Irish Bend to assist General Taylor, bringing news the remaining Confederate forces from Fort Bisland were close behind. Under cover of woods, Taylor positioned Gray's regiment quietly to the left overlapping Grover's right. General Taylor needed to ensure a safe retreat for his main army, and thought he had one. Confident, he ordered a vigorous attack on Grover's force with the entire rebel army obeying. Out of the woods, far to the rear of the 159th NY, the enemy came charging strongly, then halted, pouring a hot volley from their muzzle loaders into the regiment. Realizing the situation was critical, Captain Dayton immediately ordered the 159th NY to fall back. The men followed the order under severe fire, falling back to the road. In their retreat, they swept over the position of the 26th Maine and the 25th Connecticut assisting these battered regiments in the retreat. The restructuring of these beaten regiments was cut short as General Grover gave orders to redeploy and take back the broken battle.[15]

At that moment, Taylor's left line came down on General Birge and his

---

13 Irwin, Richard B. *History of the 19th Army Corps*. G. P. Putnam's Sons New York, (The Knickerbocker Press 1892), ch. 11, par. 19.

14 Illustration, *The Battle of Irish Bend Louisiana*, sketch by William Hall of the 22nd Maine, Harper's Weekly, May 16, 1863 - Public domain, via Wikimedia Commons.

15 Ibid., ch.11, par. 20,21.

men. Approaching Taylor's left, coming to Birge's aid, was the 13th Connecticut regiment commanded by Lieutenant-Colonel Warner. Emerging from cover under a grove of trees, the 13th Connecticut marched into the open field. They moved steadily toward Taylor's troops near Cornay's two guns positioned in the woods. Warner's men instantly replied to the cannon volleys and small firearms that greeted their appearance. Pushing on, firing as they went, their movement occurred just in time to help their comrades.

Although Taylor's troops were bearing down vigorously on the men, the 13th Connecticut's attack drove the rebels off. They also seized two cannons, wagons and a flag presented to the Confederate Battery by the ladies of Franklin.[16] Nearly sixty Confederate prisoners fell into the hands of the 13th Connecticut, who were still advancing until Grover ordered them back to restore the brigade line of battle.[17]

As Birge withdrew his troops on the right, Dwight deployed two lines: the 6th New York and the 91st New York in front, and the 22nd Maine, 1st Louisiana and 131st New York supporting the rear. They advanced against Taylor's left flank, pushing them further into the woods. Dwight's men took seventy rebel prisoners. The resistance Dwight encountered was feeble compared to the vigor Birge was met with earlier, forcing his men back. During that engagement, Confederate lines of battle almost gained their main objective, but now the only goal was to make good on their withdrawal.

General Grover advanced the division through the woods to an open field where he could see the enemy on a knoll behind their original position. To cover their precarious position, the gunboat *Diana* was brought forward, fixed from the previous day's damage, made partly good by repairs. Her thirty-pound Parrott rifle opened a slow fire on Union troops without any great effect other than increasing Grover's caution as he advanced his men.[18]

Taylor directed Mouton to take command of the troops while he rode into Franklin and led the Confederate retreat. General Green was in command of the rearguard. His orders were to prevent Emory and Wietzel from pressing the Confederate trains and troops outside Franklin heading for Yokley Bridge

---

16 Tiemann, William F. *The 159th Regiment Infantry, New York State Volunteers*, In the War of the Rebellion 1862-1865, (Brooklyn New York 1891), p. 30.

17 Irwin, Richard B. *History of the 19th Army Corps*. G. P. Putnam's Sons New York, (The Knickerbocker Press 1892), ch. 11, par. 23.

18 Ibid., ch. 11, par. 24-28.

and the road to New Iberia.

Meanwhile, Mouton withdrew from Grover's front. During their retreat, Confederates blew up the *Diana* and damaged all their transport steamers on the Teche except their hospital boat, the *Cornie*. Now in the hands of their adversary, the Confederate sick and wounded who filled this boat would be cared for by Union hands.

By 9:30 a.m., the last of the enemy exited the town's north end. General Weitzel's men entered from the south, unsure of which direction the rebels were headed. After interrogating prisoners and questioning townspeople, Banks concluded the direction, sending Weitzel and Emory to follow. The delay allowed Confederate forces to fall back quickly and make their escape[19] as they retreated to Opelousas, by New Iberia and Vermilionville, heading for their first defensible position at Bayou Vermilion, thirty miles south of Opelousas.

General Banks decided not to pursue the Confederate army before morning. He felt nothing would be gained by marching his tired troops. The decision was made; the men went into bivouac early that afternoon.

Federal Troops Marching Through Franklin, April 15, 1863, After General Richard Taylor's Withdrawal Following The Battle of Irish Bend.[20]

---

19 Ibid., ch. 11, par. 30.

20 Illustration, *Federal Troops Marching Through Franklin, April 15, 1863, After General Richard Taylor's Withdrawal Following The Battle of Irish Bend,* Courtesy of Morgan City Archives, 501 Federal Ave, Morgan City, La.

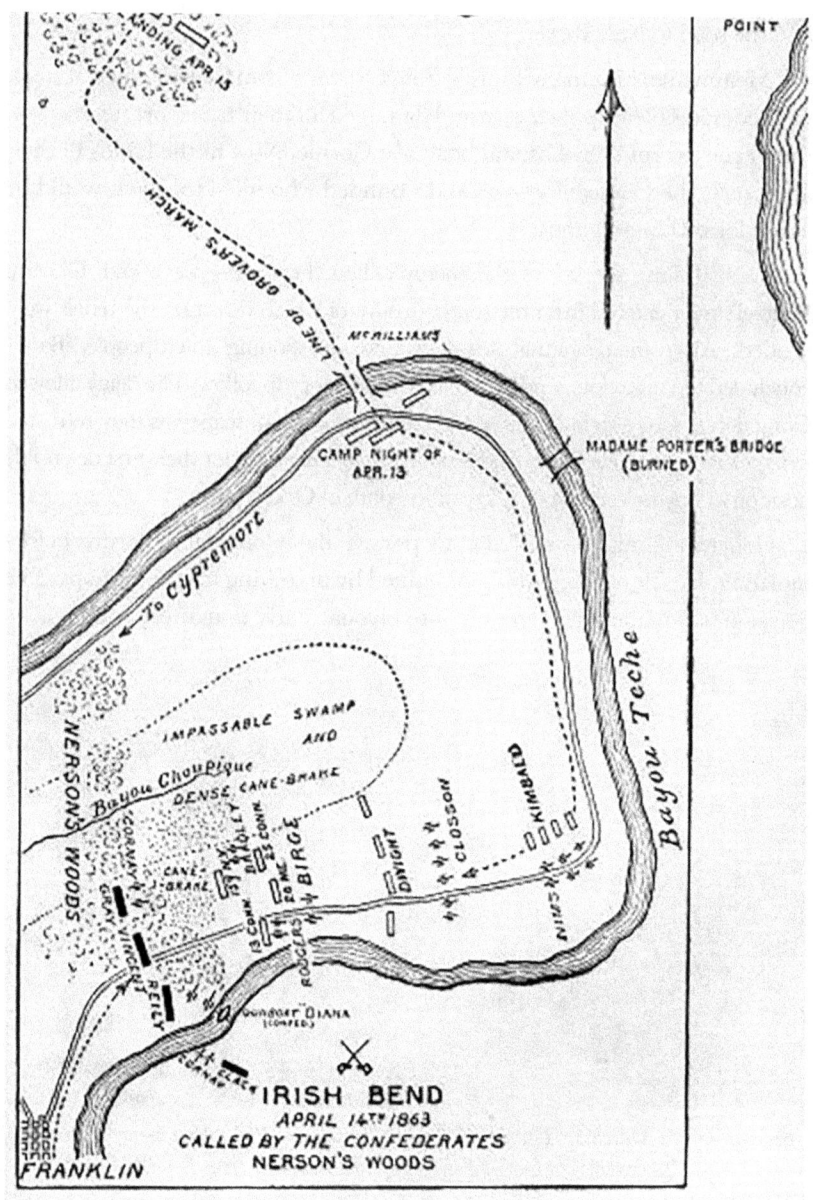

Battle at Irish Bend.
(Map from History of the Nineteenth Army Corps, 1892)

Battle at Irish Bend [21]

---

21 *History of the 19th Army Corps, 1892*

A UNION VICTORY AT GREAT COST

Victory went to the Union army, unfortunately, the loss of brave soldiers was great in numbers: six officers and forty-three men killed, seventeen officers and two hundred fifty-seven men wounded and thirty men missing, in all three hundred fifty-three. Of this total, Dwight's brigade lost six men killed and thirty wounded. Birge's brigade lost forty-six men killed and two hundred-twenty wounded, with an additional forty-nine missing, in all two hundred ninety-six.

The 159th NY took heavy losses, the nominal lists show four officers and fifteen men killed, five officers and seventy-three men wounded, and twenty men captured or missing—in all, one hundred-fifteen. But the regiment suffered more severely than the figures indicate. Besides mourning the death of the gallant and promising Lieutenant Draper, Colonel Molineux received a grievous wound that deprived the regiment of one of the best Colonels for weeks.[22]

OFFICERS LOSSES AND WOUNDED
Killed

Lieutenant-Colonel Gilbert A. Draper
Adjutant Robert D. Lathrop
1st Lieutenant John W. Manley Jr, Co. A

1st Lieutenant Wm. Plunkett Co. K
Second-Lieutenant Charles P. Price Co. K
Second-Lieutenant Byron Lockwood

Wounded

Colonel Edward L. Molineux
Captain Wells O. Pettit, Co. H

Second-Lieutenant William F. Tiemann, Co. A

COMPANY A, LOSSES AND WOUNDED
Killed

Private Patrick Connery
Private James Reynolds
Private John Kelly

Private Joseph Snyder
Private Robert L. Kipp

Wounded

1st Sergeant Edward Tynan
Corporal John Higgins, mortally
Private Solomon Maurer
Private William Brenzel
Private Richard Mosier

Private Thomas Daley
Private John G. Tator
Private John Dennis
Private George Finney
Private Warren Winslow, mortally

---

22 Irwin, Richard B. *History of the 19th Army Corps*. G. P. Putnam's Sons New York, (The Knickerbocker Press 1892), ch.11, par. 36.

Private Wm. H. Hollenbeck
Private Thomas Atkins
Private Thomas Ward

Private Patrick Keegan
Private Charles I. Winans

COMPANY B, LOSSES AND WOUNDED
Wounded

1st Sergeant Frank P. Gavin
Private Thomas Corson
Private Smith H. Lewis
Private Stephen French

Private Fredrick Siegler
Private Balthazar Wurtz
Private John Keron

COMPANY C, LOSSES AND WOUNDED
Killed

Private James Houghtaling
Private Daniel Riley

Private James Morrison

Wounded

Wm. J. Shufeldt
Private Ambrose Coon, mortally[23]
Private John Rilsing, mortally

Private Wm. Calkins, mortally
Private Christian Schnack
Private Mortimer Rockefeller

COMPANY D, LOSSES AND WOUNDED
Killed

Private Charles Hulfas

Wounded

Sergeant Isaac L. Rose
Private Frank W. Kisters

Private Lewis Messonsole
Private Adam Schuck

COMPANY E, LOSES AND WOUNDED
Killed

Private Richard Boice
Private Henry D. Wolf

Private Peter Silvernail

Wounded

1st Sergeant Samuel B. Macy
Private James Burns, mortally.
Private James Decker
Private Jesse Miller
Private James Doran, mortally
Private Chas. H.G. Peterson

Private Wm. H. Hart
Private Francis R. Syre

---

23 Ambrose Coon, Private 159th C0. C: Killed in action Irish Bend, age 16, Taghkanic NY, left behind parents and seven siblings (his age at enlistment 15).

COMPANY F, LOSSES AND WOUNDED
Killed
Corporal John G. Laws
Private Henry Eaton

Private Zebulon V. Flowers

Wounded
Privat John Doyle
Private Edward J. Mackey
Private Bartholomew Doser
Private Terence Mackey

Private Richard Keron
Private Thomas White
Private Aaron Miller

COMPANY G, LOSSES AND WOUNDED
Killed
Private John Murphy

Private James W. Sharon

Wounded
Corporal John Devlin
Private Cornelius Stickles
Privates Leonard Smith

Private Michael Fortin
Sergeant Wesley Tanner, mortally

COMPANY H, KILLED AND WOUNDED
Wounded
Corporal John Neefus, mortally
1st Sergeant Wm. J. Kennedy
Private Washington Adams
Private Michael Murtha

Private James Bennett
Private Bryan Hopkins
Private George Rodan

COMPANY I, LOSSES AND WOUNDED
Killed
1st Sergeant Mark Baker
Corporal Wm. H. Haws
Sergeant Theodore Bohrer

Wounded – Corporal Joseph O. Reed
Private Alvarus Coon
Private Jacob Coon

COMPANY K, LOSSES AND WOUNDED
Killed
Corporal Andrew Asbell
Private George Carr

Private David Miller

Wounded
Sergeant John Day
Private John Coughlan
Private Henry Hahn
Private Joseph Corcoran
Private James Kelly

Private John M. Kewan
Private John Emmons

The wounded men were transported to hospitals in New Orleans, since there were limited medical accommodations in the field.[24]

THE SCABBARD

When Colonel Molineux was injured in battle, his scabbard and belt were cut from him and discarded on the battlefield. His 2nd Lieutenant, Tiemann, stayed close as Molineux was being carried away from the fighting. Taking possession of Molineux's sword, Tiemann was struck by a musket ball and knocked to the ground. Although injured and carried off the field, Tiemann held onto the sword. [25]

After the two officers were safely out of harm's way, the scabbard and belt were discovered by Confederate soldiers who presented the items to their commander, General Taylor. Under a flag of truce, Taylor returned the items to the Union line.[26]

Colonel Molineux being carried off the field at the Battle of Irish Bend, Artist unknown was a soldier with the 159th NYSV. [27]

---

24 Tiemann, William F. *The 159th Regiment Infantry, New York State Volunteers*, In the War of the Rebellion 1862 -1865, (Brooklyn New York 1891), pp. 31-35.

25 Young Sanders Center, Franklin LA, Letter from William F. Tiemann dated April 18, 1863.

26 *Genealogical and Biographical of the Molineux Families*, by Nellie Zada Rice Molineux & Edward Leslie Molineux.

27 *Colonel Molineux being carried off the field at the Battle of Irish Bend, Artist unknown was a soldier with the 159th N.Y.S.V.*, E. L. Molineux Photo Collection, NYS Military Museum, Property of Will Molineux.

Colonel Molineux described the battle scene in his diary:

"I received an order to charge bayonets and drive the rebels out. I obeyed but not being supported properly was outflanked and met with tremendous crossfire which cut us down in scores. Still the boys followed me but by the time we had reached within fifty feet of the woods I found in looking back that the men had been slaughtered by the two fires so dreadfully that the line was a broken mass without unity. I tried to reform them, but it was only murder and then I ordered them to lie down between rows of sugar mounds and fire. This we did for some five minutes. It was then I was wounded by a Mississippi rifle ball. I was cheering the boys and was raising on my feet to see how we could best protect ourselves from the rebels who outflanked us. Hardly had I raised on my knees when whack! And I tumbled to my face. The ball entered my open mouth tearing away the gums and teeth on the left side, scraped the jaw and out of the cheek."[28]

Colonel Molineux returned to New York spending time healing his wound. While he was there, he did an interview which was shared with a local Hudson newspaper:

we are indebted to our neighbor of Gazette (newspaper) for the following--The 159th Regiment, Colonel Edward L. Molineux, of the 159th Regiment, favored us with a call last week. He was badly wounded at the severe engagement at Irish Bend, Louisiana, on the 14th day of April, when so many gallant young men of Columbia County fell. While leading his regiment on to a bayonet charge, and uttering a word of command, a rifle ball entered his mouth, taking off one half of the upper jaw, and passed out at the center of his left cheek. His escape from death was miraculous, but bullets sometimes take mysterious freaks. We are happy to announce that the Colonel has recovered from his injury and left on Friday to join his regiment. He is a fine looking, athletic young man, of a slight frame but large courage, and he will be sure to win high distinction or an honorable grave. [29]

---

28 James I Robertson Jr Civil War Sesquicentennial Legacy Collection at The Library of Virginia – Edward L. Molineux collection, *War Stories 1863*.

29 NYS Military Museum & Veterans Research Center, Unit History Project, *159th NYS Regt. Newspaper Clippings, paper unknown*.

THE OPPONENT

General Banks celebrated his triumph at Irish Bend against his formidable opponent Confederate General Richard (Dick) Taylor, a Louisiana boy from St. Charles Parish. His father was the twelfth President of the United States, Zachary Taylor. He was also the former brother-in-law of Confederate President Jefferson Davis.

Taylor, a powerful plantation owner and Louisiana state senator, was known as a feisty character, an "army brat" with no interest in a military career. He attended Harvard University and graduated from Yale in 1845. He managed a sugar plantation for his Father, married a southern heiress and settled into life as an affluent planter. He was thirty-five and a member of the Louisiana State Senate when the state seceded. He joined the Confederacy with little military experience and took command of the 9th Louisiana Infantry regiment in July 1861. Proving himself an able combat commander, he was promoted to Brigadier General on October 21, 1861, Major General on July 28, 1862, and Lieutenant General to rank on April 8, 1864, becoming one of only three non-West Point graduates to do so.[30] In 1879, Richard Taylor wrote "Destruction and Reconstruction," his Civil War memoir, based on his experiences as a regimental officer of the Confederate Army. The memoir was published one week before his death in New York City on April 12, 1879.

FORAGING AND PLUNDERING

On April 15, 1863, the 159th NY regiment marched along the Bayou Teche waterway fifteen miles to McGuire's plantation where they camped for the night. The following day, they marched ten miles to the town of New Iberia meeting several of their comrades captured during the engagement, then liberated on parole. The next morning, the men started a twenty-mile march reaching Vermilion Bay and establishing camp. While encamped, one of the soldiers was found dead—he was shot while drawing water from a local well.[31]

A New Orleans correspondent for the Boston Traveler sent the Hudson Gazette the following story:

---

30 Young Sanders Center, Franklin LA, *General Richard (Dick) Taylor, article* from the Civil War Times Jan. 1985, v. 23, no. 9.

31 Tiemann, William F. *The 159th Regiment Infantry, New York State Volunteers*, In the War of the Rebellion 1862 -1865, (Brooklyn New York 1891), p. 36.

The assassination of a Federal soldier. As we reached Vermilionville Bayou on Friday, one of our soldiers, Corporal Appleton W. Rackett,[32] of Co A, 159th New York Regiment, went up to a well for the purpose of filling his canteen with fresh water. He was alone and nothing was seen of him for several hours, until soldiers belonging to other regiments went up to the well for water, and found the Corporal lying dead upon the ground, when Dr. C. A. Robertson, surgeon of the 159th New York regiment came up, and examining the wound he became convinced that the soldier had been shot from the adjoining house. He knocked at the door and was met by a French Creole, who pretended he could not speak no English. The surgeon inquired if he knew anything of the soldier's death, and he replied he had not. The Doctor determined to search the premises, and upon examination he discovered a shotgun, bullets and powder, the gun evidently having been used within a short time. He also found a complete Rebel uniform belonging to an orderly sergeant. The window fronting the well was open. And the evidence was so strong that the man was placed under arrest. The soldiers were so enraged upon learning the particulars of the murder that they gutted the house and set it on fire. A blackened pile of smoking ruins now marks the former dwelling of this heinous traitor.

We presume the destruction of this murderer's house will be regarded by disloyal men as another piece of "wanton vandalism" on the part of our soldiers, calculated to repress the growth of the Union sentiment among our "Southern brethren. [33]

Knowing the enemy was not in the immediate region following the battle, regiments took advantage of foraging for food. Throughout the region, men searched for provisions including beef, poultry, pork, and dried goods. One hoped Union soldiers would collect only what was necessary for a few fresh meals (because men cannot survive on a regular diet of hardtack and

---

32 Appleton W. Rackett, Corporal 159th Co. D: Age 26, killed April 17, 1863. From Southold in Suffolk, New York.

33 NYS Military Museum & Veterans Research Center, Unit History Project, 159th NYS Regt. Newspaper Clippings - *New Orleans correspondent of the Boston Traveler the assassination of a federal soldier.*

dried rancid beef), leaving plenty for civilians residing in the surrounding communities. But it was war, and often the men went hungry due to deficient regimental stores. Command gave the order to collect provisions, and the men followed, possibly without consideration for the poor residents also trying to survive the harshness of war.

A local distinguished Louisiana gentleman, of "high character," gave this graphic eyewitness account of foraging by the Union army:

"licentious march of the motley crowd," in the country through which the Union line passed, "The road, was filled with an indiscriminate mass of armed men on horseback and on foot, carts, wagons, cannon and caissons, rolling along in most tumultuous disorder, while to the right and to the left, joining the mass, and detaching from it, singly and in groups, were hundreds going empty handed and returning laden. Disregarding the lanes and pathways, they broke through fields and enclosures, spreading in every direction that promised plunder or attracted curiosity. Country carts, horses, mules and oxen, followed by negro men, women, and even children, (who were pressed into service to carry the plunder,) laden with every conceivable object, were approaching and mingling in the mass from every side. The most whimsical scenes presented themselves, at every step, horses and even gentle oxen, were pulled, pushed, and beaten along towards this seething current, with pigs, sheep, geese, ducks, and chickens swinging from their backs, fluttering, squealing, and quacking, while the burthened animals, in bewildered amazement, were endeavoring to escape from their persecutors. These' scenes, repeated at every step on my way to Carencro, was only varied on my return, by the diminished objects of plunder left for those that came after."[34]

On April 19, 1863, four companies of the 159th NY detached from the brigade with orders to collect horses and cattle. Successfully collecting about 5,000 head of cattle, along with a number of horses and mules, the men drove the animals back to Berwick. Along the way, they stopped at Fort Bisland in Franklin, leveling the enemy's earthworks before arriving in Berwick City on April 29, 1863. Immediately upon their arrival, orders were given to

---

34 Young Sanders Center, Franklin LA: *Official Report relative to the Conduct of Federal Troops in Western Louisiana during the Invasion of 1863 and 1864*. Compiled from Sworn Testimony, Under Direction of Governor Henry W. Allen, Shreveport, April 1865, p. 9.

escort a wagon train to Vermilion Bay.[35]

The 159th NY started for Opelousas the next morning, but John was not with them. He was ordered by Doctor Briggs to be left behind and recuperate from his illness at Elvira hospital in Berwick City. After experiencing six battles, feeling weak and sick, John took a moment to write home from his hospital bed.

Letter #14 - April 29, 1863

La. Brashear City

Since writing my last letter I thought you might feel uneasy about me. I was very sick and became exhausted by marching and got diarrhea and it must be something it run me down very fast and now I am very coe in flesh and very weak. I have been in the wagon every day since on the march this morning, Doctor Briggs said he would leave me in the Hospital at Brashear City to recoup up. I don't want you to feel worried about me, I don't think I am dangerously sick I sent $30 to you by express and said it would be to Churchtown, but if it doesn't come there by the time you receive this, then you go to the express office in Hudson and inquire there for it. I sent it with Lieut. Colonel Draper before we left Bayou Boeuf, and he sent it to New Orleans and he was killed and I have no receipt for it. I think it will all come safe as all the rest sent the same way and got no receipt. You need not say to the officers of the express that you have not received the receipt, let me know if you receive it. I have been in six battles and am still safe yet. The first you could not call a battle, at Baton Rouge. The second was on the Clinton Roads, we were within four miles from Port Hudson, as you heard our business was to destroy a bridge and we did it. The third was at Irish Bend, when we landed the bombs and bullets flew pretty well, they heaved a dozen bombs right over the vessel I was on. I expected to see one come right through the ship every minute. When they were fighting, I got down and ate my breakfast as if nothing was going on. They shot the Lieut. Col. of

---

35 Tiemann, William F. *The 159th Regiment Infantry, New York State Volunteers*, In the War of the Rebellion 1862-1865, (Brooklyn New York 1891), p. 36.

the 1st Louisiana Regiment right through the leg. The fourth was at Franklin the same day the bombs and shells flew, they wounded several of our men and killed one of Billy Wilsons men, his insides all out. The fifth was the next day the battle I wrote about in my other letter. After that our Regiment was thrown in the rear, it being cut all to pieces. So, we became the rear guard, before that we were the advance guard, two days after that our front worked over the Rebels and had quite a brush and as I was going along, I saw a large number of horses and one Rebel lay along the road dead. The Rebels got ahead and crossed a bridge and burnt it, and then we had to fix it and then they turned, and we had another fight. They killed two or three of our men and wounded several of our Regiment. Six Companies went on with the Army, four Companies were ordered to go back to Brashear City and drive all the cows, horses and mules, and take all the cotton and clean the country, and so we did. You will see in the papers what we have been doing, this expedition has done a great deal to finish this rebellion. The prisoners too, we have captured a large number, and they say they are done, we strip the country and leave them to starve and they were done. I think the war has to be done now. Direct your letters to Bank Expedition, 159th Regt. N.Y.V., New Orleans or elsewhere, La. The weather is very warm, blackberries are ripe and by the thousands. Corn is waste high, we will have young potatoes in a short time. Write soon, trust God that he might guide and protect me and bring me back again to greet you in my native town.

Yours Xe, John W. Mambert [36]

TECHE REGION STRIPPED CLEAN

For weeks following the battles at Franklin and Irish Bend, it was said the Teche region was stripped clean of cotton, sugar, and cattle—the confiscated commodities valued in the millions.

The region was described as "diminished" and "desolate" in a published report released by the Governor of Louisiana investigating the atrocities of the Union army.

---

36 *Letter #14: April 29, 1863, La. Brashear City.* John W. Mambert letters from the Civil War.

"the absence of the domestic animals through St. Landry and Lafayette, where the broad parries sweep down to the road, may yet be been a few cattle that have wandered in by the way side; but along the Teche, animal life diminished at every step, until, below Franklin, even the most necessary domestic animals disappeared. For miles nothing could be seen but the vulture brooding, from some scattered tree top, over the desolate scene; or the hawk, flying low, in search out his prey, over the tangled thickets usurping the once cultivated fields."[37]

On the other hand, in a letter from Captain Tiemann to his father, he described the same country with "corn and beef in abundance."

"They talk about starving out the rebels. Why! It is a thing utterly impossible. They have corn & beef in abundance. Our regiment drove in over 10,000 head of cattle & say there is still over 100,000 head still running over the country. Wherever we go it is the same. The corn cribs are full & the cattle still running over the country. Sugar & molasses they have thousands of barrels of."[38]

The two conflicting opinions couldn't be further apart.

BILLY WILSON'S MEN

John mentioned Billy Wilson's men, who were the 6th Infantry NYS Volunteer Infantry Regiment recruited and organized by Colonel William Wilson on Staten Island, NY. They were accepted as a regiment by the State on May 22, 1861, and mustered into service of the United States for two years.

---

37 Young Sanders Center, Franklin LA: *Official Report relative to the Conduct of Federal Troops in Western Louisiana during the Invasion of 1863 and 1864.* Compiled from Sworn Testimony, Under Direction of Governor Henry W. Allen, Shreveport, April 1865, p. 12.
38 Young Sanders Center, Franklin LA, *Letter from William F. Tiemann dated May 9, 1863.*

CHAPTER THREE

Nicknamed "The (Billy) Wilson Zouaves" at Tammany Hall they took the oath of fidelity to the flag on April 24, 1861. Colonel Wilson was among the first to offer his services to the government when war broke out. He recruited a regiment of nearly twelve hundred men from the rowdy and criminal classes of New York City. The regiment was formally mustered in at the old Tammany Hall. There, with the men arranged around the room and the officers in the center, the Colonel with a sword in one hand and the American flag in the other, led the men into swearing to 'support the flag and never to flinch from its path through blood or death.' The Zouaves, A few days later the Zouaves left for the South." —Frank Leslie 1896 [39]

The 6th NY became part of Grover's 4th Division on April 1, 1863, and joined them in Brashear, Louisiana, prior to the transport's departure for Irish Bend. They fought alongside the 159th NY throughout the region, also assisting the 159th companies assigned to collect horses and cattle, driving them back to Berwick City. The regiment remained with the division until June 10, 1863. Returning to New York City, they were mustered out of service on June 25, 1863.

---

39 *The Billy Wilson Zouaves*, artist Frank Leslie.

Before their departure from the Department of the Gulf, the following special order was issued:

HEADQUARTERS, 1ST BRIGADE, GROVER'S DIVISION,
ALEXANDRIA, LA., May 14th, 1863.

[Special Orders, No 43.]

Lieutenant Colonel Cassidy, the Officers, non-commissioned Officers and Privates of the 6th Regiment, New York Volunteers: The commanding general of the 1st brigade cannot allow the 6th regiment to leave the department of the Gulf and the service of the United States without conveying to them his high appreciation of their conduct as men and their valor as soldiers during the present movement. Since the landing of this command at Irish Bend, La., on the 13th of April; until the arrival at Alexandria on the 8th of May, 1863, an interval in which the regiment endured the hardships of severe marching under an almost tropical sun, and during which they encountered the enemy three times, sustaining well their reputation for endurance and bravery. The members of the 6th regiment, officers and men, carry with them the earnest desire of the commanding general of the 1st brigade for their future welfare and happiness, mingled with a regret that the Government should have lost the services of this regiment, though the time has arrived for its members to enjoy their well merited repose.

By command of WILLIAM. DWIGHT, Brig. Gen'l.
Commanding First Brigade[40]

LEFT BEHIND TO RECUPERATE

Contaminated water supplies were concerning, and a major issue when it came to a soldier's maintaining proper health. Lack of sanitation was the principal reason for the ongoing problem. John, still sick, next writes to his family:

"I was taken with diarrhea on the march by contaminated water and then was

---

40 *6th NYS Infantry Regiment Civil War Historical Sketch*, NYS Military Museum & Veterans Research Center.

taken with vomiting and lost flesh in 3 days that you would hardly know me."

His letter gives a glimpse of what goes through the mind of a soldier sickened by tainted water. He had already watched many of his comrades die around him with the same symptoms.

Letter #15 - May 3, 1863

Berwick City, La.

I thought I would write this morning and let you know how I get along; I am getting better. I am in the Hospital at the City above mentioned instead of Brashear City. Brashear City lays on the east side of the river, the same as Hudson, and Berwick City on the west side, the same as Athens. There is a railroad that runs from New Orleans to Brashear City, 83 miles. I have been very sick; I was taken with diarrhea on the march by contaminated water and then was taken with vomiting and lost flesh in 3 days that you would hardly know me. When Doctor Briggs[41] saw me, he said to me you're on the sick list, you must go to the hospital. Just as you say I said and he went off and was gone awhile, and then he sent two men with me to carry my things. The other sick ones and me went to a large and convenient building standing half over the river two stories high with a narrow dock on the river side say 16 feet, a place to fish when I get so I can. So, we were put overhead in the upper story, there is about 30 beds from bedstands for one to lay on, they gave me a very good bed, right by a window in the southeast corner. Where I can lay and look right at the river and see the boats run. The Doctor looked at me and he told me I was exhausted; he said my case was not that bad. He thought rest and a good diet would soon bring me up again, but yesterday morning he seemed to think there was a little more to matter with me than that. For I found I wasn't quite so well as I had been, so he changed the medicine and this morning I feel quite easy, I made up my own bed and that I wasn't able to do before.

---

41 Caleb C. Briggs, MD, Assistant Surgeon of the 159th regiment: Age 36, doctor from Martindale in Claverack, NY. Lived with wife & brother-in-law.

But I must confess that the Doctor is a perfect gentleman, I don't mean our Regiment Doctor the Hospital Doctor. I have the very best of attendance and I can ask no better, the nurses are perfect gentlemen, and eating I have enough of everything, such as beef soup, beef T, toast-bread, custard pudding and everything, tea, or coffee and as much as you want. I could not be taken better care of at home. We have mosquito bars to put over the beds which the nurses put over every night. In the morning, they bring soap, a towel and wash dish and after that they come around with the medicine and then breakfast. I am doing very well, and I think I shall soon be able to join the Regiment again. They went up the Red River and have raised the old Nick with the Rebel. One thing I forgot to mention in my letter about that seventh battle was that they dropped a shell about 10 to 15 paces behind me, about the time I was a going to turn back and get in the ditch, but happily it did not explode. I don't know what the consequences would have been if it had, I picked it up after the battle was over. We fought several battles after that of which I have not written, and when I left to go back I heard they fought several more afterwards. After that battle it was a sight to see the wounded and dead. Some wounded in the head, arms, body, leg and neck, Ed Mc Lean[42] said he did not feel afraid until a ball passed between his adams apple, and his chin and the wind blew as you would blow with a bellow and squealed like a rooster, then he said his eyes stuck out; he came out all safe. John H. Allen[43] came out all safe but since he is sick, he was left in Newtown or New Iberia as it is called. Little John Kipps[44] son is dead, he was shot through the knee, and they had to amputate his leg and he died. I helped carry him to the hospital from the field. Rich Raught[45] went off with Ez.

---

42 Edward McLean, 159th regiment Co. E: Age 25, died at hospital in Thibodaux Oct. 7, 1863. A farm laborer for Charles Jones in Smokey Hollow, Claverack NY.

43 John H. Allen, Private 159th regiment Co. C: Age 32, promoted to Corporal. Lived in Taghkanic, NY with his wife and four children ranging in age from 1 to 7 years old.

44 Robert Kipp, Private 159th regiment Co. A: Age 19, died of wounds received at Irish Bend L.A. April 14, 1863. Farm laborer from Greenport, NY. Lived and worked for his father "little John Kipp" who he left behind along with his mother and four siblings.

45 Richard Raught, Private 159th regiment Co. E: Age 40, deserted at Baton Rouge L.A. Jan 25, 1863. From Claverack, NY.

& Eph. Stickles and have not heard from them since, only that one of the Stickles had been shot and the other hung as spies, but I don't credit the report; they maybe home now. It is very warm here we have to pay $5 a piece for eggs here, I have got $6 with me, I shall get me a bottle of wine in a short time if I can. I want you all to write and direct your letters to New Orleans, 159th Regiment, Bank Expedition, or elsewhere. I don't know how long I will be here, or you might direct them here, I heard there was a letter for me before I got here and papers, but I suppose they are gone to the Regiment, and I don't know when I will get them. Tell Jobe I read his letters, he gives me more news than all the rest. Tell Walt to write and tell Permilia I read her letters and Marthas too. I got the letter with the stamped envelope; they were detained at New Orleans, they got wet in coming from New York. Maria, you must write, I have heard nothing from you yet. You must go to Hudson to the express office and see about that money, if it doesn't come to Churchtown, write, and let me know. I write just as I am, I don't want you to think I make it right to ease you. Write soon.

Yours John W. Mambert [46]

THE SANITARY COMMISSION

Disease caused by unsanitary conditions should have been the number one concern for soldiers; after all, it was the number one killer. Before the war ended, more than 400,000 men died from one of the many illnesses commonly spread in the camps.

Joseph Janvier Woodward, M.D., Surgeon U.S. Army, documented breaches of sanitation in camps:

"In great armies in time of war personal cleanliness is often nonexistent. The men are unwashed, their clothes filthy, bodies full of vermin, and heaps of garbage lie about." "Especially needed was policing the latrines. The trench is too shallow not the requisite five feet deep, daily covering with dirt is entirely neglected." "Large numbers of the men will not use the sinks latrines but instead every clump of bushes and every fence border." It is

---

46 *Letter #15: May 3, 1863, Berwick City, La.* John W. Mambert letters from the Civil War.

impossible to step outside the encampment without having "both eye and nostril continually offended." [47]

An army sanitary program was desperately needed but was dependent upon the cooperation of officers and surgeons to enforce it—this came hard. All too many line officers combined complacent ignorance with a lazy tolerance of their men's dirty habits. Soldiers fell ill to diseases caused by bacteria that flourished and spread in these unsanitary conditions. Diarrhea, dysentery, typhoid fever, and food poisoning from Salmonella were among the consequences of unsanitary practices.

Finally on June 9, 1861, Order No 3 of the Secretary of War established the Sanitary Commission. It was decided that the Commission, along with a designated Surgeon of the Federal Government, "will review the principles and practices connected with the inspection of recruits and enlisted men, the sanitary condition of the volunteers to the means of preserving and restoring the health and of securing the general comfort and efficiency of troops to the proper provision of cooks, nurses and hospitals." [48]

These changes within the military did not happen quickly, sometimes not at all. In 1861, John Strong Newberry, M.D. of the U.S. Sanitary Commission reported on "contamination of the soldiers' water supply" caused, in part, by the construction of army latrines "in the vicinity which slopes down to the stream from which all water in the camp is obtained." Clearly, unhealthy habits caused unsanitary conditions and illness to run rampant among the troops.[49]

It was unlikely that surgeons were aware of these ill-effects to soldiers; they were too busy with twice daily surgeon calls, examining soldiers reporting to sick bay. Surgeons were overwhelmed, and no matter how much help they sought or received, it was never enough. Through it all, the Sanitary Commission was persistent and did make progress, most notably in the care of wounded and sick men. But the diet for soldiers in the field was still a problem. The commission wanted fresh fruits and vegetables included in army rations.

---

47 Lemuel Shattuck reported on a general plan for the promotion of general and public health devised, prepared and recommended by the Commissioners appointed under a resolution of the Legislature of Massachusetts, relating to a sanitary survey of the State (1850). This report is one of the fundamental documents in public health in the United States.

48 History of the United States Sanitary Commission, being *The General Report of its Work During the War of the Rebellion*, by Charles J. Stille. p. 532.

49 Ibid., Lemuel Shattuck report.

(Unfortunately, my research shows their attempts failed.) An example of the Sanitary Commission's success in following-through is found in the conditions under which John was placed to recuperate. Five days after the battle at Irish Bend, the Sanitary Commission established a hospital in Berwick City.

The next day, hundreds of wounded men from the battlefield were transferred to Berwick City hospital using flatboats, skiffs and small steamers. The wounded men covered with blood and black soot from battle arrived at a much different facility than what the Sanitary Commission secured only hours prior.

The buildings chosen were deserted, bordering the Atchafalaya River. There were storefronts below and tenements above, also a large adjoining building. Empty, the rooms were gloomy and covered with dust and cobwebs, and totaled about 1,500 yards of open filthy floor space. The buildings were transformed into a clean, sanitary and comfortable hospital by the U.S. Sanitary Commission.[50]

Upon arrival of the wounded, the nursing staff washed the injured soldiers, giving them places to rest in iron beds neatly arranged in rows with good mattresses, clean sheets and pillows. The nursing staff then cleaned wounds, dressing them with clean bandages and lint dressings. The men were provided with clean shirts, drawers and towels. Flowers in vases were placed on tables with pitchers of ice water accompanied by wholesome food. The injured men could sleep comfortably under mosquito nets.

THE NURSES

In August 1862, an order was issued by the War Department requiring all able-bodied enlisted men doing hospital duty to return to their regiments. Before this order, most nursing was provided by soldiers, (either musicians or the wounded), sick men helping to care for those not capable. From this point on, Washington employed civilian nurses to take their place. The men acting as nurses were needed back in their regiments, but even more important was the realization that the wounded had a better chance of survival under the care of skilled medical staff.[51]

---

50 Ibid., pp. 428-429.

51 United States Sanitary Commission. Western Department & Newberry, J. S. (1871) The
  U. S. Sanitary Commission in the valley of the Mississippi, during the war of the rebellion,
  -1866. Cleveland, Fairbanks, Benedict & Co., printers. [Pdf] Retrieved from the Library of
  Congress, https://www.loc.gov/item/01010332/. pp. 51-52.

The Sanitary Commission replaced the unqualified with well-trained, and experienced men, competent nurses who were paid a moderate compensation.[52]

An observation of the USSC nurses reported by Reverend Sloan noted:

"Those who have labored in this noble cause have found that far more is done than talking, distributing publications, and praying. They have had to nurse, dress wounds, strip off filthy garments, wash from helpless soldiers the blood and dust of hard fights and hard marches; cleanse them of vermin, and put upon them clean and comfortable clothing; dig graves for the dead; lift and open boxes; make wearisome visits on foot; sleep on the ground, or floor, or bags, or boxes, and often work from daylight until midnight, or all night long, with little to eat except dry bread or crackers, and meat without cooking." [53]

THE ADJUTANT'S FATHER

On May 4, 1863, four companies of the 159th NY received orders to rejoin their brigade ninety-five miles north at Barre's Landing on Bayou Courtableau. John remained in the hospital.

The next day, the companies marched toward the main army two days in advance of them. The men marched seventy-seven miles on roads covered with six inches of dust in extreme heat. Totally exhausted, their feet blistered from the long march. Knowing his men needed rest, General Grover halted the troops, allowing them to remain camped through the next day.

The following day, the men had a short march to Well's Plantation where they rested once again. This time, they were surrounded by vast, beautiful fields, acres of corn one foot high, with dense woods in the background. They saw little cotton or sugar cane because Confederate President Davis instructed plantation masters: "plant corn for the needs of the people." [54]

The 159th NY remained in camp for three days. On the last day, they received a visit from Captain Gideon Lathrop of Stockport, a well-known steamship Captain who navigated on the Hudson River. The captain, recognized as "the

---

52 History of the United States Sanitary Commission, being The General Report of its Work During the War of the Rebellion, by Charles J. Stille. p. 258.

53 Moss, L. (1868) Annals of the United States Christian commission. Philadelphia, J. B. Lippincott & co. [Pdf] Retrieved from the Library of Congress, https://www.loc.gov/item/02018786/. p. 364.

54 Young Sanders Center, Franklin LA, *Letter from William F. Tiemann dated May 9, 1863.*

Adjutant's father," was greeted with expressions of sympathy by the men.

Captain Lathrop arrived in Louisiana soon after the battle of Irish Bend with hopes of procuring the body of his son Robert. D. Lathrop. While visiting the regiment, he discovered Robert was buried along with three fellow officers, Lieutenant Colonel Draper, Lieutenant Byron Lockwood, and Lieutenant Manley. They were all interred on the grounds of Mr. C. T. Carlin in Franklin, twenty-eight miles from Brashier City. An officer furnished him with a diagram of the premises and the precise location of the graves. The captain also learned that Robert Kipp, Joseph Snyder and James Reynolds were interred at the forts in Brashier City: buried in separate coffins, but all under one grave marker.

His visit with the regiment was cut short. On May 14, 1863, orders were issued to move down the Mississippi River and proceed to their next destination. Unfortunately, battle conditions near Franklin made it impossible for Captain Lathrop to retrieve his son's body. But he hoped information obtained would help friends reclaim the body of the brave officer at a later date.[55]

First Lieutenant Byron L. Lockwood[56]

---

55 NYS Military Museum & Veterans Research Center, Unit History Project, 159th NYS Regt. Newspaper Clippings, *Return of R. D. Lathrop.*

56 *First-Lieutenant Byron L. Lockwood*, photo, United States Army Heritage and Education Center, Carlisle, PA.

Lieutenant Colonel Gilbert A. Draper [57]

The battle at Irish Bend reduced the 159th NY numbers considerably, including the loss of eight officers. Then, news reached the regiment of another loss: First Lieutenant Bradley, sick in the hospital, had died.[58]

First-Lieutenant Bradley[59]

---

57 *Lieutenant Colonel Gilbert A. Draper*, photo, United States Army Heritage and Education Center, Carlisle, PA.

58 Tiemann, William F. *The 159th Regiment Infantry, New York State Volunteers*, In the War of the Rebellion 1862 -1865, (Brooklyn New York 1891), p. 37.

59 *First-Lieutenant Bradley*, photo, United States Army Heritage and Education Center, Carlisle, PA.

CHAPTER THREE

THE PROCLAMATION

John shared with his family thoughts on President Lincoln's proclamation, believing it would bring an end to the Confederacy. He wrote:

"There is but one way to put down the rebellion, to stand up for the President's proclamation, as long as slavery exists there will be abolition and as long as there is slavery and abolition there will be war."

Knowing no good would come to the country if slavery were allowed to continue, he also believed enlisting the freedmen would help bring the conflict to an end—and it could be his ticket home.

Letter #16 - May 16, 1863

Louisiana - Berwick City

To Mises or Mrs. Maria Mambert and Family

Dear Wife, I thought I would write a few lines as plain as I can so you can read it yourself. I am getting along very good; I don't know how long I will stay here in the Hospital. There is a camp right across the river in Brashear City where they send us from here to get strength, and from there to the Regiment. I feel middling good, only very weak, my leg bends under my body. My appetite is so strong that I can eat anything, I can see that I have more than I can do to guard against it. I get enough to eat; I bought a bottle of wine, but it is good for nothing I give $1 for it. It wasn't worth 25 cents, your wine would be worth between $3 to $5 per bottle here. I dreamed last night that I was at home and how glad I was. I thought I kissed Jobe and Walt. I thought you were not at home. I thought the neighbors came to see me, but before I saw you, I woke up and found it to be a dream. I hope and trust to God that I may soon see home again, the prospect is somewhat encouraging, General Banks told us that in 60 days from the 4th day of April we would be pretty near ready to go home. I think there is a prospect of it for our expedition is doing much to kill it. We strip the country where we go, I have talked to Rebels and they say they are done in, one said he wished we would leave the country, but he wanted us to take all the Negroes with us. He said he did not know what would become of them, he wished they was all out of the country. They always were a curse, but they had the institution and they got rid of it and the country is ruined now. He said that there was a large majority for the Union, but their head men made false returns and the majority could not help themselves, he said the south was gone and ruined. I told him that if they had represented themselves in Congress before the 1st of January 1863 they would have been protected. He said their Representatives would not do it and so it was. I told him that their Representatives said all the Representatives they need was Stonewall Jackson[60], Lee Davis[61] and Beauregard[62], I told him if that was all then they must suffer the consequences. Yesterday there was 500 or 600 Negroes came here from Texas, they said the Texans were shooting all the Negroes they could get sight of and they were all fleeing from the country. After the poor Negroes have made their farms and dollars and made them all they ever had, now they shoot them down like crows which pulls up the corn. There is but one way to put down the rebellion, to stand up for the Presidents proclamation, as long as slavery exists there will be abolition and as long as there is slavery and abolition there will be war, and if you protect slavery you will never have permanent peace because abolition has just as good a right to say that slaves aught to be free as the rest have to say there aught to be slaves, and so you see there will always be quarreling. So if there is no slavery then there can be no abolition and if there is non abolition or slavery there will be nothing to fight about, so you see there would be no war about that so wipe slavery out and war will cease. Arm every Negro and make a soldier of him and make him fight and let the white soldier go home, there is 4 times as many Negroes in this State as Whites, and

---

60 Thomas Jonathan *"Stonewall"* Jackson:(January 21, 1824 – May 10, 1863) served as a Confederate general (1861–1863) during the American Civil War, and became one of the best-known Confederate commanders after General Robert E. Lee.

61 Jefferson Lee Davis: (June 3, 1808 – December 6, 1889) was an American politician who served as the president of the Confederate States from 1861 to 1865. As a member of the Democratic Party, he represented Mississippi in the United States Senate and the House of Representatives before the American Civil War. He previously served as the United States Secretary of War from 1853 to 1857.

62 Pierre Gustave Toutant-Beauregard: (May 28, 1818 – February 20, 1893) was an American military officer who was the first prominent general of the Confederate States Army during the American Civil War. He became the first brigadier general in the Confederate States Army.

if these Negroes were all armed, I don't see what would hinder them to keep the secessionists down. We soon expect to take Port Hudson and Vicksburg if it is not already taken. If we want more soldiers, we expect you will try to send them, and we will make as many Regiment of Blacks as we can here to do the drudgery and digging. We left the Negro Regiment at Baton Rouge and the Rebels thought they could easily take it, so they tried, but I tell you the Negroes heisted them, they fought like tigers. They knew what the result of been if they had been licked. The Negroes say if you don't believe we can fight just let us try, we will fight for our liberty as long as we can raise a hand. We have heard good news from Virginia that Fredericksburg was taken by our forces. I hope the next news will be that Richmond and Charlette are in our possession and then I think I should soon be at home again. Rebellion is pretty near played out here. I was informed that our Regiment would be back this way in a short time. You will hear good news from us on our expedition, we are doing well. The sick in the hospital it is full they die off very fast, one died right at my feet. They told him he was dying, he said do you think so, they said yes and that he would not live till morning, he laid a few moments and then said "well that is rough", and then gave orders what to do with his things. This was about 1:00 o'clock at night and about daylight he died. Lieut. Bradley[63] from Hudson is dead, he was a stout rugged man, he died in the Hospital, they are buried decent who die in the Hospital, they have a good coffin, plain white wood no paint about it but good and stout. And they are all buried in a row and a board like a tombstone got up by their head with their name, Regiment and age. These boards as a matter don't stand long so if anyone has friends die here in the hospital, they can fetch them if they think proper. They can get metallic cases in New Orleans and put them on a ship, I don't think they charge very much, they are airtight. This country is level the rivers are higher than the land when it is high water, I have seen it 3 feet above the ground, the rivers are all banked up like the ground around a ditch and when

---

63 Wesley Bradley, 2nd Lieutenant 159th Regiment Co. A: Age 32, promoted to 1st Lieutenant. Co. E. Died in New Orleans Hospital, May 10, 1863. From Hudson NY., left behind a wife with three children, ranging in age from 4 to 6 years old.

you walk from the river it looks as if you were going downhill. This is all I will write today. You can send your letters, one at least to Brashear City Hospital. I will put a card in here now, but you must not put confidence in my receiving it for I don't know where I may be, but it may follow the Regiment. So, you must not put anything valuable in the letter so there will be any loss if I don't get it.

John W. Mambert [64]

INTERRING THE DEAD

Wherever the military encamped, whether during battle engagements or extended quarters, a suitable burial location was selected for the dead to be interred: this also applied to any hospitals, field or otherwise. Comrades held a ceremony for the departed; they would fall in and follow the soldier's remains preceded by fife and drum to the chosen resting place. The fife, John or maybe Eli, played a dirge with the drum accompaniment. A simple coffin constructed of pine was used. As the survivors marched to the solemn notes of the dirge, many tears were shed in silence. Once the coffin was placed in its resting spot, a detail would fire a salute over the grave. Chaplain Kipp prayed, and the graveside ceremony ended. The procession line reformed, marching back in quick time. The firing detail marched with arms reversed and the fife and drum played a lively tune such as, "The girl I left behind me." A wooden marker was placed at the top of the grave.[65] Sadly, the 159th NY repeated this agonizing ceremony far too often, given the number of men who succumbed to illness, or the brave who perished in battle.

LINCOLN'S ULTIMATUM

John had a number of conversations with Confederate prisoners. In his discussions, he noted the South's ruin due to their "head men" lying and misrepresenting the majority of the people appears to be a time-honored tradition by government representatives. John referenced the date January

---

64 *Letter #16: May 16, 1863, Louisiana - Berwick City, To Family.* John W. Mambert letters from the Civil War.

65 *The Third New Hampshire and All About It*, by D. Eldredge, Captain 3rdNH Volunteer Infantry, Boston, Mass. Press of E.B. Stillings and Company, 1893.

1, 1863, when President Lincoln issued the Emancipation Proclamation as the nation was approaching its third year of bloody war. The proclamation declared "that all persons held as slaves" within the rebellious states "are, and henceforward shall be free."

Despite this expansive wording, the Emancipation Proclamation was limited in many ways. It applied only to states that seceded from the United States, leaving slavery untouched in states bordering the loyal Union. It also expressly exempted parts of the Confederacy already under Northern control. Most important, the freedom it promised depended upon the Union winning the war.

Although the Emancipation Proclamation did not end slavery in the nation, it captured the hearts and imagination of millions of Americans and fundamentally transformed the character of the war into a war for freedom. It added moral force to the Union cause and strengthened the Union both militarily and politically. As a milestone along the road to slavery's final destruction, the Emancipation Proclamation assumed a place among the great documents of human freedom.[66]

The proclamation became a guide for the Union army, a declaration of freedom taking effect as the army continued advancing into Confederate territory. Confederate states affected by the proclamation were: Arkansas, Louisiana, Tennessee, Mississippi, Alabama, Georgia, Florida, South Carolina, North Carolina, Virginia, and Texas.

In John's letter, he mentions how Negroes were fleeing Texas in large numbers and coming into Union controlled Louisiana. When the Emancipation Proclamation became effective, it meant nothing to the South. The Federal government had no power over the Confederacy. But it did mean something to enslaved people: hope. Word of the declaration traveled throughout the Confederate states by decree and word of mouth, although it could not be enforced without Union presence. The Union army found Texas especially difficult to occupy, leaving it unscathed, and the treatment of slaves remained relatively unaffected. In fact, the state was considered a refuge for planters (with their slaves) fleeing Louisiana and Mississippi.

The Union army had two options to occupy Texas: First, by the state's southern coastline. It was far from east Texas counties where about 200,000 slaves existed. Second, entrance by the interior from the east through

---

66 The National Archives. *The Emancipation Proclamation*. Archives.gov/

Louisiana—the most noteworthy attempt would be General Banks's Red River campaign failure.

Throughout the Confederacy, including Texas, once slaves learned of the proclamation, many made a run for freedom. Regrettably, many Black persons were murdered, lynched, and harassed by Texans. Slave patrols scoured the countryside looking for the runaway Blacks; if caught, they were beaten and sometimes killed.

The Emancipation Proclamation's international importance was far greater. The Union's suspension on the world's largest source of cotton became a general calamity. Confederate officials hoped the English and French governments would intervene in the conflict, but with the struggle turning into a crusade against slavery, it made European intervention impossible. The Emancipation Proclamation did more than lift the war to the level of a crusade for human freedom, it brought substantial practical results allowing the Union to recruit Black soldiers. The Black population responded in considerable numbers when invited to join the Union army. By the end of the war, almost 200,000 Black soldiers and sailors fought for the Union and their freedom, enabling the liberated to become liberators.

Also, the Confederacy understood there was no turning back the clock. The Emancipation Proclamation made the promise that the Civil War would change the United States forever. The proclamation dealt a deathblow to slavery in the United States. As a temporary war measure, the proclamation needed to be codified into law with an amendment to the Constitution. Congress officially outlawed slavery with the passage of the 13th Amendment on January 31, 1865. [67]

In June 1865, as the war slowly came to an end, two thousand Federal soldiers of the 13th Army Corps arrived in Galveston, Texas. Two days later on June 19, 1865, they made the announcement—all slaves were hereby freed.

---

67 Britannica, The Editors of Encyclopedia. *"Emancipation Proclamation."* Encyclopedia Britannica, 20 Feb. 2023, https://www.britannica.com/event/Emancipation-Proclamation. Accessed 21 March 2023.

Chapter Four

"My mind is more settled and composed. I think I am pretty well climate, since having that dreadful shock on Ship Island"

John remained at Elvira hospital in Berwick City recuperating while the 159th NY continued onward to the next engagement. It would be three months before his return to the regiment. John was hoping to reunite with his comrades sooner. He didn't want to miss out on the excitement of, "raising the old Nick with the Rebels." Instead, he was hospitalized longer than expected. John's life as a convalescent was not as dull as one might expect. He later described this time as a "living hell," much like life for his regiment at Port Hudson.

PORT HUDSON

General Banks received information that Confederate troops were moving from Port Hudson to support forces defending Vicksburg. With this unexpected opportunity, he began planning an invasion of the garrison before the troops heading for Vicksburg could return. Banks divided his army into two columns: one group of three divisions approaching Port Hudson from Alexandria, and a second column of two divisions advancing north from Baton Rouge.

General Banks delayed organizing his troops for a week in preparation for the attack on the garrison. His untimely delay allowed General Gardner, commander of the Confederate defenders, time needed to reinforce garrison lines while moving additional artillery into position.

Marching under Banks's orders, the 159th NY reached the shores of the Atchafalaya River on May 17, 1863, where they set up camp. The following afternoon, they were transported across the river on large flatboats to Simmesport. Marching a short distance from the river's edge, they went into camp. On May 21, 1863, the regiment packed their items, broke camp, and marched back to the river where they embarked on the transport Empire Parish.

The next day, they arrived in Bayou Sara, on the east bank of the Mississippi River, fourteen miles above Port Hudson. The 159th NY commenced an advance to the front line, supporting a section of artillery strategically placed on the road heading toward Port Hudson. The regiment remained there until their brigade caught up: the two columns combined created a force of 30,000 men. Banks's army surrounded the Confederate force defending Port Hudson. The Union Troops outnumbered the enemy four to one.

On May 25, 1863, the advance resumed at 11:30 a.m. Soldiers marched to the outer rifle pits of the garrison, and action commenced with the regiment skirmishers pressing forward on the rebel position.

The enemy constructed their works to cover the entire bend in the Mississippi River at Port Hudson; it extended three miles with heavy batteries strategically placed protecting it. Inland, the garrison's fortifications[1] were about seven miles long encircling the rear of Port Hudson.[2] The Union positioned mortar[3] and gunboats above and below the enemy defenses. North of the enemy's works, on their right, was General Weitzel's division, to his left Grover's division and General Dwight's division. To the left of the works were the divisions of generals Auger and T.W. Sherman.

As the Union army approached, the enemy sharply opened fire. The 159th NY, along with the 13th and 25th Connecticut regiments, advanced through the woods, staying on the right of Bayou Sara Road: doing so while the enemy was strongly opposing them.

Shells bursting overhead were tearing away great limbs on trees as the solid shot ricocheted over their heads while pressing the enemy positioned in their outer line. The regiment remained under intense fire all day and were thoroughly tired. When nightfall finally arrived, there was a much-needed temporary halt.

---

1 Fortifications: Make a defensive position stronger. For example, high mounds of earth to protect cannon, or spiky breastworks to slow an enemy charge. Fortifications may be man-made structures or part of a natural terrain. Man-made fortifications could be permanent (mortar or stone) or temporary (wood and soil). Natural fortifications could include waterways, forest, hills and mountains, swamps, and marshes.

2 Tiemann, William F. *The 159th Regiment Infantry, New York State Volunteers*, In the War of the Rebellion 1862 -1865, (Brooklyn New York 1891), p. 38.

3 Mortar: An unrifled artillery gun which was designed to launch shells over walls and enemy fortifications. The most famous Civil War mortar is the "Dictator," a mortar which was mounted on a railroad car and used during the siege of Petersburg. With its 13-inch bore, it was capable of launching two hundred-pound shells.

CHAPTER FOUR

One soldier was killed from the 159th NY while several were wounded during this first exchange against the well-defended Confederate garrison.

The following day, the 159th NY marched to the front supporting the skirmishers; the musketry was immense and deafening. The men advanced on the garrison as rapidly as possible, finding it difficult to dislodge the enemy from their well-defended works. For the next eighteen days, the Union force kept a ceaseless, harassing fire upon the garrison night and day, giving the enemy no rest or sleep.

"Forward!" was the word as the army made their assault. The 159th NY, in the extreme advance as pickets, had enemy sharpshooters keeping a constant fire on their position. Advancing, they climbed a perpendicular slope, reaching just under the parapet while the enemy fire fell like hail upon them. The Color Bearer died with the staff shot from his hands, but it was secured and carried again. The 159th NY captured a Confederate captain along with eight sharpshooters in an ambush outside the works. This accomplishment felt trivial considering the regiment sustained losses of twenty-one men killed and thirty-eight wounded that day.[4]

On June 13, 1863, another assault began at 11:15 a.m. with a constant shelling of the Confederate works lasting one hour. When the guns silenced, Banks sent a message to the enemy commander demanding he surrender the garrison. Gardner's reply was, "My duty requires me to defend this position, and therefore I decline to surrender."[5] Banks then continued the bombardment which lasted throughout the night.

The next day, Banks gave orders to his division commanders for what was supposed to be a simultaneous three-pronged infantry attack. Unfortunately, the lack of an agreed-upon plan, along with heavy fog, disorganized the attack as it was beginning. Before the other Union divisions advanced, Grover's column struck the Confederate line at "Fort Desperate."[6] Attacking

---

4 Duffy, Edward B., *History of the 159th Regiment, NYSV* (New York 1890), p. 15.

5 Irwin, Richard B. *History of the 19th Army Corps.* G. P. Putnam's Sons New York, (The Knickerbocker Press 1892), ch.17, par.12.

6 Fort Desperate: The name given to the Confederate position sitting at the top of an exposed ridge on the northeastern corner of the Port Hudson defensive line by the men serving inside it; primarily the officers and men of the 15th Arkansas Infantry commanded by Colonel Benjamin W. Johnson. They withstood the ferocious Union attacks of May 27th and June 14th, along with continuing artillery barrages and sniper fire without giving an inch during the entire 48-day siege.

the same daunting terrain prior, the enemy enhanced their defense, and again stopped the assault outside their works. The next attack occurred on the center of the works after the first attack—also in failure. The final assault was made on the southern end of the line at dusk. Failure was also the result.

The failed attacks resulted in more dead and wounded Union soldiers. Their casualties totaled 1,792 compared to the Confederate loss of forty-seven. After this botched chaotic movement against the enemy works, Banks decided future actions on Port Hudson would be reduced to bombardment and siege.[7]

Shortly after the siege began, the following letter was received and published by a Hudson, NY, newspaper:

> We make a few extracts from a letter received last week from one of the noble wounded of the 159th Regiment, N. Y. V. Hoping that the perusal of these extracts will awaken all to the realization of the hardships and trials through which our own and other regiments have passed, and will still undergo, during the siege of, or attacks upon Port Hudson:
>
> NEW ORLEANS, June 1863, "One has no idea how difficult it is for wounds to heal in this climate. Those who have been wounded at Port Hudson, are dying off very rapidly; this is even the case with officers who receive the best care. I heard to-day, that there were many wounded down at the Levee, in an awful condition, their wounds not having been dressed in two days, were infested with maggots and worms, and their suffering intense and excruciating. It is awful, but such is the fate of war! There are many coming down from Port Hudson every day, affected with fevers and every kind of disease. There are over eight thousand sick and wounded at New Orleans, besides those at Baton Rouge, and in the different Hospitals elsewhere. Quite a large army!"
>
> In view of this distressing representation of affairs, we would again call the attention of all our readers to the box which the Ladies of the Relief Society, hope, if possible, to send this week, to the sick and wounded of the 159th Regiment, anything which the exigencies of t h e occasion suggest will be gratefully received on Wednesday

---

[7] Tiemann, William F. *The 159th Regiment Infantry, New York State Volunteers*, In the War of the Rebellion 1862 -1865, (Brooklyn New York 1891), p. 47.

afternoon, at the Reformed Dutch Church. We would also call the attention of the ladies, to the Society under the auspices of which the box is to be sent. It meets Wednesday of each week, and there is always work sufficient for any number who may desire to attend.

The dawn of better days seems breaking. The sky looks brighter, but to ensure this brightness, the light of many homes has gone out in darkness, and bleeding and suffering ones still linger, in need of those attentions which benevolent and patriotic women can furnish. Let not the call of our brave defenders fall unheeded on the ears of those, who in peaceful security, enjoy the blessings vouchsafed them by the Government, to the sustenance of which they have so cheerfully and so gloriously consecrated their energies—even their lives; published by Francis H. Webb.[8]

The ladies of Columbia County, NY, organized themselves into associations for the purpose of providing much needed comforts for the local soldiers serving far from home. Their work began from the infancy of the war and throughout the entire conflict. With untiring diligence, they provided food, clothing, and luxuries for the men in the hospitals.[9]

BRAVE YOUNG MEN OF THE STORMING PARTY

When the siege on Port Hudson began on the heels of the bloody repulse of the previous day, General Banks issued an order congratulating his troops upon the steady advance made on the enemy's works, expressing his confidence in an immediate and triumphant issue of the contest:

"We are at all points on the threshold of his fortifications." The order continues: "Only one more advance and they are ours!"

For the last duty that victory imposes, the commanding General summons the bold men of the corps to the organization of a storming column consisting of 1,000 men. Banks stated they will:

"vindicate the flag of the Union and the memory of its defenders who

---

8 NYS Military Museum & Veterans Research Center, Unit History Project, *159th NYS Regt. Newspaper Clippings published by Francis H. Webb.*

9 Wendell, C., *Columbia County NY in the Civil War (3rd Annual Report of the Bureau of Military Statistics of the State of NY*, New York State Military Museum & Veterans Research Center.

have fallen! Let them come forward!" Officers who lead the column of victory in this last assault may be assured of the just recognition of their services by promotion. Every officer and soldier who shares in its perils and its glory shall receive a medal to commemorate the first great success of the campaign of 1863 for the freedom of the Mississippi, his name will be placed in General Orders upon the Roll of Honor."

Colonel Birge of the 13th Connecticut straightaway volunteered to lead the storming column. So prevalent was the feeling of confidence in Birge, that within three days they had all men needed for the engagement. The officers, and men, of the newly formed storming column quietly prepared themselves for the serious work expected of them. Those having anything to leave made their wills in the customary manner sanctioned by the army, and all confided to the hands of comrades the last words for their families and friends.[10]

Colonel Molineux, in a letter to the Committee of Correspondence of the War Fund Association stated:

"It is with great pride that I inform you the regiment not only fully sustained its reputation for firmness in face of the enemy but won new honors. When volunteers were called for, the forlorn hope to lead the storming columns organized for the assault," He further wrote, "the regiment furnished more by far than most other organizations, and as I think it may be of interest to your Committee, I furnish a list of the names of the volunteers." [11]

The list of names provided by Colonel Molineux is below, including additional information on the brave men willing to storm the walls:

Robert McD. Hart, Captain Co. F – age 22. Lived in Brooklyn, N.Y. Later killed in action at Cedar Creek on October 19, 1864.

Duncan Richmond, First Lieutenant, Co. H – age 21. Lived in Brooklyn, N.Y. Later killed in action at Cedar Creek on October 19, 1864.

Alfred Greenleaf, Second Lieutenant, Co. B – age 24. Lived in Brooklyn, N.Y. with his parents and five siblings. Worked as a train conductor prior to war.

---

10 Irwin, Richard B. *History of the 19th Army Corps*. G. P. Putnam's Sons New York, (The Knickerbocker Press 1892), ch.18, par.10.

11 O.R. *Colonel Molineux, letter written to Committee of Correspondence of the War Fund Association*, Series I – Vol. XXVI – Part 1.

CHAPTER FOUR

Hugh McElravey, Private Co. B - age 38. Brooklyn, N.Y.

Amos Hark (Hauge), Sergeant Co. B – age 19. Brooklyn, N.Y.

George W. Hatfield, Corporal Co. B - age 18. Brooklyn, N.Y.

John Taylor, Private Co. B – age 18. Brooklyn N.Y. At the close of war in confinement by sentence G.C.M.

Gilbert S. Gullen, Sergeant, Co. F – age 21. Brooklyn, N.Y. Would later be promoted to N.C.S. as Sergeant Major.

Bartholomew Doser, Corporal Co. F – age 32. Brooklyn, N.Y. Would later die as a P.O.W. in Salisbury N.C.

Thomas Bergen, First Sergeant, Co. K – age 34. Lived in Brooklyn, N.Y with his wife and three children. Promoted to Commissary Sergeant shortly after.

Michael Hogan, Private 159th Co. C – age 19 (actual age of 16 at enlistment). Lived in Stuyvesant, N.Y. with his parents and six siblings.

Edgar (Ed) Hollenbeck, 1st Sergeant 159th Co. C - age 17. Lived in Hudson, N.Y. with his parents and seven siblings.

Christian Schnack, Private 159th Co. C – age 20. Born in Bavaria, farm laborer, lived in Claverack, N.Y. with a relative and his family.

James Perkins, Private 159th Co. E – age 18. Later in the war he would be killed in action at the Battle at Cedar Creek V.A., he was an only child who lived with his father in Hudson, N.Y.

John Thorp, Private 159th Co. E – age 18 (likely 16). Lived in Austerlitz, NY with his parents and four siblings.

James Brazier, Jr., Private 159th Co. I – age 18 (actually 17). Father was in the same company, born in Scotland, lived in Poughkeepsie, N.Y. with his parents and four siblings. When the war ended Junior was absent from mustering out due to being in confinement for desertion.

George W. Schofield, Private 159th Co. I – age 21. Laborer lived in Fishkill, N.Y. with his parents and one sibling.

Upon close review of the brave volunteers, with exception of two men in their thirties, youth appears to be a prerequisite for storming the enemy's garrison. Possibly a little persuasion by more senior soldiers, letting the young men become the heroes: or perhaps, the challenge and promise made by General Banks to share in its glory and receive a medal. Young men filled with intense enthusiasm wanted to participate and receive the great recognition of the crucial appeal. It should be acknowledged, the 128th NYS Volunteers, also from Hudson, N.Y., and consisting of Columbia and Dutchess County men, provided thirty-two volunteers for the storming party. When volunteers were summoned to join the storming party, the 159th NY had 140 soldiers and ten officers fit for duty.[12]

LOUISIANA HELL

At the close of June 1863, losses to Banks's army at Port Hudson exceeded 4,000 men, with just as many in hospitals, sick from exhaustion or diseases. There was no reserve to draw from: the last were in the trenches. The effective strength of all men at arms never exceeded 17,000. Of them, less than 12,000 could be regarded as available for any duty directly connected with the siege. Every day, the numbers dwindled. Men fell under fire from enemy sharpshooters, succumbed to the deadly climate, or collapsed from exhaustion by continuous labor and deprivation.[13]

The Confederates defending Port Hudson could not compensate for the loss of personnel resulting from starvation, disease (particularly scurvy), dysentery, malaria, sniping, shell fragments, sunstroke, and desertion. The use of mule meat and rats as rations could not maintain the health of the enemy soldiers left standing, not to mention a further drain on their morale.

The June heat in Louisiana was unbearable. Soldiers fortunate enough to catch a few hours of rest in the dense shade of surrounding forests, agonized over their next tour of duty.

In the trenches, the sun beat upon the parched clay generating heat like a furnace. The still air was stifling and the steam from the tropical showers was severe—even the healthiest, strongest men became sun struck daily just

---

12 *The 159th Infantry Unit, nicknamed Second Dutchess and Columbia Regiment.* NYS Military Museum and Veterans Research Center.

13 Irwin, Richard B. *History of the 19th Army Corps.* G. P. Putnam's Sons New York, (The Knickerbocker Press 1892), ch.18, par.16

CHAPTER FOUR

walking a few yards in this burning zone. The siege extended over a wide front, pressing severely on the Union army already far too weak for such an undertaking in any climate, but especially summer in Louisiana.

Constantly on duty, either digging trenches or tunnels, as guards, sharpshooters or on outpost service, the men got little rest. As the number of men available for duty grew smaller, and the physical strength in the ranks wasted daily, the work fell heavier on the remaining men.

During the siege, "nine-month men" whose terms expired, or were about to, became dissatisfied with their situation and unwilling to perform any duties involving danger. There was great embarrassment and trouble caused by the conduct of those troops. One regiment, the 4th Massachusetts initiated an open mutiny.[14]

Toward the siege's end, the effective force barely exceeded 8,000 men. As the number continued to decline, every officer and man still fighting in the trenches might have gone on the sick report. Their pride dwindling under the harsh conditions, it may have been mere duty that held the remaining men to their posts.[15]

Lieutenant Tiemann became sick during the siege. Unable to perform his duties, he was placed in a makeshift hospital in the woods just outside the enemy works, along with a few other sick and recuperating officers. Tiemann spent his time writing letters to family members. In one to his sister, he wrote about an amusing back and forth between the adversaries:

"The 75th N.Y. regt. lay on the right of ours in the rifle pits, so close to the rebs. that they could talk to them. One night one of the rebs. sung out, "I say, Yanks!! Where's Hooker?" One of the 75th immediately replied, "gone to Stonewall Jackson's funeral." This reply was satisfactory & there were no more questions asked. Our boys would sing the National Airs, to which the rebs. would reply by singing Secesh airs! You would hear our boys offering to trade hard tack (army biscuit) for com bread & telling the rebs. to come over & get them. Such is war. Shoot at one another one minute & laugh & joke together the next."[16]

---

14 O.R. Department of the Gulf, *General Banks, General Report*, Series I – Vol. XXVI – Part 1, pg.14, par 4.

15 Irwin, Richard B. *History of the 19th Army Corps*. G. P. Putnam's Sons New York, (The Knickerbocker Press 1892), ch.18, par.18.

16 W.F. Tiemann Collection at The N.Y.S. Military Museum, *Letter June 26, 1863, Port Hudson*.

On May 27, 1863 prior to Tiemann being on the sick list, he found himself in a particularly dangerous situation with soldiers of the 159th NY, and additional supporting regiments. For several hours the men lay close to the breastworks in the position they gained. Unable to move they found themselves trapped between Confederates' crossfire coming from Fort Desperate, and Commissary Hill—until, as reported by Lieutenant Tiemann:

"At this junction what was supposed to be a flag of truce was raised, and the rebels, thinking it a signal for a cessation of hostilities, ceased firing along the line, during which the regiment sought and secured a safer position not quite so near the works..."[17]

The Union commander never authorized a truce, and the soldiers did not follow proper military procedures for one. The following day, General Banks formally apologized to the garrison's Confederate commander General Gardner, for their error.[18]

ENDING THE SIEGE

General Banks, in strategizing to find a method to end the siege, gave orders to construct three mines underneath the enemy works, two of them directed against Priest Cap[19] and one under Citadel.[20] Once the mines were excavated, chambers at the end would be loaded with powder, then exploded under the Confederate works—destroying them, and blowing gaps in the trench lines. Afterward, an infantry assault launched with hopes of overrunning the entire fortification.

On July 7, 1863, the shaft for the mine under Priest Cap was complete, the chamber tunneled, charged with 1,200 hundred pounds of gunpowder, and the mine entrance tamped with sandbags. The mine under Citadel was

---

17 Tiemann, William F. *The 159th Regiment Infantry, New York State Volunteers*, In the War of the Rebellion 1862-1865, (Brooklyn New York 1891), pp. 41-42.

18 Port Hudson State Historic Site, *Lieutenant Tiemann "flag of Truce,"* historic plaque located near Fort Desperate.

19 Priest Cap: Port Hudson's interior of the Confederate stronghold known as "Priest Cap," one of the two main forts. It had sharpened stakes to repel attackers who breach the first line of fortifications. Priest Cap was located on the northeastern side of the lines, slightly south and east of Fort Desperate.

20 Citadel: The second main fort at Port Hudson. These fortifications, parapets, ditches, and gun emplacements were mutually supporting; an advance on one position invited fire from those adjoining it.

CHAPTER FOUR

already complete, charged with 1,500 pounds of gunpowder, the entrance tamped shut. Since warm rain and heavy thunderstorms were frequent, it was thought best not to trust a damp fuse to explode the mines. Daybreak on July 9, 1863, was set as the time for the simultaneous explosion of the mines, followed by one last rush of Union forces through the gaps created.[21]

While Banks and his army waited patiently for the storming party's explosive entrance into the Confederate garrison, a different event was moving rapidly toward ending the siege. In the early morning of July 7, 1863, the gunboat *General Price*, part of the Union ram fleet, came down river bringing great news: Vicksburg surrendered to General Grant on July 4, 1863. An aide-de-camp made his way to the General of the Trenches bearing the brief announcement. The General ordered a note written, then wrapped securely around a clod of clay, the closest thing to a stone in the lowlands of Louisiana, then tossed into the trench of the enemy's works.[22] When General Gardner received the note, he quickly dispatched a letter in response. It was the first of many notes exchanged with General Banks. Gardner wrote:

"Having defended this position so long as I deem my duty requires, I am willing to surrender to you and will appoint a commission of three officers to meet a similar commission appointed by yourself at nine o'clock this morning for the purpose of agreeing upon and drawing up the terms of surrender and for that purpose I ask a cessation of hostilities."[23]

On July 9, 1863, the terms of surrender were agreed upon. Confederate defenders laid down their weapons, ending the forty-eight days of continuous fighting. It was the longest siege in U.S. military history. An unconditional surrender was negotiated, which included additional arrangements, the most important: getting medical attention and food for the starving garrison. Gardner at once approved the articles, finalized by the signature of Banks.

A few minutes later, the long wagon train, loaded with provisions and standing for hours on Plains Store Road, was signaled to go forward into the garrison. Cheers from the Confederates welcoming the wagons through the "now inviting"

---

21 Irwin, Richard B. *History of the 19th Army Corps*. G. P. Putnam's Sons New York, (The Knickerbocker Press 1892), ch.18, par. 34.
22 Ibid., ch.18, par. 36.
23 Ibid., ch.18, par. 40.

sally port,[24] were perhaps not as loud as the cheers from the Union men surrounding the garrison when they received the tremendous news from Vicksburg.[25]

The "Volunteer Stormers" commanded by Colonel Birge, marched into Port Hudson's garrison. Confederate soldiers formed a line facing their adversary, piling their arms and colors[26] in front: This completed the surrender. From that moment, the men, and officers of the two armies mingled freely. There were many issues discussed among the opposing soldiers, but most centered around the desire to go home and be with their loved ones.

Over 6,000 prisoners along with small arms and artillery fell into Union hands. After the surrender, 5,935 enlisted Confederate soldiers received parole, and the 405 officers were sent as prisoners to New Orleans and Vicksburg.

The 159th NY regiment's total losses at the siege were seventy-three: fifteen men and one officer killed. Fifty-five men wounded (four mortally) and two men missing. The following list is as thorough as could be created.

COMPANY A, LOSSES AND WOUNDED
Killed
Col. Corp. Jonathan J. Race

Wounded
Corporal Silas W. Peary
Corporal George Maurer
Corporal Jacob Sagendorph
Private Thomas Atkins

Private George Howes
Private Patrick Keegan
Private John Tator

COMPANY B, LOSES AND WOUNDED
Wounded
Corporal William Roberts
Private John Joy
Private Alex F. Irving

Private Wurtz Balthazar
Private Thomas McCartney, mortally

---

24 Sally Port: a secure, controlled entry way to a fortification or prison. The entrance is usually protected by some means, such as a fixed wall on the outside, parallel to the door, which must be circumvented to enter and prevents direct enemy fire from a distance. It may include two sets of doors that can be barred independently to further delay enemy penetration.

25 Ibid., ch.18, par. 48.

26 Colors: A flag identifying a regiment or army. The "Color Bearer" was the soldier who carried the flag in battle, which was considered a great honor.

COMPANY C, LOSSES AND WOUNDED
Wounded
Captain Charles Lewis
Sergeant Michael Hogan
Sergeant Samuel A. Norman
Col. Sergeant Wm. H. Spanburgh
Private Joseph Peterson

Private John Schermerhorn
Private Myron Staats
Private William Tator
Private Cornelius Van Valkenburgh

COMPANY D, LOSES AND WOUNDED
Wounded
Corporal Alanson Pearsall
Sergeant John F. Jennings
Private Michael Furness

Private David McKinley
Private John Hannophy
Private Thomas Newins, mortally

COMPANY E, LOSES AND WOUNDED
Killed
Corporal Jacob H. Christman
Private Wm. H. Proper27

Private John Maxwell
Private Wm. Pugh

Wounded
Private John W. Myers
Private Robert Proper
Private Robert McCracken

Private Robert Race
Private Patrick J. O'Brien

COMPANY F, LOSES AND WOUNDED
Killed
Private John McCauley

Wounded
Corporal John H. Ferguson, mortally
Sergeant Sam'l C. Tompkins
Private John R. Brokee

Private John O'Mara
Private Wm. Callaghan

Company G, Losses and Wounded
Killed
Private John Gallagher

Private Harvey Pultz

Wounded
Corporal Andrew Goshia
Corporal Silas W. Richmond
Private John C. Bollinger

Private Thomas Shea
Private John Lynch

---

27 William H. Proper, Private 159th Regiment Co. E: Age 22, killed in action Port Hudson, La May 27, 1863. Lived in Taghkanic, NY. with his parents and five younger siblings.

COMPANY H, LOSES AND WOUNDED

Killed

Sergeant Thomas E. Cowell
Corporal Wm. Uggla
Private James McCormick
Private Charles Rossiter
Private John Daley

Wounded

Private Thomas Barrett
Private Davis Jacobs
Private Peter Miller
Private Edward McGreen
Private Edward Pettinger
Private Niel Petersen

COMPANY I, LOSES AND WOUNDED

Killed

Private William Coon

Wounded

Corporal James s. Brazier
Corporal Edward Cosgrove
Private Fredrick Bogardus
Private Wm. Kellerhouse
Private Edward Vail

COMPANY K, LOSES AND WOUNDED

Killed

Private Edward Bridges

Wounded

Corporal Francis L. Vandergaw
Private John Brush
Private Albert C. Trumbull, mortally

Missing

Corporal John Cook Co. I
Private Augustus Garholt Co. K 28

---

28 Tiemann, William F. *The 159th Regiment Infantry, New York State Volunteers*, In the War of the Rebellion 1862 -1865, (Brooklyn New York 1891), pp. 59-62.

CHAPTER FOUR

Siege at Port Hudson [29]

BATTLE AT BRASHEAR CITY

Union forces, heavily invested with men and equipment at Port Hudson during the siege, created an opportunity for the enemy. Confederate forces planned an attack eighty miles south of Brashear City. In an attempt to divert

---

29 Sneden, Robert Knox. Siege of Port Hudson. [S.I., to 1865, 1863] Map. Retrieved from the Library of Congress,www.loc.gov/item/gvhs01.vhs00139/.

Union attention away from Port Hudson, Confederate General Taylor was planning an offensive movement in the Union controlled Lafourche Parish. The large number of Union troops participating in the siege weakened other Union strongholds throughout the region. If Taylor could reoccupy Lafourche Parish, he could threaten New Orleans, forcing Banks to divert his army from Port Hudson to protect the Union occupied Crescent City.

Taylor's plan called for a two-pronged invasion of the parish, starting with his primary mission: seize control of Brashear City, and capture the large military stores within the village. By late May, the Union occupied city was overflowing with spoils of war. Gathered during Banks's Teche campaign, items other than collected commodities, included arms, ammunition, uniforms, medical supplies, and other military necessities. Banks ordered the depot of military stores be broken up and moved to Algiers and New Orleans, but for unknown reasons his orders were never followed.

Brashear City had a small Union force prepared, they thought, to meet a Confederate force if one should ever attack. Colonel Holmes of the 23rd Connecticut regiment commanded the troops there. He also posted soldiers along the railroad leading from the east into Brashear City. It was believed, if an attack by the enemy should occur, it would happen there.

On June 1, 1863, a small Confederate force commanded by General Taylor attacked the hospital in Berwick City, on the bay opposite the Union forces in Brashear City. Company K of the 23rd Connecticut instantly embarked on a small steamer lying at the village wharf, soon followed by three more companies: G, I and C. The force was under the command of Captain Crofut. They advanced quickly on the hospital, driving off the enemy. After the engagement, Crofut's men protected those involved in the removal of sick and wounded, along with government property, transporting them over the bay to Brashear City.[30]

John, along with the other soldiers recuperating at the Berwick hospital, plus the sick already there, took up residence in the convalescent camp in Brashear City. Located near Star Fort, the camp had rows of tents specifically for the hundreds of garrisoned convalescing soldiers. The men were gaining strength to eventually return to their regiments.

---

30 Sherman, Andrew M. (Andrew Magoun), 1844 – 1921; *In the Lowlands of Louisiana in 1863*, pp. 21-22.

Brashear City was situated a few feet higher than the maze of rivers, lakes and swamps stretching for miles in all directions. To the east was a single-track rail line which offered a tenuous link to New Orleans. Berwick Bay to the west created a formidable barrier against enemy attacks.

The Union occupied two dirt forts at Brashear City: Fort Buchanan on the north, (opposite the entrance of the Atchafalaya into the bay) and Star Fort, south of the rail depot, near the junction of Bayou Boeuf and the bay.[31] Besides a partial Indiana regiment numbering about thirty men, there were four companies of the 23rd Connecticut, two companies of the 176th New York, and one company of the 42nd Massachusetts. In all, less than 400 effective soldiers were positioned to protect Brashear City.

Brashear City[32]

The sketch above shows the city receiving commodities and stores for the military. In the fort positioned behind the stockade are rows of tents and cannons. The steamboat loaded with bales of cotton, most likely is heading for New Orleans.

General Mouton, along with Taylor, set out to attack and capture the Union garrison at Brashear City by way of Teche country. Colonel James P. Major, commander of a Texas cavalry brigade, traveled down the west bank of the Mississippi to Donaldsonville, positioning his men to cut off any attempt of a Union retreat. Major Hunter, who commanded a section of Confederate

---

31 Finch, Boyd L., Tucson Arizona. *Surprise at Brashear City Sherod Hunter's Sugar Cooler Cavalry.* Young Sanders Center, Franklin, La.
32 Illustration, *Brashear City, La.* Young Sanders Center, Franklin LA.

infantry, would play a vital role in the capture of the Union garrison.

Brashear City's defenders could have been prepared for the enemy attack about to come, had a report from a Union scout been delivered to the appropriate officer. The scout observed the enemy crossing Lake Palourde in boats behind Brashear City the previous night. Although he reported the movement to Lieutenant-Colonel Duganne, (who forwarded the information to Major Anthony, the commander at Brashear City) there is no evidence that Anthony actually took the warning seriously.[33]

On the evening of June 22, 1863, the enemy was busy on the west shore of Berwick Bay, opposite Brashear City, moving 500 soldiers into the woods and deserted buildings along the banks. Hoping to move quietly and undetected, they left their horses at camp. With brute manpower, they stealthily transported the enemy's four-gun Valverde Battery[34] to the waterfront.

The next morning, June 23, 1863, the sun rose for the Union garrison along with the impact of Confederate shells—catching the Yankees off guard. Reveille was sounded for the men in camp by the guns of the Valverde Battery. General Green's artillery shot forty to fifty rounds before any response from the Union cannons occurred.

The first shots toward the enemy's location came from the steamer Hollyhock, anchored in the river a short distance below Green's position. Initially, the gunboat advanced toward the enemy's line, but it quickly retreated about one mile downriver after receiving a few well directed shots from the rebel battery: At that range, the Hollyhock opened on Green's men with little effect.[35]

The Union Navy, now drawn south, left no vessel to occupy the water north of Brashear City. The opposing general felt the Union gunboat

---

33 Duganne, A. J. H. *Camps and Prisons, Twenty Months in the Department of the Gulf.* (New York J.P. Robens, 1865), p. 135.

34 During the Texas invasion of New Mexico, soldiers of Confederate Brigadier General Henry H. Sibley's brigade captured five guns, three six-pounders and two twelve-pound howitzers, at the battle of Valverde, New Mexico Territory, on February 21, 1862. Volunteers from three cavalry regiments then organized a battery with the trophy cannon under the leadership of Captain Joseph Draper Sayers. The battery fought numerous battles and skirmishes in Louisiana. It was notable for the capture of the Union gunboat *Diana* in March 1863. In June, the unit served in the battle of Brashear City.

35 Noel, Theo. *A Campaign from Sante Fe to Mississippi,* p54. Hunter to Mouton, June 26 and Green to Mouton, June 30, 1863, O.R., Series I Vol. 26, Part 1, pp. 224,225.

"shamefully retreated early in the action."[36] Meanwhile, the enemy's solid shot crashed through the depot, while other balls struck the wharf landing in the convalescent camp. The Union battery positioned a twenty-four-pound cannon from Fort Buchanan next to the depot, opposite the enemy's guns. The fire from Green's light artillery across the bay was no match for the heavy gun. When the big cannon went into action, the Valverde Battery across the channel was forced to move their guns out of reach.

Union soldiers rushed to the waterfront, opening fire with their smooth-bore muskets, but at the long distance their accuracy was not effective. During all the turmoil on the bay, there was no attention paid by the Union to the swamp located at the rear of the city, but the alert enemy took advantage of the neglect and quietly snuck into the rear of the Union garrison.

While Union defenders were concentrating their attention on Green's infantry in their front, behind them, Major Sherod Hunter was leading his men through the woods 400 yards from the rail depot. His intentions were: Flank the road between the depot and Fort Buchanan. He ordered his soldiers of the 2nd Texas-Arizona to follow him in a bayonet charge across the open ground toward the fort. At the same time, Major Blair led the Louisianans and the remaining Texans past an orange grove, across the tracks toward Star Fort, south of the rail depot. Their plan of both columns closing in simultaneously on the rail facilities in the village center was successful.

When the Confederate attack from the rear began, the rebels paid little attention to whether or not the enemy had arms. Unarmed convalescents scattered in all directions becoming targets, as were the armed Union troops who attempted a defense. Confusion ensued, intensified by the gunpowder smoke hanging like a low fog. Rebel columns led by Lieutenant-Colonel Clough swept down upon the convalescent camp, making short work of the few who ventured opposition.

Somewhat able-bodied convalescents could have fought, but they had neither arms, ammunition, nor officers to direct them. As unarmed men were shot, clusters of soldiers raised white handkerchiefs to surrender, others, in a feeble attempt, continued firing. Clough's men proceeded to capture all the sick, letting none go unnoticed, including those in the toilet.

John, among the unarmed convalescents, became a prisoner of the

---

36 O.R. *Mouton to Surget, letter written, July 4, 1863*, Series I, Vol .26, Partl, p. 215.

Confederate colonel.

Surprised by the enemy, hundreds of unorganized Union troops rallied in an attempt to fight back, taking the gun at the waterfront and leveling it on Hunter's column of men: but to no avail.

According to a Texas newsman, Hunter's soldiers seized the fort within twenty minutes and:

"dispersed the garrison, torn down the stars and stripes and hoisted the bonnie blue flag on the ramparts."[37]

Resistance by Union troops was over by 11 a.m. Some soldiers escaped, a few found boats and desperately rowed away. One fourteen-year-old Massachusetts fifer was fleeing in a small skiff "yelling like an Indian," while others fled afoot through the dangerous swamps.

At Fort Buchanan, Major Hunter asked the defeated commander, Major Anthony, to surrender Brashear City, his army, the convalescents, runaways, and navy. While issuing his demands, Anthony glanced in the direction of the bay watching the *Hollyhocks* black smoke streaming backward.

The gunboat with its engines running at full speed was withdrawing further from the conflict heading down river. Disgusted at the sight of such blatant cowardice, Anthony remarked forcibly to his captor:

"he hoped the disloyal gun-boat be caught by the Texan along with the wretched coward who commands her."[38]

Brashear City was now in Confederate hands. They possessed over 1,300 prisoners and an estimated $2,000,000 in commissary, and quartermaster stores. They also received eleven siege guns and 2,500 small arms.

Meanwhile, Hunter's victorious cavalrymen took to eating, drinking, and looting. Generals Taylor and Green obtained a Cajun canoe and paddled across the bay meeting up with Hunter. Two flatboats and several skiffs were found along the waterfront to ferry reinforcements including the cavalry; their horses swam alongside. When Taylor arrived, he found high-spirited

---

37 Rampart: A large earthen mound used to shield the inside of a fortified position from artillery fire and infantry assault. Occasionally ramparts might be constructed of other materials, such as sandbags.

38 Duganne, A. J. H. *Camps and Prisons, Twenty Months in the Department of the Gulf.* (New York J.P. Robens, 1865), p153.

chaos among his troops in Brashear City.[39]

The report of the capture of Brashear City from Confederate Major Sherod Hunter, Baylor's Texas Cavalry, Commanding Mosquito Fleet, reads:

BRASHEAR CITY, June 26, 1863

GENERAL:

I have the honor to report to you the result of the expedition placed under my command by your order June 20. In obedience to your order, I embarked my command, 325 strong, on the evening of June 22, at the mouth of Bayou Teche, in forty-eight skiffs and flats, collected for that purpose. Proceeding up the Atchafalaya into Grand Lake, I halted, and muffled oars and again struck, and, after a steady pull of about eight hours, reached the shore in the rear of Brashear City. Here, owing to the swampy nature of the country, we were delayed some time in finding a landing place; but at length succeeded, and about sunrise commenced to disembark my troops, the men wading out in water from 2 to 3 feet deep to the shore, shoving their boats into deep water as they left them. Thus cutting off all means of retreat, we could only fight and win. We were again delayed here a short time in finding a road, but succeeded at length in finding a trail that led us by a circuitous route through a palmetto swamp, some 2 miles across, through which I could only move in single file. About 5:30 we reached open ground in the rear of and in full view of Brashear City, about 800 yards distant. I here halted the command, and, after resting a few minutes, again moved on, under cover of a skirt of timber, until within 400 yards of the enemy's position, where I formed my men in order of battle. Finding myself discovered by the enemy, I determined to charge at once, and dividing my command into two columns, ordered the left (composed of Captains [J.P.] Clough, of [Thomas] Green's regiment, [Fifth Texas Cavalry]; [W. A.] McDade, of Waller's battalion; [J.T.] Hamilton, of [L.C.] Roundtree's battalion, and [J.D.] Blair, of Second Louisiana Cavalry) to charge the fort and camp below and to the left of the depot, and the right (composed of Captains [James H.] Price, [D. C.] Carrington, and [R.P.]

---

39 Taylor, Richard. *Destruction & Reconstruction, Personal Experiences of the Late War*, (New York 1879), p141.

Boyce, all of [G.W.] Baylor's Texas cavalry) to charge the fort and the sugar-house above and on the right of the depot; both columns to concentrate at the railroad buildings, at which point the enemy were posted in force and under good cover, each column having nearly the same distance to move, and would arrive simultaneously at the point of concentration. Everything being in readiness, the command was given, and the troops moved on with a yell. Being in full view, we were subjected to a heavy fire from the forts above and below, the gun at the sugar-house, and the gunboats below town, but, owing to the rapidity of our movements, it had but little effect. The forts made but a feeble resistance, and each column pressed on to the point of concentration, carrying everything before them. At the depot the fighting was severe, but of short duration, the enemy surrendering the town. My loss is 3 killed and 18 wounded; that of the enemy, 46 killed, 40 wounded, and about 1,300 prisoners. We have captured eleven 24 and 32 pounder siege guns; 2500 stand of small-arms (Enfield and Burnside rifles), and immense quantities of quartermaster's, commissary, and ordnance stores, some 2,000 negroes, and between 200 and 300 wagons and tents. I cannot speak too highly of the gallantry and good conduct of the officers and men under my command. All did their whole duty, and deserve alike equal credit from our country for our glorious and signal victory. I am, sir, very respectfully, your obedient servant,

SHEROD HUNTER, Major,

Baylor's (Texas) Cavalry, Commanding Mosquito Fleet. Brig. Gen. Alfred Mouton, Commanding South Red River [40]

---

40 O.R. Series I-Volume XXVI: Battle of Brashear City 23 June 1863, *Report of Maj. Sherod Hunter, Baylor's (Texas) Cavalry, Commanding Mosquito Fleet, of the capture of Brashear City.*

CHAPTER FOUR

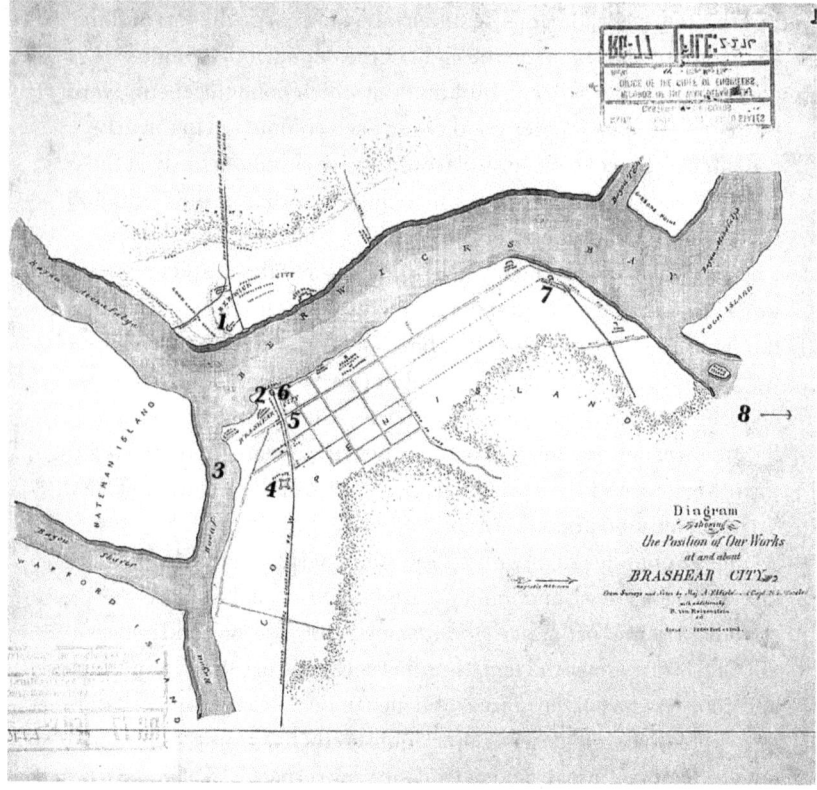

Position in Brashear and Berwick Cities [41]

1. Location of Confederate General Green, the Valverde Battery and 500 rebel soldiers.
2. Location of Union cannons and soldiers on the waterfront.
3. Location of Union gunboat Hollyhock and the coward for a commander.
4. Star Fort located south of the rail depot near the junction of Bayou Boeuf and the bay.
5. Convalescent camp John was transported to after the rebel attack in Berwick City.
6. The battle where Green and Major Blair attacked and captured the Union garrison.
7. Fort Buchanan on the north opposite the entrance of the Atchafalaya into the bay.
8. Direction of Lake Palourde where Major Hunter crossed with the 2nd Texas-Arizona.

---

41 Map, *Positions at Brashear & Berwick Cities*. Editorial illustration by Garrett Kipp.

A CONFEDERATE PRISONER

After Sherod Hunter's triumph in Brashear City, the Texan sent troops a short distance to the railroad crossing in Bayou Boeuf. Union Lieutenant-Colonel Duganne, commanding the 176th New York Volunteers, was preparing defenses while waiting for the enemy he knew was heading in his direction. Once attacked, Duganne's defense rapidly failed against the enemy; he along with his men, also became prisoners of Taylor's army.

Duganne later alleged Lieutenant Stickney (the commanding officer in charge of the Brashear City forts and hospital) failed to make proper use of the hundreds of convalescent soldiers located there—if he had, the result may have been different.[42]

John, with the rest of the recuperating sick and wounded at Brashear City, had no idea their situation could have been different if orders from General Emory were obeyed. At the beginning of June, C.W. Worden of the 23rd Connecticut, received a telegraph from Emory: an order to evacuate the convalescent because of rumors General Green could attack.

The order, initially ignored, was obeyed at a later date when some sick were transported to New Orleans. Only 600 of the 1,500 convalescents moved to Crescent City. The remaining could thank Worden for their new position as prisoner.

Brashear City was in the hands of the Texans, along with the stores, guns and a Union prisoner count of over 1,500 adding Duganne's men. Ordered to line up, each prisoner received three pounds of wheat flour, the only food they received while in enemy hands, even though Confederates captured enough hardtack, salt-horse, and other rations to supply an army for three to four weeks.

The captured soldiers at Brashear City and Bayou Boeuf received paroles by Confederate commanders on June 25-26, 1863. The following is an example of the paroles given to the Union privates:

Headquarters C S Forces, South of Red River

Brashear City, LA, June 25th, 1863

---

42 Duganne, A. J. H. *Camps and Prisons, Twenty Months in the Department of the Gulf.* (New York J.P. Robens, 1865), p. 151.

CHAPTER FOUR

I, Private A M Sherman, Co F, 23rd Regt Conn Vol, do solemnly swear and pledge this, my parole of Honor, that I will not take up arms against the Confederate States, or their allies, nor any manner whatsoever aid, assist, or abet the Government of the United States, during the existing war, until regularly and duly exchanged.

A M Sherman [43]

Unfortunately, the Union officers captured were on their way to Camp Ford, a Confederate prison in Tyler Texas.

Under the guard of Confederate soldiers, prisoners took their haversack, blanket, and small amount of wheat flour on their ninety-mile march toward the Union line in Algiers.

The marching line of prisoners extended for several miles. Emaciated, still weak from recent illnesses, many of the men were in no condition to keep up the steady pace.

Half the distance between Brashear City and Algiers, the men marched on the railroad. The southern sun of June beat down directly on them. A dense forest on either side of the track made it impossible for air to stir. The heat was intense. It was an unbearable situation for these weak and feeble men. As the march continued, the Confederates did not give prisoners any additional food. As a consequence, it was forage or go hungry. The men had a little coffee and sugar in their haversacks. Occasionally they would stop, build a fire, and boil coffee in the familiar and indispensable tin can. A few ears of sweet corn plucked from adjacent fields, then toasted over the coffee fire was a great treat for men surviving on so little.[44]

Quenching their parched throats as they made their way, they often had to push aside heavy thick green scum accumulating atop stagnant pools of swamp water along both sides of the rail track. One prisoner marching along the tracks by the name of Sherman wrote:

"every mouthful of which contained poisonous matter...alligators were numerous all along the railroad, and some were of such dimensions that we

---

43 Sherman, Andrew M. (Andrew Magoun), 1844 – 1921; *In the Lowlands of Louisiana in 1863*, p. 31

44 Ibid., p. 33.

did not care, in our defenseless condition, to disturb them."[45]

On July 3, 1863, Confederate General Mouton sent a flag of truce from Des Allemands; he wished to deliver 1,200 prisoners to General Emory. In a letter dated the same day, Emory informed General Banks of the situation, stating a large number of men were being delivered, "where they come from I do not know." He also stated, the men returned used such seditious language, the commanding officer put them in confinement. Banks responded two days later that he could not understand why the prisoners would use such language. Perhaps if Banks or Emory had experienced what those men had endured, they would not have been so perplexed.[46]

SORTING THE PAROLES

Once the paroled prisoners returned to the Union army, the task of where to send the sick, injured, and deflated soldiers was at hand. Of the 1,200 paroled, less than half returned to their regiments immediately. The remaining 782 displaced soldiers—the majority convalescents, John included, needed to be processed. Under normal circumstances, New Orleans would have been the location; however, with an impending attack on the Crescent City by Dick Taylor's men (especially after their two recent victories) that option was improbable.

On July 8, 1863, General Emory sent a message to Commodore Morris regarding the unprotected conditions in New Orleans:

"what I am most concerned about is the scatted disposition of our stores and armament and the demoralization from the presence of 1,200 or 1,500 paroled prisoners and some 5,000 noncombatants, sick and wounded."[47]

Due to the desperate situation in Crescent City, paroles were loaded on the transport steamer St Mary and sent down river to Ship Island.

A portion of the journey wound through a narrow canal, with the steamer frequently colliding against the banks of the river, nearly taking the men off their feet. Entering into Lake Pontchartrain, they made a brief landing, re-embarking on a different sidewheel steamer: the Sallie Robinson.

---

45 Ibid., p32; Finch, "Surprise at Brashear City," pp. 427-428.
46 O.R. *Emory to Banks July 3, 1863*, letter written, Series I–Vol. XXVI–Part 1, pp. 49-50.
47 O.R. Union & Confederate Navies. *General Emory to Commodore Morris July 8, 1863*, message written. p. 336.

CHAPTER FOUR

After a delightful trip through Lake Pontchartrain, John with the others, arrived on Ship Island, Mississippi, on July 11, 1863.[48] The fort served briefly as a reception center for the paroled Union prisoners.[49] The men were held on the island while being processed which was dreadful since Island conditions were unpleasant.

Private Sherman, a soldier with the 23rd Connecticut regiment Co. F, (positioned in Brashear City doing provost duty) fought the attacking Confederate force. Now a paroled prisoner stuck on Ship Island, he described the men's experience in the following letter:

Ship Island Gulf of Mexico, July 28, 1863

Dear,

When I tell you this island on which we have been encamped since the first part of the month consist almost entirely of fine white sand with scarcely a tree for shade, and with only here and there a patch of grass you cannot doubt the propriety of applying the word "barren" to our present quarters. In this sand our tents are pitched, and on this sand, with a mere blanket for a bed we be, and sleep as best we can, with the various insects that minister to our discomforts. Our shoes are never free from the irritating presence of sand. You may find it difficult to believe me when I say from about 10:30 am till about 1:30 pm the sand is so hot from the sun's rays that an attempt on our part to walk in it in bare feet as some of the acclimated natives do, will prove so painful as to deter one from a second attempt.

The comfortable nights which we invaluably have offset to a considerable degree, other inconveniences we suffer.

Every steamer that lands at the wharf is eagerly watched by the boys in the ott-disappointed hope that it is the one to take us to the land of trees and shrubbery, and grass, and to our regimental comrades who are strangely endeared to us.

Yours sincerely,
A M Sherman [50]

---

48 Sherman, Andrew M. (Andrew Magoun), 1844 – 1921; *In the Lowlands of Louisiana in 1863*, p37.

49 Bearss, Edwin C., *Historic Research Study Ship Island*.

50 Sherman, Andrew M. (Andrew Magoun), 1844–1921; *In the Lowlands of Louisiana in 1863*, p. 36.

Taylor's Confederate force fled Lafourche Parish once they learned Vicksburg and Port Hudson surrendered to Grant and Banks. Thousands of Union soldiers no longer needed at Port Hudson, would soon be hunting Taylor's troops down: Lafourche Parish was no longer a safe place for them.

Because the threat of enemy attack was no longer a concern in New Orleans, after surviving seventeen days in their "barren quarters," the convalescents on Ship Island were released from "sand hell" on July 28, 1863. John and the other prisoners from Brashear City were transported to New Orleans where they could be returned to their respected regiments throughout the region.

BATTLE AT COX PLANTATION

The 159th NY received orders to move at daybreak on July 10, 1863. Each man received five days' rations and 100 rounds of ammunition. The next day, the regiment proceeded down river on board the steamer Iberville. At 10 p.m. that day, the 159th NY arrived in Donaldsonville with orders to protect the town and fort. The men immediately disembarked and marched to the front as pickets. They retained the position, with a section of artillery, and reinforcements by the 25th Connecticut on July 12, 1863.[51] The enemy was nearby, holding a position just behind the town.

The following morning, General Grover moved out an advance guard on either side of the bayou. He desired additional space for the troops and new fields to forage their animals.

On the right bank of the bayou marched Colonel N.A.M. Dudley, commander of the 30th Massachusetts, along with two sections of the 6th Massachusetts battery commanded by Lieutenant John Phelps, Colonel Charles Paine, who led the 3rd brigade (which included the 159th NY infantry) supported by the 1st Maine battery commanded by 1st Lieutenant Haley.

On the left bank was Colonel Joseph Morgan, temporarily commanding Grover's division. The two lines moved simultaneously on opposite sides of the bank.

Moving forward slowly, they advanced six miles when they found themselves upon the estate of a planter by the name of Cox. The enemy commanded by General Green was also there, but hidden by narrow fields

---

51 Tiemann, William F. *The 159th Regiment Infantry, New York State Volunteers*, In the War of the Rebellion 1862 -1865, (Brooklyn New York 1891), pp. 55-57.

of tall corn, dense thickets of willows, and deep ditches common with sugar plantations. Green decided to take the initiative, and with his entire force attacked Dudley's and Morgan's columns. Driving them in their front, he caught the Union advance completely off guard.

Morgan managed his brigade poorly and soon suffered for it, falling into a tangled plight forcing him to withdraw his troops. His departure fully exposed the flank of Dudley, who was making a good fight against the fierce onset of Green's Texans in his front. Morgan's mismanagement caused an assault on Dudley's troops from both banks of the bayou, forcing him back one mile. Observing Dudley's retreat, Colonel Paine came forward in support with his men.

Grover, also observing, brought forward the remaining brigade, throwing them forward to cover the retreat. But as the Union force was coming forward, the Texans did not stay to fight—they fled.

The 159th NY began pursuing the enemy, halting it at Brashear City. The regiment was not strong enough in force to safely pursue the enemy further. They continued their retreat to a position west of the Atchafalaya River. At camp, the 159th NY posted on the right and in front, remaining on picket duty all night, closely listening and watching for enemy movements.

General Grover's losses at the Cox plantation engagement were two officers and fifty-four men killed, seven officers and 200 men wounded, three officers and 183 men captured or missing: in all 465. Adding to the disappointment of the deficient performance at the hands of an inferior force, the 1st Maine and 6th Massachusetts's batteries both lost a gun.

Green reported his loss as three killed and thirty men wounded, including six mortally. He stated the Union loss was a "little less than one-thousand." In reality, over 500 of Green's men were killed or wounded, of whom 200 were left on the field by their fleeing comrades; an additional 250 became Union prisoners.

After the engagement, Colonel Morgan was charged before a general court martial with charges of misbehavior before the enemy and drunkenness on duty. Found guilty of both charges, he was sentenced to be cashiered and disqualified from holding any office of employment under the government of the United States. General Banks disapproved of the proceeding's findings and sentence, he believed the evidence was too conflicting.[52]

---

52 Irwin, Richard B. *History of the 19th Army Corps*. G. P. Putnam's Sons New York, (The Knickerbocker Press 1892), ch.19, par. 36.

159th New York Volunteer Infantry at Donaldsonville, Louisiana.[53]

Camp of The 159th New York Volunteers, Donaldsonville, Louisiana [54]

---

53 Dunham, Marshall Sgt. 159th NYSV Photograph Album, ca 1861-1865. *Photo of soldiers.* Louisiana digital library, LSU Libraries Special Collection.

54 Ibid., *Photo of Camp.*

REUNITED WITH THE REGIMENT

On July 15, 1863, Colonel Molineux, though not fully recovered from his gunshot wound, rejoined the regiment after a three-month absence. Back in command, there was great rejoicing among the men. The weather at Fort Kearney was intensely hot and humid so duties were kept to a necessity and as light as possible. The men appreciated the rest after the fatigue of the lengthy siege at Port Hudson. They especially enjoyed feasting on muskmelons and watermelons harvested from a field not far from camp.

On July 29, 1863, the 159th NY, with two batteries, was transported to the east bank of the Mississippi River. They were detailed to safeguard a baggage train of 100 wagons coming from Port Hudson. The detachment marched seventy-three miles to Carrollton, Mississippi arriving August 2, 1863. They encamped there, naming it "Fort Kearney" after the gallant General Philip Kearney who recently laid down his life for his country.

The first week of August 1863, John finally reunited with his comrades, arriving at Kearney shortly after his departure from New Orleans. The last three months were no picnic for him, the last month proving to be the most dreadful of all. He had some wretched stories to tell his friends.

On July 30, 1863, Captain William H. Sliter of Company G resigned and was discharged.[55]

FAITH

John had many objections to the mistreatment of soldiers: abuse, rotten food, sickness, and the ongoing effort to get his pay corrected. All these issues led up to numerous conversations and negotiations with different hometown lawyers in an effort to get him released from the army.

But now, unexpectedly, he has decided to complete his term of service. Did John believe the rebellion was finally ending or was it his faith?

In his letters, it is apparent that John had a strong trust in God. There is no doubt from his writings he believed God answered prayers. At times, he felt that the support he received to stay alive was an answer to his prayer.

---

55 Tiemann, William F. *The 159th Regiment Infantry, New York State Volunteers*, In the War of the Rebellion 1862 -1865, (Brooklyn New York 1891), pp. 57-58.

In his next letter, John finally sounds comfortable, accepting his fate and completing his duty as a soldier, believing he will be delivered home safely to his family. Recovered here is a partial letter from John while at Fort Kearney, LA:

Letter #17 – August 6, 1863

the conclusion that the Quartermaster might pay me twenty-one per month, but I had been mustered for $13 per month and was paid up to the 1st of January 1863, and since I was mustered with the noncommissioned staff but never have been paid, but I expect $21 per month under these circumstances. I can tell when I get paid, and they said I should have the back pay from the 1st of Nov. 1862. So, I am waiting for that on Monday next week to get mustered again, I think I will get the $21 per month. Before Post left, I said to him "you told me the wages for Fife Major was $41 per month" said he to me "it is", I told him It was down to $21 per month," he made no reply. This is the way Post and Mart Best treated me. When Post sees the elephant, he runs and so did Mart, and finally all that could get away went, and I was left all alone to take care of myself, but all things will come right. I have one who takes care of me and one who will never leave nor forsake me and I trust by his goodness toward me he will bring me back home again. I want you not to pay anybody any money on my account, and null all agreements concerning me, if you have made any for, I expect to stay until my time is out or I am honorably discharged, and I don't want you to feel uneasy about me. I feel quite good now and I think that rebellion is about done, at least about here. I don't think we shall see another battle, but this is not certain. I write this letter so that you may know just how I have been deceived and please don't let anyone have any copy of my letters. The letters I have received since I have been here is 15, I think I shall destroy them. I have not received a newspaper. Postage stamps I think about 14 or 15, so you need not send more postage stamps yet. For the present Camp Kearney lays about 4 or 5 and I don't know but 6 miles above New Orleans, it is a beautiful place, and we have a nice camp, the road runs right through it. Everything looks promising for this war to end soon, only enlist enough soldiers, the war will soon be over. Our camp swarms with locusts, they are almost twice as large as ours, but I think the same thing. You must write, tell Permilia I like to read her letters and all the rest. The name for the foal I cannot give, if it was a boy, I would call it as you think proper. I can't write much news in this letter but trust in God and he will answer prayers. I feel I have been sustained in answer to prayer. You may direct your letter, 159th Regiment N.Y. Vols., New Orleans Fort Kearney, Banks Expedition

Yours, John W. Mambert [56]

DESERTERS

On August 7, 1863, Colonel Molineux had his fill of desertion in the regiment. He ordered the Commandants of each company to provide him with a list of all deserters, including their hometown, and any knowledge of where the runaways might be. They were to deliver the list (along with any additional information they may have assisted in their apprehension) to Captain Waltermire.[57] The regiment had more than its share of deserters, most recently more than 100. With so many men on the sick list, Molineux could certainly use another 100 men to keep his regiment strong.

On August 15, 1863, the Nineteenth Corps went under reorganization. Major General W. B. Franklin assigned the command. The brigade comprised the 13th Connecticut, 19th New York, mustered in from Cayuga County, 91st New York, mustered in from Albany County, (this regiment also included soldiers from northern Columbia County) the 131st New York, mustered from Herkimer and Otsego Counties, and the 159th New York. They were constituted as the First of the Fourth Division, designated as "Grover's" and the division assigned to the district "Defense of New Orleans."[58]

---

56 *Letter #17: August 6, 1863,* A partial letter from Fort Kearney, LA. John W. Mambert letters from the Civil War.

57 G.O. *Molineux to Regiment commandants on deserters.* Regimental Company Books for the 159th NYSV, National Archives Washington DC.

58 Tiemann, William F. *The 159th Regiment Infantry, New York State Volunteers,* In the War of the Rebellion 1862 -1865, (Brooklyn New York 1891), p. 58.

'PROVO DUTY' IN THIBODAUX

The 159th NY was in the 1st brigade, under the command of Colonel Birge. The regiment remained at Fort Kearney until August 29, 1863, when they were ordered to garrison the district west of the Mississippi River extending to Brashear City. The 159th NY ferried over the river on the transport Laurel Hill to Algiers, directly opposite New Orleans. On September 1, 1863, the regiment loaded onto rail cars and then transported to Terrebonne where they proceeded to march three miles to Thibodaux. The 13th Connecticut, and the 159th NY, were assigned to garrison the town, performing provost and picket duty.

The regiment encamped on the outskirts of town on the front lawn of Madame Guion's plantation along with the 13th Connecticut. Lieutenant-Colonel Burt assumed command of the 159th NY at this time. Although a devout secessionist, the regiment found her to be a kind, true-hearted woman, a lady of the highest respectability and moral worth.[59]

The camp faced Bayou Lafourche and was named Camp Hubbard in honor of Weitzel's late assistant Adjutant General John B. Hubbard, killed during the siege of Port Hudson.

Because of the brigade's total loss of equipment during the capture of Brashear City, new tents were issued to the whole command along with cooking utensils, books, and what was most needed: new clothes for every soldier. For the first time in months, the men enjoyed themselves. With great energy they performed their camp duties, each soldier striving to present his best appearance at dress parade and guard mount.[60]

To pave their company streets, the men used bricks from the ruins of an adjoining house. Tents housing the men of each regiment were laid out "with military precision and kept with neatness and care." A narrow ditch was dug around the outside of each tent to carry off water that accumulated after a heavy rain. Company tents faced one another on their assigned street. With all the tents raised, the arrangement resembled a "canvas village." The men made their tents as comfortable as possible inside, erecting each with a double row of floorboards raised two feet. This allowed room for elevated bunks, far better than sleeping on the ground. To the rear of the company

---

59 Sprague, Homer B. (Homer Baxter), b.1829, *History of the 13th Infantry Regiment of Connecticut Volunteers, during the Great Rebellion,* (Hartford Conn. 1867), p. 180.
60 Beecher, Harris H., *Record of the 114th Regiment N.Y.S.V.* (Norwich, N.Y.,1866), p. 237.

tents, also arranged in line, were the company officers; still farther back were the tents of field officers and staff. [61]

After the men found large earthen jars in an old sugar mill, they carried them to camp, placing them in the ground as water coolers. By the aid of what little lumber the boys could find, cook shanties were constructed on the banks of the bayou, and sentry stations were built around the camp to protect the sentinels from the weather. An old building behind the camp was repaired and transformed into a hospital.

The camp's sutler, a citizen of New Orleans, opened an establishment filled with all necessities and luxuries a soldier could think of. Between the enjoyment of ease, amusement and light duties, the days at Camp Hubbard passed pleasantly and swiftly. [62]

Camp Hubbard in Thibodaux Louisiana, encampment for the 159th NYS Volunteers and the 13th Connecticut Volunteers. [63]

The 159th NY camped on the right, while the 13th Connecticut positioned on the left of Madam Guion's front yard. The house was owned by Brigadier-General Francis T. Nicholls, a Confederate. After the war, Nicholls became

---

61 Tiemann, William F. *The 159th Regiment Infantry, New York State Volunteers*, In the War of the Rebellion 1862-1865, (Brooklyn New York 1891), p. 63.

62 Beecher, Harris H., *Record of the 114th Regiment N.Y.S.V.* (Norwich, N.Y.,1866), p. 238.

63 From the collection of E. L. Molineux, *photo encampment for the 159th NYS Volunteers and the 13th Connecticut Volunteers*, Thibodaux, LA 1864, NYS Military Museum and Veteran Research Center.

two-time governor of Louisiana. Today, the house is nestled in a residential community—it still remains in the family.

Duties for the regiment were extremely strict because of not having enough officers available to assist with guard and picket responsibilities. The men were constantly engaged in company and regimental drills, achieving a high state of proficiency. Every Sunday morning, company inspections took place. All arms and accouterments were to be in perfect order. It took only a few days for the grass underfoot to be stamped into hard ground from the constant traffic of foot soldiers. Clothing was expected to be clean and in good repair as well as the tents. Company streets and parade display areas were to be neatly kept. There was little variety in camp life, soldiers rose in the morning at the tap of the drum, moving through the day by its signals to eat, work, and sleep. The drum corps of the 159th NY, thoroughly drilled by Drum-Major Dayton and assistant Drum-Sergeant Dunham, became superior to any other regiments.[64]

Dress Parade, 159th New York Volunteers at Thibodaux, LA in 1863 [65]

Orders below were for the soldiers of the 159th NY while garrisoned in the town of Thibodaux, LA. It reveals the typical day in the life of a foot soldier.

---

64 Tiemann, William F. *The 159th Regiment Infantry, New York State Volunteers*, In the War of the Rebellion 1862 -1865, (Brooklyn New York 1891), pp. 62-63.

65 From the collection of E. L. Molineux, *photo Dress Parade*, Thibodaux, LA 1864, NYS Military Museum and Veteran Research Center.

Thibodaux September 21st 1863

General Order
No 15

The following hours of service will be observed by the command until further orders

Reveille	5:30 A.M.
Breakfast	6:00
Surgeons Call	6:45
Guard Mounting 1st Call	8:00
Adjutant's Call	8:30
Drill Call	9:00
Recall	9:30
Non Commissioned Officers Call	10:00
Dinner Call	12:00
1st Sergeant Call	12:30 P.M.
Drefs Parade 1st Call	5:00
Assembled	5:15
Adjutant's Call	5:30
Retreat	Sunset
Supper	6:00
Drill Call	6:30
Recall	7:00
Tatoo	8:30
Taps	9:00

All enlisted men at drill must wear a blouse and accouterments, Commissioned Officers a blouse or jacket and side arms

(Signed)
By order of
Geo W Hufsey
Edward L. Molineux
Lt & Acting Adjutant
Col Commanding 159th NY [66]

Because Thibodaux (located next to the regiment camp) was one of the largest and finest towns in Louisiana, Molineux instituted a system

---

66 G.O. *Molineux to Regiment hours of service*. Regimental Company Books for the 159th N.Y.S.V., National Archives Washington DC.

of passes allowing soldiers to visit the town's businesses, homes, and surrounding plantations to obtain milk, hoecakes, or fruit. A few men were granted permission allowing them to visit New Orleans, the crown jewel of Louisiana.[67] Other than being granted a pass, not much excitement occurred to vary the monotony of camp life. Sick and wounded soldiers from Port Hudson returned to the regiment's hospital to recuperate. Thibodaux proved to be a pleasant spot to occupy for a short spell before the toils of another campaign. Camp Hubbard was an extremely quiet encampment.[68]

THE PLANTER'S ARISTOCRACY

John's time in the South exposed him to the ways of the "planter's" aristocracy. He saw the devastation it created in the South, the unfairness to the average citizen, and the poor treatment they received, including the mistreatment of the Black population. During his conversations with either Confederate prisoners or local citizens, John's eyes were opened to the dismantling of democracy by unjust rulers: the privileged minority in the Confederate states.

In John's next letter, he describes slavery as a "cancer" threatening the life of the nation, submerging the poor into the depths of destitution while the riches of the wealthy increase, fueling the aristocracy.

Letter #18 - September 11, 1863

To Maria M. And family, preserve this letter, private letter

Thibodaux, LA (called Tibado)

To my wife and family, grateful to Almighty God this morning that I am well, and I think I am enjoying better health than I have since I have been in Louisiana. I feel good I have been gaining flesh, I have a light cough. My mind is more settled and composed. I think I am pretty well climate, since having that dreadful shock on Ship Island. I have been gaining slowly, I was weighed yesterday, I weighed 144

---

67 Beecher, Harris H., *Record of the 114th Regiment N.Y.S.V.* (Norwich, N.Y.,1866), p. 238.
68 Pellet, Elias P., *History of the 114th Regimental New York State Volunteers*, (Norwich, N.Y. 1866), pp. 143-144.

lbs. When I went aboard the vessel in New York I weighed 169 lbs, a decrease of 25lbs. I suppose I have been as low as 128 or 130. I want you to feel perfectly content about me, I think by the help of God and your prayers I shall be home one of these days. I don't want you to do anything as it regard paying out money to any one for the purpose of trying to get me out of this department. When the proper time comes, if I live, I will be at home, so rest easy. I suppose you would like some money; I think we will be paid by the first of next month, then we won't be until Nov., about the middle, but I think sooner. I will send it to the Hudson Express office to be left there until you call for it. You did not write how you got the 1st, whether you got it at Hudson Express office, or at Albert's or where? I don't want you to let anyone know how much I send; it may be more or less. I will send a letter at the same time stating what you must do. The next day after I wrote my last letter, we were ordered from Camp Kearny to this place, Thibodaux, 65 miles southwest of New Orleans. We, or our Regiment, are doing provo duty i.e. guarding the city. The 13th Conn. is with us, everything seems to be still at present and the rebellion is about done. There was a report of a battle just beyond Brashear City a day ago, but how true I can't say. When you write after this direct your letters all in this way, until I order it different. John W. Mambert, 159th Reg. NY Vols. Co. G, Banks Expedition, New Orleans, or elsewhere La. And then the letter will follow the Regt. no matter where it is even if it is in Texas.

I have not received a letter since I wrote my last letter from Camp Kearny. I suppose you have sent one to Ship Island. Perhaps I may never see it unless you put the no. of the Regt. on it. I want you to be careful what you write on account of if the letter should fall into other hands, they would know all your desires which might prove injurious to me. The letters which I have received I have burnt, all except one, so as to be sure that no one could get hold of them. I am doing duty now. The General said and issued an order which he received from the War Department that the Government of the U.S. did not recognize more paroles which was given by Dick Taylor the Rebel General, saying that they are or were not given according to the Cartel of Exchange made by us and the Confederate State,

July 23, 1862. The matter had been deliberated upon in Washington and the matter was all understood, but our Government and theirs that the paroles were null, and void and we would if taken again be protected. And hence it was that we were ordered back to our Regiment to report for Duty, this is the State of things now. I won't write much only to let you know that I am well, but tomorrow I might be dead, you can't tell in this country today what tomorrow will bring forth. I can't tell how long we will stay here; we may leave before night. I want you to write and let me know who is drafted from in our town and about how I am glad to have them come and help cut down the rebellion. I state before closing that I see that slavery is a greater curse to the nation than I had anticipated. At first, I have seen enough and the evils of it. I am now more appalled by it than ever, it is like a cancer, it has threatened the life of our nation. It is an aristocratic principal, it beats back education, it keeps free labor in the factories, it gives them one vote to every five Negroes they own, which us northern men ought to have for every five horses to be equal. It makes the poor, poorer and the rich so rich that it inhales into them aristocracy. I have seen enough of it. I must close, remember me in prayer and take good care of yourselves and trust in God. Write soon all the news you can.

John W. Mambert, Principal Musician [69]

Enhancing the South's prosperity was the invention of the cotton gin creating enormous wealth for a very small minority. Rising up the social scale was nonexistent for the millions of slaves producing the majority of the South's wealth. Poor southern Whites, who may have envisioned improving their socioeconomic status by becoming slave owners never saw their vision achieved, because the wealthy southern planters had economic and political control over lower-class citizens.

The most crucial tool the planters had in maintaining the wealth and hierarchy of the South was the election. Controlled by those who had wealth and money, this equated to votes, allowing rich planters to maintain dominance over the poor majority. The planters had political strength, silencing those who

---

69 *Letter #18: September 11, 1863, To family, Thibodaux, LA.* John W. Mambert letters from the Civil War.

opposed them, and influencing the poorer class. The poor felt their vote did not matter, and elections represented the planters' interests. Elections were passive in design, suppressing and not encouraging the lower class to vote, instituting methods such as a poll tax paid prior to voting.[70]

PRISONER EXCHANGE

Prisoner exchange agreements were nonexistent between the Union and Confederate governments at the beginning of the war. Based on reports of the poor conditions in Confederate prisons, the public wanted this agreement to happen.

There were attempts to reach an agreement, but President Lincoln initially remained opposed to a general prisoner exchange. He was, however, a skillful politician, sensitive to the rumblings of public opinion. The President, and his Secretary of War, Edwin Stanton, realized due to public outcry that they could no longer hold off creating an agreement. Stanton gave the assignment to Major General John Adams Dix, commander at Fort Monroe. Confederate General Robert E. Lee sent Major General Daniel Harvey Hill as the Confederate negotiator.[71] Following a series of meetings at Haxall's Landing on the James River, the two generals reached an agreement known as the Dix-Hill Cartel exchange of Confederate and Union Army soldiers.

This was the agreement John mentioned in his letter. Soldiers of equivalent rank were exchanged on a one-to-one value, corporals and sergeants were worth two privates, lieutenants were worth four privates and so on. As rank increased, so did the corresponding number of soldiers released.

John and the other men in the regiment, as paroled prisoners from Brashear City, believed they were obliged to follow the orders stated on the parole they signed, pledging to not take up arms against the Confederate States. But now under the Dix-Hill agreement, there was no exchange provided for prisoners paroled in Brashear City. The prisoners marched and were released at the Union line. The technicality may have been in the lack of a negotiated exchange of prisoners; therefore, not recognized as a formal exchange.

Colonel Molineux, frustrated by the parole's nonparticipation, issued the

---

70 Merritt, Keri Leigh, *Masterless Men, Poor Whites and Slavery in the Antebellum South*, (Cambridge University Press 2017)

71 *New York Times*, June 23, 1862.

following order on September 5, 1863:

Commandants of Companies will inform all the so-called Paroled Prisoners, wagoner's & musicians that have been ordered to their companies for duty, that they will be severely punished if they persist in refusing to "do duty." Commandants of companies will be held responsible for the fulfillment of this order.[72]

Once John learned he would be protected if captured again, he returned to duty.

SHIP ISLAND

Controlled by the Confederacy at the start of the rebellion, Ship Island was quickly transferred back to Union hands in September 1861. The Confederates did not have enough soldiers to properly garrison the island and were forced to evacuate after a twenty-minute cannon exchange with the *USS Massachusetts*, a Union gunboat.

The Navy wanted to move on to New Orleans, and President Lincoln was anxious for a fleet to move up the Mississippi River, opening the transportation route. This would allow supplies to be delivered to forces on the western banks and commodities to start flowing back to New Orleans. General Butler was assigned the task of capturing New Orleans, but he was to keep the movement confidential. He decided Ship Island would be a good base of operations, since it was equally effective for movement against either Mobile or New Orleans.[73]

In December 1861, Butler sent two infantry regiments and a battery of light artillery to the island, keeping it secure. By April 1862, the Navy commenced its flotilla up the Mississippi River, taking control of the waterway which allowed Butler to move his troops into New Orleans to take command of the city.

Shortly after Butler's arrival in New Orleans, he needed to create order within the community. He decided Ship Island would make a suitable prison, one to be feared by the people of New Orleans.

---

72 G.O. *Paroles back to service,* September 5, 1863. Regimental Company Books for the 159th NYSV, National Archives Washington DC.

73 Butler, B. F. (Benjamin Franklin). (1892). *Autobiography and personal reminiscences of Major-General Benj. F. Butler: Butler's book : a review of his legal, political, and military career.* Boston [Mass.]: A.M. Thayer & Co. p. 324.

Those accused of crimes or agitators appeared before Butler himself. He demonstrated little mercy, as in the case of Mr. Andrew:

Butler: "You are charged with having exhibited a breastpin, claiming it was made of the thigh bone of a Yankee. Did you exhibit such a pin?"

Mr. Andrew: "Yes, sir I was wearing it."

Butler: "Did you say that it was made of the thigh bone of a Yankee?"

Mr. Andrew: "Yes, but that was not true, General."

Butler: "Then you added lying to your other accomplishments in trying to disgrace the army of your country. I will sentence you to hard labor on Ship Island for two years, and you will be removed in execution of this sentence."[74]

For soldiers, life on the island during the Civil War was boring, uncomfortable, and often a deadly experience. An officer with the Sanitary Commission, Dr. George A. Blake, reported on the island's health conditions:

"The wretched condition of Ship Island, a barren, desolate sand-spit, left free for the most part to alligators and such reptiles as abound in the swamps and lagoons of that region; the painful and variable climate; the sufferings of the men from diarrhea, influenza, and rheumatism; the badness of the food, which was of salt meat (no fresh meat being issued); the badness of the water, and the wretched system of cooking, made the presence of the Sanitary Commission not undesirable."[75]

Many regiments were garrisoned on Ship Island, stationed there anywhere from a week up to three months, except for a Union regiment of African Americans, the 2nd Louisiana Native Guard, whose unfortunate orders kept them on the desolate island for a three-year duty. By the end of the war, 232 Union soldiers died and were buried there.

---

74 Ibid., pp. 513-514.
75 The United States Sanitary Commission: *A Sketch of Its Purposes and Its Works.* Boston, Mass., 1863. p. 201.

Chapter Five

"I AM GLAD TO HAVE SUCH ENCOURAGING LETTERS FROM YOU THAT YOU TRUST IN YOUR GOD AND MY GOD THAT HE WILL BLESS AND PROTECT ME AND BRING ME BACK AGAIN TO ENJOY YOUR SOCIETY."

STEALING AMONG THE MEN

While at Camp Hubbard, men in the 159th NY felt it necessary to take their muskets apart and clean them. One would expect a soldier to keep his weapon clean and effectively working. Unfortunately, musket parts were getting lost, and it was becoming a challenge for the quartermaster to replenish the parts. Molineux ordered the commanders of each company to inventory all missing parts and charge the expense to the individual. Commanders were also to caution the soldier on their liability for arms and ordinance—their pay could be stopped should their returns not be satisfactory.

And there was a larger problem; certain men in the regiment were replacing their broken equipment, and arms with working equipment belonging to other soldiers—without their permission. It is likely those individuals who lost parts also feared a loss in pay.

Colonel Molineux, upon learning about the situation, issued an order to the company officers stating:

"any person detected in thieving will be paraded through the camp three times on the day following his actions with his hands tied and the placard "Thief" around his neck; he will also be made to stand in front of the battalion and guard at the next evening parade." [1]

While reviewing newspapers from back home, John notices many friends being drafted, some with whom he has financial partnerships. His thoughts travel back home to his family, their welfare, and concerns over their financial stability while he is away.

---

1 G.O. *Molineux to Regiment on punishment*. Regimental Company Books for the 159th NYSV, National Archives Washington, DC.

CHAPTER FIVE

Letter #19 - October 5, 1863

Thibodaux

To Miss John W. Mambert

I am well and I hope you are. I feel better now than I have since I have been south, thanks be to God for his goodness and mercy toward me. I feel that prayer has sustained me. I received your letter dated Sept. 20th, I was glad to hear from you. Don't feel discouraged for you do not know what trouble is. I think this war will soon be over and I hope and trust in God that I may return to my native home again. I think my enlisting is perhaps all for the best. I had a paper this week before I got this one from you today and I saw that Harm[2] was drafted and all the rest.

Some die including Kline[3] from our town, and others, we are at Thibodaux and fixing for winter quarters. The 159th and the 13th Connecticut Regt., we get our mail regularly, every week and you can depend on it. I expect to be paid next week and you must look well to your letters and don't worry about me, I am doing very well, don't let every little thing trouble you. If anyone wants to know how I get along tell them to go and see, and if they would go and defend their homes as I did it would be more honor to them. You must not let anyone see or read my letters, if you and John read them that is enough. I can't write much tonight. The weather has been very warm and the last of Sept 29th and 30th it rained, and since that it has been cold nights, and last night there was a little frost. You wrote that Mag and her man was down. If you could get the money on that note of Andrew Decker[4] and Rockefeller, you could pay her $200, and if I have good luck, I will send you some money to pay all the rest of them you would have some left and the interest next

---

2 Harmon V. Mambert: Age 26, J.W.M.'s younger brother. A butcher in the hamlet of Churchtown, NY. Married with a five-year-old daughter.

3 Pleasant Kline, Corporal 159th Regiment Co. I: Age 19, died at Thibodaux, La. of disease on September 28, 1863. A farm laborer for the family farm in Taghkanic, NY. Left behind his parents and eight siblings ranging in age from four to seventeen years old.

4 Andrew Decker: Age 40, draft records, however not listed in N.Y.S. regiments. Farmer in Taghkanic, NY. Widower with eleven-year-old son.

spring. I saw that Andrew Decker and Walter Decker[5] were drafted and Walter Miller[6], let me know if it is old mortal, and two of Uncle Abrams[7] boys Henry[8] and Holly Gardner[9]. This is private and I don't want anyone to read it, if anyone does except you or John, I want them to stop at once. I want John to try to buy back the place of Andrew Decker if he can get it for $2900, if Andrew will take back that mortgage of Besse's and if he will buy it and then move there in the spring, and I want you to see that he pays the interest to Elsoffer. I want you to put some money in the bank for that purpose, when I send you money so you can get it, and if Andrew Decker is going into the service you must get him to let Solomon Avery[10] sign that note, and then it will be safe. If Bess could pay you the $850 you could take up the Elseffer mortgage and then all would be right. Tell Jobe and Walt that I thank them for their letters, I am glad to read them and the rest too. I will write more the next time.

Trust in God John W. Mambert [11]

THE DRAFT, BOUNTIES AND RIOTS

On March 3,1863, President Lincoln signed into law the Enrollment Act of 1863, also known as the Civil War Draft Act. The law enabled the government to draft men into the military for the first time in the United States. It passed at a time when the volunteer system to supply troops failed and the North suffered a series of military defeats, creating a critical need

---

5 Walter Decker: Age 27, draft record, however not listed in any NYS regiments. Farm laborer in Taghkanic, NY. Lives and works for his parents, has one sibling.

6 Walter Miller: Signed draft records, however, not listed in any NYS regiments, (2) Walters were drafted – Walter A. age 31 from Germantown, NY, and Walter J. age 41 from Taghkanic NY. Both signed the registry at the same time.

7 Abrams Gardner (Uncle): Age 62, a railroad flagman from Hollowville located in Greenport, NY.

8 Henry Gardner: Age 26, draft records, however not listed in any NYS regiments. A painter from Greenport, NY.

9 Holly Gardner: Age 23, draft records, however not listed in any NYS regiments. A saloon tender from Greenport, NY.

10 Solomon Avery: Age 50, a farmer in West Taghkanic NY. Lives with wife and four children.

11 *Letter #19, October 5, 1863, Thibodaux, To Miss John W. Mambert.* John W. Mambert letters from the Civil War.

for more soldiers. The Enrollment Act was designed to stimulate volunteer enlistments; the draft itself would be employed only where a district failed to meet its quota of volunteers.

The law mandated enrollment of all male citizens and would-be citizens between the ages of twenty to forty-five. A federal agent was assigned to each congressional district with the task of enlisting the number of new soldiers required. Once the agent received the number, the task of filling the quota through both drafted men and volunteers fell upon the states.

To avoid the dreaded stigma of the draft, communities worked hard to fill their quotas with enlisted volunteers; those men received a bounty of at least $100. State bounties were not the only ones available. The Federal government and local districts offered bounties. Due to the offered bounties, the number of soldiers who fought as conscripted men was fairly small.

In reality, the draft had a social impact on the men. It revealed racial and class divisions which, at times, created resistance.

NEW YORK CITY CIVIL WAR DRAFT RIOT

The most serious opposition to the draft act happened in New York City. The city contained a substantial number of residents who were sympathetic to the South, among them a large part of the Irish community.

The Irish population was worried that if the North won the war and slavery was abolished, they would be outcompeted for jobs by the newly freed Black population. Further opposition came from the fact that it was possible for the wealthy to effectively buy their way out of conscription by substituting themselves with other men, those of the lower class who would fight in their place.

New York Governor Horatio Seymour could do nothing but obey the order to institute the draft. On July 4, he gave a speech which criticized the practice of conscription as unconstitutional.[12] He claimed the report of his state's deficiency in volunteers was not correct, boldly intimating that excessive quotas were saddled upon districts known to be predominantly Democratic. His speech inflamed many within the city of New York. Grumbling citizens soon provoked a violent storm.

---

12 Van Pelt, Daniel, *Leslie's History of the Greater New York*, Volume I, p. 413.

Republican Mayor George Opdyke surmised there would be trouble when the drafts began. He opposed Seymour's proposal to withdraw all militias from the city.

The Governor simply replied,

"the city would be safe enough under the protection of its own police force."

The draft was appointed for July 11, 1863. Soon after the first draftees were called, a pistol shot was heard; the signal for attack. A volley of paving stones was fired into the drafting office knocking down officials, upsetting inkstands, smashing chairs and tables. The crowd rushed in, destroying the drafting machine, wrecking everything in the room, then setting the house on fire.[13] Rich citizens were singled out for attack, as were Black people and any businesses that employed them. Many Black residents were beaten, some lynched.

The Secretary of War ordered all New York regiments back to New York. On the evening of July 15, the Tenth and Fifty-Sixth regiments arrived. Soon after came the Seventh, Eighth, Seventy-Fourth, One Hundred-Sixty Second and the Twenty-Sixth Michigan. Destruction continued to rage in the city upon their arrival.[14]

On the fourth day of unrest, Governor Seymour, from the steps of City Hall, issued a proclamation commanding the crowd of rioters to abstain from violence, informing them he urged the Federal Government to suspend the draft. Meanwhile, the Common Council passed an ordinance appropriating $2,500,000 toward individual payments of $600 to each substitute draftee (replacements for men who bought their way out of service—paying other men to be drafted in their place). The mayor "very properly" vetoed the measure, regarded by the populace as a victory vindicating the riot and all its horrors, rather than condemning it.

When the riots were finally quelled, between 1,000 and 1,200 persons had been killed, an unknown number wounded, and $2,000,000 worth of property was destroyed.[15]

Though riots occurred in other cities, the act was complied with—because of the bounties. While it was not perfect, it did ensure the Union Army remained mostly volunteers throughout the Civil War.

---

13 Ibid., Volume I, p. 414.
14 Ibid., Volume I, p. 418.
15 Ibid., Volume I, p. 419.

JOHN'S HOMETOWN CONTROVERSY

Columbia County, New York, was not free of issues when considering the draft, as John's hometown of Taghkanic became a hotbed of controversy over conscription. The following newspaper article from the Hudson Gazette ran on September 14, 1863, uncovering political concerns:

> In Taghkanic, suspicion was felt that the enrollment of that town for the draft had not been fairly executed, and N. S. Post, Esq., was delegated by the citizens to procure a copy of the enrollment list from the Provost Marshal at Poughkeepsie, with whom it had been deposited. Upon examination, it was found that forty-four names liable to draft had been omitted from the roll, and of this number, five-sixths are Republicans. Taghkanic last fall cast 333 votes, of which 248 were Democratic and only 85 Republican – making a difference of three to one in favor of Democrats. With this fact in view, it is not surprising that the discovery gave rise to considerable speculation and created some excitement in the town. The matter will be investigated. The assertion that "forty-four names liable to draft" in the town of Taghkanic, had been "omitted from the roll" is utterly untrue. Post it seems, went to Poughkeepsie and made a copy of the enrollment of his town, comprising the first class alone, and then came back to Taghkanic with an immense "mare's nest." The people of his town were summoned together, and the enrolling officer, Christopher Miller, Esq., one of the most esteemed and upright citizens of Columbia County, denounced in the strongest manner, Mr. Millers friends at once called upon Mr. Whiting B. Sheldon and Mr. Samuel L. Myers, two prominent and influential Democrats of the town, and with them proceeded to Poughkeepsie on Tuesday last. They made a thorough examination of the enrollment, and found every man of Post's forty-four on the list.[16]

The Hudson Gazette followed up with the subsequent article on September 15, 1863:

---

16 *Hudson Gazette ran on September 14, 1863.* Unit History Project, 159th NYS Regt. NYS Military Museum & Veterans Research Center.

The Draft Completed - The process of drafting in this district was begun at Poughkeepsie on Monday of last week and completed on Friday. The proceedings were uninterrupted. No disturbances occurred. The names were published as fast as drawn, and the utmost good nature prevailed. We give a full list of conscripts from this County in another part of our paper today. Those who have not yet received official notice of their election have the privilege of enlisting, in which case they are entitled to the state bounty. The examination of such conscripts as believe themselves physically exempt was to commerce at Poughkeepsie yesterday morning at 9 o'clock. Strenuous exertions have been and are being made in the city and in some parts of the County to raise money for an indiscriminate exemption of all drafted men. We are sorry to say that some of the proceedings have this object in view have been in no wise creditable to the loyalty or manhood of those concerned.[17]

In John's next letter home, he asks about how the young men in his family are faring regarding the draft. He is also curious about the number of men from his hometown who paid not to go to war, specifically asking about his younger brother, Harmon.

Letter #20 - October 15, 1863

You must read this letter by the pages, I made a mistake in the pages

A private letter to Mrs. John W. Mambert & John Mambert

Thibodaux

To my dear wife and family, I wrote a letter dated Oct. The 13th on the receipt of your and John's letter. I wrote that I was not very well, but I have been gaining since. I feel quite well today, the fever has left me, and I am getting along. I have been paid today $168 and the $11 back pay was not reckoned in by mistake of the officer, but they say I shall have it next pay day, which I expect will be paid in less than 4 months. I will send you by express $150, you will find it in Hudson. I don't know how much you owe, but if you can pay Mag

---

17 Ibid., *The Hudson Gazette article on September 15, 1863.*

$100. You better do so. That will give you $7 less interest to pay her another spring which will give you another barrel of flour. If I live you will soon have some more money. Your rent you must pay next April from your interest of Silvester Bortle and Andrew Decker, and if you can get that note paid by Andrew Decker, then pay Mag. If not let Andrew Decker get Solomon Avery to sign or give you a new note with himself and Solomon Avery as security. Be careful how you do business, do it right and take $100 of your interest and pay Mag that in the spring, and if Best can pay the $850 then take up the Eloeffer mortgage, and then you're all safe. And what money you have left over put out at interest on good note and security, don't use any principle if it is possible. I hope and trust in God. I may be home before that time so that Andrew Decker keeps that interest of Eloeffer up, and if he can't then you pay the interest. If Andrew offers to pay the mortgage of yours take it and take that mortgage up of Granpap Mambert and the rest put on interest. If not just let John see what Andrew will take for the place, the least, and take the mortgage back on Ves Bortle and let me know. See that the interest of Bob Bortle[18] is endorsed on his mortgage of Best for your and Ves's sake, write on the receipt of this letter. Such a thing might be that I may send for you all yet, but don't put any confidence in that. I am glad to hear that you can read my writing, only get my mind and you can tell what the words are without spelling. I am glad to have such encouraging letters from you that you trust in your God and my God that he will bless and protect me and bring me back again to enjoy your society. Trust in God, let me know what Harm is a going to do, pay for ninety-nine chances to die and but one to live? We expect to stay here all winter, but after all that we may be called away tomorrow. If you write let me know how many have paid and don't go to war, let me know what Andrew Decker does, whether he pays or not. Tell Bert that I am glad he escaped the draft, tell him I thank God for that, and that John is not drafted, if John can't see out of his right eye, he may be exempt. I think of poor Rams Mambert, I hope he will escape the draft. Let

---

18 Robert "Bob" Bortle – Age 44 a farm laborer in East Taghkanic, NY. Lives with and works for Sylvester Bortle.

me know what Father thinks of Harmon being drafted. I think it may be all for the best that I am here, one year is gone, only two years more at the longer, but I may be home sooner. Send me that notebook without delay for Tommy doesn't like to let me see his. The nights are cold here and the days are pleasant. Our Regiment is pretty well. If they ask how I get along tell them good. Give my love to Father and the Old Lady, and to Old Daddy Sheldon, and Granny Sheldon, and the rest of our folks. Tell Jobe and Walt to look good after things and go to school, and you look good to the girls, and tell John I think of him often.

Yours, John W. Mambert [19]

The Columbia County draft results totaled 1,163 men enrolled; Walter and Andrew Decker, Walter Miller, Henry, and Holly Gardner, as well Harmon Mambert were all listed on the enrollment.

MOVEMENT IN THE RANKS

Charles A. Robertson, surgeon for the 159th NY, resigned on November 2, 1863. First Assistant-Surgeon William Y. Provost from Brooklyn, New York, was promoted to replace him. Colonel Molineux became acting Provost Marshal General of all the forces in the field and under the command of Major-General Franklin. In his new position, Molineux was responsible for development and execution of policing, including deserters. He also kept records on the physical condition of recruits and army personal casualties. Shortly after becoming provost marshal, Molineux was with his orderly Martin Garner of Co. C, when Garner was captured by the enemy near Carrion Crow Bayou along with the Colonel's gray horse. Soon after, the Colonel escaped.[20]

---

19 *Letter #20, October 15, 1863, Thibodaux, A private letter to Mrs. John W. Mambert & John Mambert.* John W. Mambert letters from the Civil War.
20 Tiemann, William F. *The 159th Regiment Infantry, New York State Volunteers*, In the War of the Rebellion 1862-1865, (Brooklyn New York 1891), p. 64.

CHAPTER FIVE

BOUNTY BROKERS

In a letter to John, his son (John junior) mentioned purchasing the Rossman farm. Located on Hoyle Road, it was a short distance from the family's home. The farm belonged to John Rossman, who lived there with his wife, Harriet, two young sons and a daughter (all under five years of age).

John shared his thoughts on purchasing the farm in his next letter home.

Letter #21 – November 5, 1863

La. Thibodaux

To Mrs. John W. Mambert

I am well and enjoying good health. Today the mail came in and no letter for me, and so I was mad. And in spite of that I will send you $30. which you will find at the Hudson Express office. I will see if that will send me a letter or not. I was paid 2 months pay today, you will please accept it, the spite I merely write as a joke. I receive letters every other mail, that is every 2 weeks, then I receive 2 letters. I received the letter stating you received the $150, now there is $30 more on the way. I was paid $42, if it is possible to save $100, try and save it. The next time I may not send quite as much, I shall have to settle my clothing bill. The government allows $42 per year, and I think I have overrun some, so what I have overrun I will be deducting. I think we will be paid again in the middle of Feb or somewhere about that time. About that Farm of Rossman, I don't know hardly what to say about buying it, I think we are making money pretty well and if we keep going on a few years we can buy a farm and pay right down for it, and then we can buy another if we run in debt for the whole, still do as you think proper. If you think it can be paid for you may try, things are high now, but these prices won't last long. I think that the war will soon be over now, the prospect looks favorable, and the news is good. I think by next spring I will be home if I have good luck. I hope you will be content, and rest satisfied until I return, which I trust in God I may. The weather here is nice, warm days and cool nights and sometimes a little frost, just enough to make the tender leaves drop. The trees are all green leaves on them except the fig and orange trees and a few others, they show as though they had been frost bitten. There are plenty of mosquitoes and flies, so you judge how the weather is. Doctor Dedrick gets along well, tell Philip Dedrick[21]. I wish you were all here to spend the winter, I have not seen a snowflake since I left New York State. There are plenty of sweet potatoes here, but they are not worth much to eat, they are not so sweet as they are up there. The cabbage doesn't head here unless the seed comes from the north, and then it runs out in 1 or 2 years. I mentioned in my letter before this one, that the southern people were behind. I wish you could see their buildings, sometimes the planters have a large house, but it is after the old fashion. The Negro houses are all in rotation, like a street, but it is nothing but a shell of a house, only boarded on the outside. Rough floors laid overhead, the chimneys are on each end on the outside of the house, and the house or hut is partitioned in 2 for 2 families, sometimes a stoop before that in general and the outside of the house, whitewashed. These are those beautiful cottages which N. L. P. alluded to in a letter which appeared in the Hudson Gazette. Soon after we captured Baton Rouge Mr. Post only seen those beautiful cottages from a distance, he had not seen the elephant yet, nor had he ever seen much for he was not here long enough. (letter ends)[22]

Possibly based on John's thoughts, Junior never made the purchase. But there may have been a different reason: John Rossman was drafted into the war, knowing his wife could not manage the farm and the young family alone, may have thought it best to sell.

But there is no record of Rossman serving. As more volunteers enrolled due to bounties, and the draft number reduced, Rossman did not need to report to the provost's office: therefore, his decision to sell the farm may have changed.

Getting prospective volunteers to reduce the draft quota was challenging. By now, potential recruits were well aware of the situation and the reality of

---

21 Philip S. Dedrick: Age 57 a farmer in Churchtown, NY. Lives with wife and five children.
22 *Letter #21, November 5, 1863, Thibodaux, To Mrs. John W. Mambert.* John W. Mambert letters from the Civil War.

soldier life. Many soldiers, especially those with disabilities who returned home, relayed horrifying stories of war. Individuals receiving letters from soldiers, like John's, warned them not to join because of the unbearable circumstances. But now, cash bounties enticed men to volunteer.

BOUNTY JUMPING

The money offered became so great, some men were tempted to try "bounty jumping" by repeatedly signing up. Once signed up, they collected the cash bounty, then re-enlisted at another location, repeating the process. This unethical behavior often encouraged by "bounty brokers," made "jumping" a highly profitable practice. Bounty brokers collaborated closely with provost marshal offices recruiting men for service. The brokers' relationship with government officials made it almost impossible for a volunteer to enlist without their assistance. So misinformed were the new recruits, they believed a broker needed to process their enlistment form. They also became convinced that a considerable portion of their bounties belonged to the broker.[23]

Bounty brokers also employed "runners" to explore the countryside for prospects. They opened offices in or near recruiting and provost marshal headquarters, posted convincing signs, and advertised in newspapers. In New York advertisements, incentives to join, or re-enlist became extremely enticing to help persuade new volunteers.

An individual by the name of Jacob P. Miller, having no history of military involvement, ran the following ad on December 10, 1863:

> All able bodied men between the age of 18 and 45 who have served for not less than 9 months, may reenlist as Veterans Volunteers in any one of the old regiments in the field. All such re-enlisting will receive $865--$465 in advance. All other volunteers will receive $600--$330 in advance. Those that enlist before the 5th of January, 1864, will receive $600 more than those who are drafted.[24]

I can only conclude he was a recruitment broker.

---

[23] *New York Times*, Aug. 25, 27, 1864.
[24] *Kinderhook Herald.* (Kinderhook, N.Y.) Dec 31, 1863, pg3, image 3. NYSHistoricNewspapers.org

Antebellum Cotton Plantation[25]

In John's last letter, he described the housing provided for the plantation slaves and the scene described by Nathan Post[26] to the Gazette newspaper. The painting above matches the description almost perfectly.

COMING HOME

Captain Lewis, on furlough, was heading home to Columbia County with a scheduled stop in Hudson, N.Y. He was transporting the bodies of Robert D. Lathrop, the regiment's adjutant, along with the other officers killed at the battle of Irish Bend. The bodies buried in Franklin were retrieved once it was safe to do so. Captain Gideon Lathrop got his wish to have his son's body found and returned home to New York.

FURLOUGH FAVORITISM

This next letter notes the first time John mentioned furloughs. He will bring the subject up six more times throughout his three-year military career. John may not have requested one but speculated he would not be granted a furlough based on the result of other soldiers' appeal.

---

25 *Antebellum Cotton Plantation*, painting, permission to use by Edwin L. Jackson.
26 Nathan Post, 1st Lieutenant in Co. E: Age 37, an attorney in Livingston NY. January 14, 1863, he was discharged with a disability by orders of General Banks.

CHAPTER FIVE

Letter #22 - November 22, 1863

La. Thibodaux

Dear Wife,

I am thankful to God on this Sabbath that I am well and enjoying good health, except I have a severe headache and I could not go out on inspection. I have just had my dinner, I had a good cup of T, beef soup, bread, butter, cheese and fried onions. This was a very good dinner and I feel a great deal better. But you can't make good T here on account of the water, it being rainwater. I don't expect mail until Tuesday and then I will finish off this letter and let you know how my head turns out. I wrote a letter and sent it with Capt. Lewis[27] and it is to be left at Ed Rorrabacks in Hudson, Capt. Lewis comes home with the dead bodies of R. D. Lathroop[28], Andy's[29] and Liut. Lockwood and others which were killed in the Battle of Irish Bend. I suppose there will be quite a time when they arrive in Hudson. The letter which I send with Capt. Lewis contains my wallet, a letter and 7 seals which denotes the 7 members of our family, one is me, one is you, one is John, one Parmelia, one Martha, one Jobe and the other Walt. I hope you will try to keep the whole together until I return, and then we will see further. I dream of being home very often, then I feel just as if I would be sorry that I want to come back here again. I feel just as if I should get home on a furlough[30], I would hate to leave and come back again, and so I would rather stay until I should

---

27 Charles Lewis, Private, 159th Regiment Co. C: Age 32, promoted to Captain, then Major of the 176th NYSV. Wounded at Port Hudson. Mill worker in Stockport, NY. Lived with his wife and sister.

28 Robert D. Lothrop, Adjutant 159th Regiment: Age 22, killed in action at Irish Bend, La. April 14, 1863. A farmer in Stockport, NY. Left behind his mother and four younger siblings ranging in age from 12 to 20 years old.

29 Andrew Asbell, Corporal 159th Regiment Co. K: Age 22, killed in action at Irish Bend Apr 14, 1863. From Brooklyn, NY.

30 Furlough: A leave from duty, granted by a superior officer. The furloughed soldier carried papers which described his appearance, his unit, when he left and when he was due to return. Furlough papers also contained a warning that failure to return on time would cause the soldier to be "considered a deserter."

come home to stay. There is no such thing as getting a furlough, they give none in this department. I think you will get this letter before you get the one sent with Capt. Lewis. It is quite warm here, but cool nights which have a curious effect on a person. It will cause your bones to hurt just as if your feet or hands were in cold water. If you lay with your head out from under the blanket, your head will pain you and you won't know what to do, and it will seem through and through cold, I sleep in a night cap. I sent you $150 which I suppose you have got and also my miniature which did not suit me, but he would not take it over again. You will see that I have grown old fast and that I am almost bald headed, I suppose you have got all of them by this time. You will let me know if the war news is encouraging, there is no use for me to write for the papers give the whole of the news and more, besides I would say though that we have been expecting an attack from the enemy. A few nights ago, the alarm was given, that is the long roll beat, between 12 and 1:00 o'clock at night and such getting out you never see, in less than 8 minutes they were all in line of battle, but the alarm proved to be false. We have been on the lookout ever since. They sent another brigade to reinforce us and a battery. A battery is cannons with men enough to handle them. So, we are pretty well fixed to give them a hardy welcome, but I think they know better than to come. I think if I live, I will be home in less than a year. I think this war will end, for the rebs are played out here. If it should be that I am mustered out of the service here or in New Orleans, I shall come home through Illinois or Wisconsin. To John I am glad to hear or rather see in your letter that you have saved a little money, so when I get back, I will help you to do something, I won't say at present what. I won't write anymore until I get a letter, or the mail arrives which will be Tuesday, Nov. The 24th. Today I received two letters, one dated the 1st of November, one the 9th. I was quite sick since I commenced this letter but am well now. I have written some letters which you have not got, I have sent my miniature, and you sent some letters which I have not got but I expect to get them. I wrote a letter today to Grandpa Mambert, this is the second letter I have written to him. Tell Martha to take good care of herself, I received her letter in yours, but her miniature I have not as of yet, it may be that the box in which it was sent has gone to the 128th instead of the 159th. If so, I may not get it, I may get it. I don't see why you have not got my letters; I hope you may get them. I am glad that John writes every Sunday. I think this war will end by spring. I dreamed of seeing you (Maria), the night before last in your black dress with your old, croaked back, and I have felt quite good ever since. I don't know if I can fill this letter, I hope you will get the money and I hope and pray that I may return home again. May God grant the desire of my heart and I know that it is all of yours, and trust in him who is able to bless and protect me. Remember me in your prayers, trust in God. I must close, tell the little boys if the Lord spares me, I will be home again. Tell Jobe and Walt to write, I write to you and John reads I suppose, and that is to him the same as if I wrote to him. Tell Permilia to write how she gets along, and her man and how they all do. Take good care of yourself and manage your business as good as you can. When I get home, I will relieve you of that burden, which I hope will be soon.

Yours, I must close in a hurry. Write soon.
John W. Mambert [31]

Colonel Molineux, the regiment's commander, could only grant the much sought-after furlough. Officers could apply for furloughs, more commonly called "leaves," as Captain Lewis had. Lewis received one, but it may have included negotiations including overseeing the return of deceased comrades.

Men requesting a furlough found it to be a long-drawn-out process, subject to approval from a lengthy list of field and staff officers before reaching Molineux. Receiving a response could take months and often end in disappointment. For those not receiving one, complaints increased. Foot soldiers complained that officers received the greater share of leaves, and of course, favoritism. Married men and the wealthy were higher on the list than single and poor men. John's chances of receiving a furlough from the colonel, considering their history, was questionable.

---

31 *Letter #22, November 22, 1863, La. Thibodaux, Dear Wife.* John W. Mambert letters from the Civil War.

As the war progressed, furlough promises to the men became empty. This was the case for John during the fall of 1864: two or three times he was promised by his commander that he would receive a furlough. But John never received one.

While in winter quarters of 1864, John believed he would be returning home on furlough, only to once again be disappointed when the commanding general ordered no furloughs granted. In deep southern territory, the commanders turned down furlough requests on the grounds that soldiers were too far from home to make them practical.

In June 1865, John and the regiment's new colonel had an agreeable relationship, and John could finally have his leave, only to decide against it because there was too much travel involved.

THANKSGIVING DAY, 1863

The 159th NY observed Thanksgiving Day on November 26, 1863, along with relief from guard and picket duty on that day. The cooks made roast turkey and pumpkin pie, with extra meals to satisfy the men. The regiment's compatriots of the 13th Connecticut celebrated the day in good old New England style, holding events like blindfold sack races, climbing a greased pole, chasing a greased pig and other sports of which they invited the 159th NY to join.[32]

---

32 Tiemann, William F. *The 159th Regiment Infantry, New York State Volunteers*, In the War of the Rebellion 1862-1865, (Brooklyn New York 1891), p. 64.

Miniature of John W. Mambert

John's view of slavery is strongly stated in his next letter home as, "the greatest curse" to befall the nation.

Letter #23 – December 16, 1863

La. Thibodaux

To Mrs John W. Mambert and family.

I am well and enjoying good health thanks be to God for his goodness and mercy toward me, and I hope you are enjoying the same blessings. I received your letter yesterday the 15th inst. I was glad to hear from you, you don't know how good I feel when I get a letter and papers. I have had a letter in each of the last 2 mails. To be 2 weeks without hearing from home seems to be a long time and I watch the mail office very close and when the mail comes and no letter for me then my lip hangs over my chin. I hope you will send me some postage stamps for my last goes on this letter. I knew your letter was dated wrong, you stated that Harm's little Ella[33] was dead and at rest. I am sorry to hear such bad news, but Permelia wrote about it in the last letter before this. Write to me where she was buried. John must attend to his little girl and Doctor once or twice a year for worms without fail. To John I am glad that you have plenty of work. I hope you enjoy yourself well and be contented for you don't know what blessing you enjoy if you knew what I know and what I have experienced, you would know what happiness you enjoy. I see everything is high up there but here in Dixie it is a great deal higher, butter here is 50 cents but we will have to pay more soon, I eat about 2lbs per month. I was in the village, and I asked what the price of a checked shirt was, $3.25 was the price. $6.50 for a pair of shirts, eggs .50 cents per dozen. Oranges I can buy as cheap in Hudson as you can here. There is no less money than 5cents here, but sometimes you can get 2 for 5 cents, 3 for a dime and 15 cents for an apple, a small one 5 cents, ordinary size 10 cents. Wages, there is no price, that is no one works except a few Negroes on plantations, but they offered from $40 to $50 per month. The news about the war is good and I hope it will be brought to a close soon, and I hope it will be brought to a close satisfactory. I hope it will close with slavery abolished for it is the greatest curse ever happened to our Nation. I think now of the welfare of my country than anything else, and I think the man who says, keep the Negro a slave for planters he is doing the Negro a greater kindness, he worships him as you would a nice horse-- he forgets that he is building up the aristocracy and destroying the education and prosperity of this country. I have seen enough and had experienced enough since I have been in Dixie to judge what is the matter. The south said the north was or had taken away their Negro, when the north had never taken one colored and then the aristocracy of

---

33 Ellen (Ella) Mambert – Age 5 ½, died, daughter to Hermon & Catherine Mambert, Claverack NY.

the south wages war and rebels, and what do you think some of the loyal folks say to me whom I have become acquainted with in the village. They say, that is the Reb's, that they messed it, they the Reb's aught 1st to get our people and nation in a war with some other nations and then seceded, and then they could get a southern confederacy. Now look at it, if that is the evil of slavery then I say kick it out of existence. So, you see it is now for the benefit of the Negro, it aught to be done away with and I will say nothing more about that. I hope that they will get volunteers enough and drive the Reb's out of rebellion.[34]

(Letter ends)

THE ARISTOCRACY SUPPRESSED EDUCATION

The aristocracy is defined as a group regarded as privileged or superior in a particular section of society. John's thoughts regarding the culture created by planters over the past thirty years matched that of other objectors from the North. In the words of abolitionist Henry Ward Beecher,

"In the South ignorance is an institution. They legislate for ignorance the same way we legislate for schoolhouses."[35]

To protect and preserve slavery, the aristocrats understood that preventing education among poor Whites helped stifle the spread of abolitionist ideas. "This passive ignorance undoubtedly decreased poor whites' ability to understand the more complicated arguments against slavery, and certainly precluded them from clearly formulating their own reasons to oppose the institution."

"The more I think and see of slavery the more I detest it." -Hinton Helper.[36]

Helper was an abolitionist with a twist; he was also a White supremacist

---

34 *Letter #23, December 16, 1863, La. Thibodaux, To Mrs John W. Mambert and family.* John W. Mambert letters from the Civil War.

35 "Anti-slavery lectures," New York Times, Jan. 17, 1855, 5, as cited in: Merritt, Keri Leigh, Masterless Men, Poor Whites and Slavery in the Antebellum South, (Cambridge University Press 2017), Ch. 5, p. 143.

36 Helper, The impending Crisis, quoted on 375, as cited in: Merritt, Keri Leigh, Masterless Men, Poor Whites and Slavery in the Antebellum South, (Cambridge University Press 2017), Ch. 5, pp. 144,145.

who argued that slavery hurt the economic prospects of non-slaveholders and impeded the growth of the entire region of the South.

Aristocrats, besides being obsessed with greed, also wanted censorship to keep the poor whites ignorant. The deep South's penal code carried severe penalties, slave codes. "Most of this legislation was passed in the 1830s, likely in response to the proliferation of abolitionist literature." Once the war had concluded people from the North soon realized the widespread ignorance of the poor White population in the South.

The Christian Recorder reported:

"So completely had the power of free speech been subdued as one of the necessary means for the security of slavery, that no man dared to utter a public warning against the danger of such an experiment; and the poor white classes, as ignorant as the slaves, were easily led as sheep to the slaughter."[37]

John mentioned the Southern Unionist statement about what the Reb's intended "1st to get our people and nation in a war with some other nations and then seceded," but there is no evidence the South tried to create an international war.

Documentation shows that economic warfare was inflicted on the South. Union Naval blockades prevented import and export trades, especially cotton, the South's biggest export, a vital commodity important to the Confederacy's financial state.

The South hoped foreign governments would intervene, putting pressure on the North. But that hope faded once the Emancipation Proclamation commenced.

YEAR'S END

Christmas came. The men of the 159th NY celebrated as best possible in Thibodaux. So far from home, the men wished instead they were with family, enjoying Yuletide. The soldiers were granted freedom from drills and all other responsibilities that day. They decorated the camp with what little they had: hardtack and dried bacon were used for decorations hanging from makeshift Christmas trees.

---

37 *The Christian Recorder* (PA), July 22, 1865: ibid, ch. 5, pp. 144, 145.

CHAPTER FIVE

Men attended Christmas service in town. Since Chaplain Kipp resigned in September, the regiment did not have sermons in camp. There was no special meal prepared in celebration, but men lucky enough to have received boxes from home shared the provisions with others not as blessed.

The winter was severe, unusually cold, sleet, rain and heavy frosts. There were days when a substantial amount of snow fell, uncommon weather for Louisiana. Men congregated around campfires reminiscing, drinking, and singing carols. December 29, 1864, "acting General," Colonel Molineux was relieved from the staff of Major-General Franklin. Five days later, he assumed command of Lafourche Parish relieving General Birge.[38]

At year's end, the 159th NY took a substantial hit on their overall numbers. Fifty-seven soldiers, thirteen officers included, lost their lives in battles bravely fought. Additionally, sixty-seven men died of various illnesses including Typhoid fever, and one from alcoholism. Eighty-six men were discharged, sixty-eight with disabilities. Twenty-eight men scampered north without furloughs rounding out the regiment losses with an additional 238 men.

The regiment had their share of deserters. But one officer took issue with the negative press on the topic as the following letter posted in the Gazette Newspaper shows:

> A member of the 159th Regiment, who left this city as a private, but has been promoted to a responsible position for his bravery and soldierly merits in the field, writes to us from camp as follows:
>
> Headquarters 159th Regiment, N.Y.S.V.
>
> FRIEND GAZETTE: —I notice that the Republican Newspaper, of your city, has of late become over-zealous as the friend of us soldiers, and will probably be shocked if any should vote against the Abolition measure to place the franchise of the army in the hands of a military despot.
>
> "The hypocrisy of your Abolition neighbor is too transparent. When our Regiment left your city on the 30th of October 1862 the Columbia Republican insulted us with the following language:

---

38 Tiemann, William F. *The 159th Regiment Infantry, New York State Volunteers*, In the War of the Rebellion 1862 -1865, (Brooklyn New York 1891), p. 65.

"IT IS NOTORIOUS THAT THE MEN WERE DESERTING FASTER THAN THEY WERE BEING RECRUITED—ONLY THOSE REMAINING IN CAMP WHO WERE TOO LAZY TO RUN THE GUARD."
"The base injustice of this slur has been most eloquently refuted by the gallant acts of the 159th on the field, and their heavy bill of mortality in battle."

"Your contemporary seems too young a convert to patriotism to attempt to set himself up as a dictator to those who have been patriots from birth. Will the Republican apologize for its slanders on the 159th? It may have forgotten it, but the soldiers have not! We demand justice, but despise cringing sycophancy. "Yours truly, H. T. N." [39]

ENLISTING: A FAMILY AFFAIR—THE KIPPS

Of the sixty-seven men from the 159th NY lost to illness, one was my 3rd great-grandfather **Phillip B. Kipp**, a private serving in the 159th NYS Volunteers, Co. I. He died on March 29, 1863, from chronic diarrhea. His death was likely from Typhoid while in quarantine at Barracks General Hospital, New Orleans. Phillip was fifty-five years old when he passed.

Born May 2, 1807, the regiment muster roll had his age listed as forty-four upon his enlistment on October 9, 1862. Phillip and his 2nd wife, Sally, lived in Gallatin, NY, where he was employed as a farm laborer. At the time of his death, he had nine children ranging in age from thirteen to thirty. Three of Phillip's sons—Amos, Theodore, and Silas, were recorded as enlisted men serving in the Union.

Amos Kipp (2nd great-grandfather to author) was a private with the 16th Heavy Artillery, NYS Volunteers Co. B. He enlisted at the age of twenty-eight, Kinderhook NY, on January 4, 1864. At the request of General Butler on August 9, 1864, his company volunteered to assist in cutting out the Dutch Gap Canal through the peninsula in the James River.

Amos became ill while working on the project. By November 1864, he was sent home to recuperate but was not improving in health. The following is a sworn statement from February 1865, regarding his illness:

---

39 *Gazette Newspaper*, Unit History Project, 159th NYS Regt. Newspaper Clippings. NYS Military Museum & Veterans Research Center.

I hereby certify that I have attended and prescribed for Amos Kipp from the middle of November 1864 up to the present time he having had congestion and inflammation of the lungs with repeated attacks of diarrhea, and he has not been able to travel or report during that time except the 12th of January when he reported to the US Hospital at Albany but could not get in and returned home, and his disease reappeared. I now consider him fit for travel and have arranged his transport to Albany US Hospital, if not admitted, order him to New York City US Hospital as I think he had better be examined before going farther.

Chatham Feb 1, 1865
R. H. Vedder
Columbia County

Subscribed & sworn to before me this day of Feb 1865

Rupert Ashley
Justice of the Peace

Amos was eventually discharged due to his illness and mustered out of service August 21, 1865. Personally, I am very grateful Amos survived his illness because my great grandfather Charles A. Kipp would not be born for another five years.

Amos's brother-in-law, Anthony Mixted, age twenty-four from Valatie, NY, joined the military the same day, regiment, and company as Amos.

Amos's father-in-law, Thomas Mixted, born in Dover England, was fifty-three years old (though the muster rolls show age forty-four) when he enlisted as a private with the 128th Regiment NYS Volunteers Co. E (another Columbia / Dutchess County Regiment) on September 4, 1862. Listed as having a disability, Thomas was discharged and mustered out August 2, 1864, from a New Orleans hospital, followed by a death date of August 11, 1864, from effects of the war.

Theodore Kipp (2nd great-uncle to author) was a private with the 128th Regiment NYS Volunteers Infantry Co. G. He enlisted at the age of nineteen in Ancram, NY on August 18, 1862, for three years. He transferred to the 19th Regiment Co. A, Veteran Reserve Corps on August 12, 1863.

The 19th became a new organization after the original mustered out from two years of service, later considered an Artillery Regiment. It was engaged in sixty-four battles, sieges, and skirmishes. He mustered out July 13, 1865, in Elmira, NY.

After the war Theodore met his wife Cora McDorman while he was employed as a farm laborer in Arizona, Nebraska. They married on February 7, 1885 and had two children. Theodore died on January 6, 1936 in San Diego, California. He was the last member of GAR Post No. 315 in Herman, Nebraska.

Silas Kipp (2nd great-uncle to author) was a private with the 192nd Regiment NYS Volunteers Infantry Co. D, He enlisted at the age of seventeen far from home in Schenectady, NY, on March 28, 1865. He traveled sixty miles north to join for one of two reasons: either his mother did not approve, or it was the closest enlistment location. The regiment was attached to the 3rd Brigade, 3rd Division of the Army of the Shenandoah. They mustered out at Cumberland, MD, August 28, 1865. Serving for six months, this regiment had one of the shortest tours of duty in the Union Army. But in that short time, the regiment lost twenty-six men.

After the war Silas lived with and worked for his brother Henry as a farm laborer in Ancram, NY. The last recorded information on Silas was the 1900 Federal Census, he was fifty and single.

A fourth son, **Henry Kipp** (2nd great-uncle to author), age thirty, was drafted on July 1, 1863. No records of serving were located.

Both Phillip Kipp and Thomas Mixted died from illnesses contracted during the war. Amos Kipp was discharged due to his disability caused by sickness. John W. Mambert was sick on numerous occasions throughout his military career, and most likely survived due to his close relationship with Assistant-Surgeon Briggs, who kept a close eye on him. Anthony Mixted mustered out March 11, 1864, two months after enlisting. There was no reason listed for his discharge. John W. Mambert, the Kipp's and Mixted's contributed greatly to fighting the aristocrats of the South and assisting in the termination of slavery in our nation. Half of those who enlisted paid a heavy price for the fight, returning with disabilities, some paid the ultimate price of their very lives.

CHAPTER FIVE

THE PROCLAMATION OF AMNESTY AND RECONSTRUCTION

Unifying this fractured nation would take enormous effort by the Federal Government. President Lincoln knew he had a difficult job ahead, especially after enacting the Emancipation Proclamation freeing slaves in Confederate states.

Lincoln planned for political reconstruction of the South and began laying the groundwork in early 1862: months before announcing the Emancipation Proclamation in January 1863. He had a strong desire for reconstruction in Louisiana and wished for it to happen expeditiously. Louisiana was an obvious choice, considering the Union's stronghold in the state.

On December 8, 1863, President Lincoln introduced his Proclamation of Amnesty and Reconstruction:

> Whereas in and by the Constitution of the United States, it is provided that the President "shall have power to grant reprieves and pardons for offenses against the United States, except in cases of impeachment, and
>
> Whereas a rebellion now exists whereby the loyal State governments of several States have for a long time been subverted, and many persons have committed, and are now guilty of treason against the United States, and
>
> Whereas, with reference to said rebellion and treason, laws have been enacted by Congress, declaring forfeitures, and confiscations of property, and liberation of slaves, all upon terms and conditions therein stated, and also declaring that the President was thereby authorized at any time thereafter, by proclamation, to extend to persons who may have participated in the existing rebellion, in any State or part thereof, pardon and amnesty, with such exceptions, and at such time, and on such conditions, as he may deem expedient for the public welfare, and
>
> Whereas the Congressional declaration for limited and conditional pardon, accords with well established judicial exposition of the pardoning power, under the British, and American Constitutions, and
>
> Whereas, with reference to said rebellion, the President of the

United States has issued several proclamations, with provisions in regard to the liberation of slaves, and

Whereas it is now desired by some persons heretofore engaged in said rebellion, to resume their allegiance to the United States, and to re-inaugurate loyal State governments within and for their respective States...

President Lincoln found the effort for reconstruction to be plagued by conflict between the Unionist Free State Committee officials and the military governor of Louisiana, Shepley. Frustrated, Lincoln turned to General Banks for assistance in gaining loyalty to the Union by increasing faithful voters who would become the core electors in the reestablishment of a new republican government in Louisiana.

A constitutional convention needed to be held to ratify a new constitution ending the previous from 1852. It was important to the president an anti-slavery clause was included. The convention could not be held until after the election of a new governor. A constitution including emancipation passed on May 11, 1864.

John felt the rebel effort was coming to an end. Based on his knowledge of current events in Louisiana, he shared with his family his belief that Louisiana would soon be back in the Union—as a free state.

Letter #24 - January 26, 1864

Thibodaux, La.

Mrs. John W. Mambert and family.

I am well and enjoying good health thanks be to God for his merciful kindness towards me. I received Permelia's letter last Sunday and was glad to hear from her and her man, and today I received your letter of the same date. I was glad to hear that you were all well and also that you bought that house and lot. I hope you will arrange the matter right away. I meant to tell you to buy it or I would have bought it as soon as I would of returned. If Permelia can hire out, she and Eb to one place I think that would be a good notion for the present. I wrote a letter to Granpap

Mambert but my mind was somewhat confused after writing one to you at the same time. Our Regt. is enjoying good health, as I am, I weighed 161 lbs. yesterday so you see I am on the gain. Last summer I weighed 135 and less. I am glad to hear that mother feels as smart as a cricket, she doesn't feel smarter than I do I will guarantee. I felt somewhat alarmed. Permelia wrote that she looked poor in her miniature, and I thought that she was care worn, etc. But I hope you will hope for the better in the future. Don't worry about me, I feel good. I raised the old glory in the Drum Core yesterday and had 4 of the boys tied up one hour a piece to the flag staff for disobedience of orders. The weather here is very warm, so warm that we run for the shade as you do in the summer but as soon as the sun is down it is cold and the cold strikes right to the bone as if you put your hands in cold water. I am glad you have sent your miniatures. I shall look for them in the next mail. Dad Wheeler[40] and Solon Macy[41] and a boy named Joseph Riell[42] is my tent mates and we enjoy ourselves. Sometimes Dad and Solon go off and comes pretty well stewed and then I have got trouble. This State is making all preparations to come back into the union as a free state and you see by the papers that several other states are following, I think the Rebels are about played out. I think by the 1st of July next you will see a great difference if we all live. The first of May our time is half out and time passes very fast. I will be home before you know it if nothing happens, I wrote I never would accept a promotion for I am as far in Military tactics as I am going. I feel perfectly content. I wish you was here to see this Country; it is just as level as can be and not a stone as large as a pea. I begin to like the Country better than I did, they are fixing to plant cotton, if you was down here how you would like to plow such land as this. I also wish you could see our Battery drill it would be a curiosity to you. I sometimes think I will get the fever again next summer,

---

40 John J "Dad" Wheeler, Musician 159th Regiment Co. C: Age 54, Gallatin, NY.
41 Solon Macy, Private 159th Regiment Co. E: Age 43, Carpenter, Greenport NY.
42 Joseph W. Riell, Musician 159th Regiment Co. D: Age 16, Brooklyn, Kings, NY. Lived with his parents and four siblings.

but I hope not, I had it twice last summer. I thought about Uncle Al when he said they could not raise pees here. About pay, it may be that we won't be paid until the middle of March or even April and we may be paid very soon, you must ask the Express agent every time you go to Hudson and then you won't miss. I have got a very nice uniform and I think I will send you another miniature shortly after we are paid. Mr. Kline[43] was here after the remains of son who died here last Sept. His land joins James M. Strevers in the South and I was present when his remains were taken up, it was quite a solemn sight. He was buried near Ed McLean[44]. I was going to write to his mother and then I saw someone who had written to her the day before and so I did not write. You will please write and let me know where Ramses Post Office address is and how Elisha Smith[45] and Sarah are doing? I have written one letter to her. Tell Uncle Allen[46] if he was here, he could shoot Rebels instead of foxes. I wish he was here a spell. There is plenty of wild ducks here and everything is tame here, birds are tame, there is plenty of turkey buzzards, they are as large as a turkey hen and flock like turkeys, they come by one camp 30 or 40 and I can walk right to them 20 paces, they live on carryon, they will fight in the street, there is a fine for shooting them, they don't notice them more than you notice a turkey, they are the color of a grey turkey and a head just like a turkey.

Old Andrew Brush, I don't know how he comes along I understand he is in New Orleans. Last Sunday I was to the Presbyterian church in the forenoon and in the afternoon to Catholic church. There is a Methodist church here and an Episcopal church. I was out today to see the Battery drill and the sun was so hot that my head felt bad. I hurried back to my tent. This is a great country to raise chickens,

---

43 Benjamin Kline: Age 48, father to the late Corporal Pleasant Kline. He was a farmer in Taghkanic, NY.

44 Edward McLean, 159th Regiment Co. E: Age 25, died at a hospital in Thibodaux Oct. 7, 1863. He was a farm laborer for Charles Jones in Smokey Hollow, Claverack NY.

45 Elisha Smith & Sarah Smith: Ages 32 and 28 respectively, married and live in Taghkanic, NY. Elisha was a carpenter.

46 Allen Mambert (Henry): J.W.M.'s brother. Age 40, a carpenter in East Taghkanic, NY. Married with one daughter.

dogs, alligators, snakes, lizards and a crawfish like lobsters they will come up out of the ground and make a hole like one of these dusty toads and they will slap their claws together, and it will rattle like a woodpecker, sometimes they will come right up in the tent and in some places if you go out at night it is one continual roar of clatter. I want mother to write, and tell Jobe I read his letter, let me know where he goes to school or who teaches. Tell Grandpap Sheldon I am happy to hear that he is about again, and that Granny Sheldon is getting along. You must ask John Cambel[47] how we slept together and such. I must here tell you a joke. One night as our Regt. was on the march after dark we stopped and the cook, a negro, fetched water out of a little water hold to make coffee he got on a log as he supposed and dipped up water for coffee and made it. Next morning, he went again after water, to his surprise the old log he stood on and dipped water the night before was an old dead mule lying there, I suppose a week or two, It was laughable, the coffee tasted just as good as long as we did not know it. I will close tonight, truly yours, trust in God and hope for the better in the future.

John W. Mambert
(note on side margin) I have those samples of Martha's dress. [48]

SOLDIERS BURIED FAR FROM HOME

When a soldier lost his life from disease or battle, family members back home feared the possibility loved ones so far away would be interred where they would never have an opportunity to mourn their loved one's death. The tradition of viewing the body for the last time and burying them in a family plot or local graveyard would keep families from proper closure.

However, if the family of the deceased had the financial means they could hire someone to locate, retrieve, and bring the body home. If the loved one died in one of the larger hospitals, for example New Orleans, a metallic casket could be purchased for $100 and the body could be shipped

---

47 John S. Campbell, Private 159th Regiment Co. I: Age 45, discharged from NOLA on November 23, 1863. From Hudson NY.
48 *Letter #24, January 26, 1864, Thibodaux, La., Mrs. John W. Mambert and family.* John W. Mambert letters from the Civil War.

home. Another option would be a family member retrieving the body just as Corporal Kline's father recovered his son's remains. He traveled nearly 1,500 miles deep into the South with a considerable extent of his trip through enemy territory; escaping unnoticed was a miracle.

Families without the ability financially or physically to bring home their loved ones agonized not knowing where or how their family members were interred. Many men who died on the battlefields were buried where they lay. After the 159th NY made their gallant charge at Port Hudson, under a flag of truce a burial party located their dead comrades lying close together not far from the enemy works. They were interred where they lie, along with the piece of color-staff left on the field.[49]

Shortly after the war broke out and mortality numbers continued to rise, the government realized it needed accurate records of the dead. The War Department issued G.O. #75 in September 1861 making commanders of Union forces responsible for burying their dead and keeping records of where the men were interred, recording the soldiers last moments and the location of interment.[50] Reality was under certain circumstances this order was impossible.

After the surrender at Appomattox Court House and the war slowly came to an end, government agents began a thorough search on all battlefields and hospital burial grounds, carefully locating soldier graves they copied the writings etched into rough wood headboards. Initially the government renewed old headboards as they rotted, making sure the grave was protected.[51] Eventually wood headboards were replaced with granite stones for all dead soldiers due to the long-term cost of replacing wooden markers. To further protect the soldiers' graves, in 1867 the U.S. Quartermaster General established national cemeteries and all known graves were relocated.

CLEAN WATER & NUTRITIOUS FOOD—NOT!

A soldier's survival depended on good water, and for soldiers continuously on the move, it was not always easy to find. The desire to quench one's thirst

---

49 Tiemann, William F. *The 159th Regiment Infantry, New York State Volunteers*, In the War of the Rebellion 1862 -1865, (Brooklyn New York 1891), p. 37.

50 Moss, L. (1868) *Annals of the United States Christian commission*. Philadelphia, J. B. Lippincott & co. [Pdf] Retrieved from the Library of Congress, https://www.loc.gov/item/02018786/. p. 375.

51 Ibid., p. 439.

becomes great for men dressed in wool, carrying thirty pounds of equipment while marching for miles in Louisiana's extreme heat and humidity. Water was also vital for coffee and cooking. Retrieved from streams or ponds, in many cases the waters contained deceased bodies or human waste, which contributed to illness and many subsequent deaths among soldiers. Diarrhea and dysentery were the most common reported diseases during the war with more than 1.6 million cases in the Union army alone. Diarrhea was commonly caused by living in unsanitary conditions. Records show large numbers of soldiers died from disease caused by contaminated food and water, including typhoid and dysentery.

Chronic diarrhea was often a sign of malnutrition or critical vitamin deficiencies. Regiment surgeons dealing with numerous cases of diarrhea frequently associated it with scurvy, a disease caused by a vitamin C deficiency and easily treatable with fresh produce. Regrettably, a soldier's regular diet did not include fresh fruits or vegetables, but instead consisted largely of pork, beans, coffee, and hardtack.

While there were few variations between a Union and Confederate soldier's diet, the basic nutritional value was roughly the same—poor. Eating vegetables was the most effective way to avoid the deadliest condition of the Civil War: diarrhea. Safely stated the armies of both the Union and Confederacy would have been healthier and with their regiments ready for battle, not in a hospital.

REDUCED

On January 28, 1864, an order from Captain Robert McD. Hart reduced John in rank, returning him to Company G.

Tommy Miller, a young fife from Company A, was promoted to fill John's position as First Principal Musician.[52] No explanation was found in the records stating the demotion while encamped at Thibodaux, but it may have been due to John's questioning or protesting his superiors too often, possibly the last straw. It was obvious John and his commanding officer had problems from the beginning, mostly regarding his position and pay, but there was also his Confederate parole in Brashear City where John was

---

52 G.O. *Captain Robert McD. Hart, Mambert demotion, January 28,1864*. Regimental Company Books for the 159th NYSV, National Archives, Washington DC.

reluctant in returning to service, until Molineux ordered it. This was also evident to John's friends back home, knowing he was "so tired of being there" as Attorney Hawver acknowledged.

John and the colonel did not see eye to eye, which likely led to his reduced position.

Letter #25 - January 1864

159th Regt. N.Y. Vols.

Private letter to Mrs. J.W. and John

I want you to go to Hudson and see John Welch, and tell him that he knows in what manner I enlisted and that he made out my papers, and that he told me that if I did not have the position of Fife Major then my enlistment was null and void, and that he said he would see me discharged and he would not charge me anything for it. I am only drawing pay as a common musician and never had a warrant of appointment to the office for which I volunteered, they paid me $21 per month from the 1st of Jan. 1863 to the 1st of Jan. 1864 and then without any reason I was reduced as private musician. When I volunteered to go with the 167 Regt. NYS Vols. as Fife Major, I was told by Mr. Post that the wage was $41 per month and that if I would go with the regiment, he would see that I should get the position of Fife Major and that $41 per month should be my wages. Mr. Port told me these words, that he would not deceive me and that upon his word and sacred honor it was so. I saw Col. Nelson, I spoke to him about the position for which I had enlisted. Col. Nelson told me personally that as you and the recruiting officer makes it, it shunt be altered. It has been altered and changed without a cause. I will say to John M. Welch that he is acquainted with Col. Nelson, and he is at Washington whether he can't get me discharged through the influence of Col. Nelson. That is discharged by President Lincoln or the secretary of war, and if John M. Welch succeeds in getting me discharged, I will make him a present of $50 if he doesn't succeed then I won't give him anything. This is the only way I know to be discharged and this is the last time I shall ask to be discharged, if Mr. Welch is a man of his word, I shall always remember him here after and if not a man of his word, I will be the same, I will always remember him.

John W. Mambert
Write often.

I would like it if Mother would write a few times and Jobe and Walt. Tell Permelia if she keeps house I will come and see what kind of a house she keeps. Write and let me know where Ram's post office address is and if Sarah's post office address is the same place. Tell Grandpap Mambert to build a new house and pull down the old one, not to paint the old one. To Grandpap shall I send my best wishes and to Grandma Sheldon__ girl I soon will be 44 years old and perhaps when you get this. Today is the 17th, quite cool. Write often, trust in God and hope for the better in future.

John W. Mambert [53]

LOUISIANA'S GOVERNOR'S

It was election season in Louisiana. In Thibodaux's town square, politicians held debates and speeches were made. The excitement of the occasion created much needed relief for the men of the 159th NY from their usual monotony of garrison duty. [54]

One of the candidates who addressed the people was the Honorable Michael Hahn running as a free-state candidate from the antislavery political coalition. Hahn ran against Benjamin F. Flanders who also advocated for a free state, but his platform was more radical than Hahns. Flanders wanted full civil and social equality for the Negro whereas Hahn wished to slow the process down. Hahn did advocate for an immediate end to slavery but preferred a gradual broadening of Negro rights to include Black suffrage.

In the end, Hahn's policy was preferred by the majority. He was elected governor on February 2, 1864. Administering his powers from New Orleans,

---

53 *Letter #25, January 1864, 159th Regt. N.Y. Vols., Private letter to Mrs. J.W. and John.* John W. Mambert letters from the Civil War.

54 Tiemann, William F. *The 159th Regiment Infantry, New York State Volunteers*, In the War of the Rebellion 1862 -1865, (Brooklyn New York 1891), p. 65.

Hahn fully supported President Lincoln's policies, believing they were the way to quickly end the war and unite the country.

Adding complexity to the election of a governor in Louisiana, Michael Hahn was elected only for the Union controlled section of Louisiana. Henry W. Allen, a Brigadier General for the Confederate army was elected governor of the Confederate controlled Louisiana on January 26, 1864, at the courthouse in Shreveport, Louisiana.

After Allen's election, he delivered an inaugural address that inspired the entire South when published in Confederate newspapers. With the new Confederate governor situated in Shreveport, his state legislature went to work on fulfilling the many requests he made at his inauguration address.

INTOXICATING LIQUORS, 'A CRYING EVIL'

During Allen's address, he discussed intoxicating liquors, an issue of which he was very passionate. He told his constituents:

"the General Assembly, at its last session, very wisely enacted a law prohibiting the distillation of intoxicating liquors except from fruit. This has had a most salutary effect. There is still a crying evil in our midst. We are importing daily from neighboring states large quantities of alcoholic poison. The effect of this poison upon the community is most intolerable. I need go no further than this capital, to show you the long record of crime brought on by intoxication. In the army it is worse, nine tenths of the arrest and punishments are caused from intoxicating liquors. I therefore urgently recommend that you enact a law prohibiting under severe penalty the importation or sale of intoxicating liquors in this state, except for the medicinal purposes. The fathers and mothers of this state will "rise up and call you blessed," for such a law and the good people will generally hail it with delight; for it will save many a gallant young soldier from punishment and disgrace, and in these times will give peace and quiet and security to all".[55]

The legislature's new law created shortages of alcohol for not only soldiers but also the general public. The law left distillers challenged to find new ingredients in place of the restricted ones. Resources of Southern Fields and

---

55 *Annual Message of Governor Henry Watkins Allen, To the State of Louisiana.* Printed at the Office of the Caddo Gazette, January 1865.

Forests, by Francis P. Porcher published in 1863, listed many recipes for producing alcoholic drinks. A simple example for making beer from corn was:

"Take one pint of corn and boil it until it is soft, add to it a pint of molasses and one gallon of water; shake them well together and set it by the fire, and in twenty-four hours the beer will be excellent."[56]

Excessive drinking was an issue for both the Confederate and Union armies. At a finding of a court martial, not satisfactory to General McClellan, he sent the following:

"No one evil agent so much obstructs this army in its progress to that condition which will enable it to accomplish all that true soldiers can, as the degrading vice of drunkenness. It is the cause of by far the greater part of the disorders which are examined by Court Martials. It is impossible to estimate the benefits that would accrue to the service from the adoption of a resolution on the part of the officers, to set their men an example of total abstinence from intoxicating drinks. It would be worth FIFTY THOUSAND MEN to the Armies of the United States."[57]

A great deal of whiskey was sent to army camps on both sides by relatives. Private John D. Billings wrote,

"if there was a red-letter day to found anywhere in the army life a soldier it occurred when he the recipient of a box sent to him the dear ones and friends he left enter the service."[58]

It was common practice for families of Union soldiers to send their loved one's care packages. Commanding officers realized that whiskey was being smuggled into the camps through these packages. General Butler testified before Congress that an Adams Express Company depot search yielded 150 different packages of liquor in crates and boxes on their way to his command.

Going forward, every parcel sent to a soldier had to be opened and inspected by an officer of the regiment or brigade. Billings wrote:

"It was a little annoying to have every box opened and inspected at

---

56 *Resources of the southern fields and forests*, by Francis P. Porcher (1825–1895) Publisher, Charleston: Steam-power Press of Evans & Cogswell, p. 552.

57 McClellan and his officers. *At a finding of a court martial, not satisfactory to him, Gen. McClellan sent the following just and noble rebuke and admonition.* New York http://www.loc.gov/resource/rbpe.12303900

58 Billings, John D., *Hardtack and Coffee*. Published 1888. p. 217.

brigade or regimental headquarters to assure that no intoxicating liquors were smuggled into camp"

He followed with:

"There was many a growl uttered by men who lost their little pint or quart bottle of some choice stimulating beverage, which had been confiscated from a box as contraband of war." [59]

On February 19, 1864, Colonel Molineux, still in command of Lafourche Parish, wrote to his sister the following:

> General Birge will be back by the 5th March & I shall be right glad of it for what with the elections, quarrels between the negroes & landholders, robbery by drunken soldiers (reenlisting as veterans & receiving large bounties) I don't like my governorship. I expect the people will be even more glad than myself for General Birge is much easier.

Further in the letter:

> I have had to approve the decision of a court martial which has condemned three soldiers – one black, two white – to death for desertion & a crime too revolting to name. If General Banks approves, they will be shot. Do you wonder that I would prefer being a soldier in the field than (as you call me) an acting Brigadier General of the district?[60]

WALTERMIRE TAKES COMMAND

Captain William Waltermire, Co. E, was promoted to Major and assumed all command of the 159th NY on February 20, 1864. This arrangement was highly satisfactory to the whole regiment. In the future, Waltermire would again be promoted to Lieutenant Colonel. William was a twenty-nine-year-old farm laborer when he enlisted. From Ghent, NY, he lived with his mother, Cornelia, and two younger siblings. His father George, a farmer, died when William was eighteen.

The new Major was respected and responsible, a good leader. Waltermire rose through the ranks having no formal military training.

---

59 Ibid., p. 219.
60 From the collection of E. L. Molineux, *Thibodaux Feb. 19, 1864, Letter to sister*. NYS Military Museum & Veterans Research Center.

Major Waltermire experienced his first responsibility in maintaining a strong regiment when sixteen men from Co. G deserted the regiment during the weeks of February 9 to 27, 1864. The majority of the men were from Hudson, NY, and all were transferred into the regiment only one month prior as veteran volunteer soldiers.

The men either took issue with their new commander, or they only reenlisted for the lucrative bounty money being paid for men with their qualifications. Likely, it was the latter, with plans to make a beeline for home the first chance they got.

THE COMMISH

Louisiana's Confederate Governor appointed a commission to gather testimony concerning the conduct of the enemy during their brief and disgraceful occupancy of West Louisiana. He wanted this information for publication and historical records, a careful, accurate, authentic statement of the atrocities and barbarities committed by Federal Officers, troops, and their camp followers during their invasion of Western Louisiana.[61]

Understandably, the Confederate leadership in Louisiana was outraged by the actions of its opponent, but should they have been surprised by the atrocities of war? The Louisiana legislature was at the forefront of secession from the Union, administering their first defiant act by taking over the federal forts in their state. Such a decision to revolt does not come without consequences. But blame had to be placed on someone, and in Louisiana, it was unlikely to fall on the aristocrats who raised the Pelican flag over their Capital.

The Commission's report opened with the following:

"Before this fair land had been wasted, and the labor of years destroyed, the planter's spacious mansion was surrounded by fluids of waving corn and cane, and overlooked broad prairies animated with flocks and herds and checkered with farms of cotton, whose trim and careful culture recalled the husbandry of the patient Hollander. Around the planters' dwellings were seen the numerous out-buildings used for agricultural purposes, and the negro cottages, always enlivened by groups of happy children. When the labors of the day were over, the scene was ever animated by the loud laugh, the rude sports, and the merry faces,

---

61 Young Sanders Center, Franklin LA: *Official Report relative to the Conduct of Federal Troops in Western Louisiana during the Invasion of 1863 and 1864*. Compiled from Sworn Testimony, Under Direction of Governor Henry W. Allen, Shreveport, (April 1865), p. 3.

indicating the happiness of the returning laborers. In the midst of these evidences of contentment, the planter enjoyed a more elevated pleasure, in communion with his family, in literary pursuits, or in the entertainment of his friends, —his highest social enjoyment consisting in administering the rite of hospitality under his roof. The master and the slave were alike happy, in their respective vocations. Such a condition naturally suggests the reflection, that the system which has produced them could only be in harmony with the wise designs of a beneficent Providence." [62]

"The planter enjoyed a more elevated pleasure"—absolutely, but describing the laborers as content and happy was a far cry from reality. If the slaves were so content to furnish free labor, be sold or traded and separated from family, punished by the whip, have no opportunities for education, to raise a family and provide for loved ones, why, at first opportunity, did they join the Union army, revolting against their masters and the "happy vocation" they so enjoyed?

Though the report was created to record the atrocities and barbarities committed, one might also view the commission's report as being crafted to make the case Louisiana needed to remain as it was—a slave state. Henry Watkins Allen himself was a planter; he knew that elevating the Negro to a free man would reduce his own ability to amass wealth and affect his position in society.

DISGRACEFUL, REVOLTING CONDUCT

The Union army tried its best to maintain orderly and lawful troops, but war can also bring out the worst in men, especially men who craved violence. One such situation took place after the trial of twelve soldiers of the 159th NY before a Provost Court in New Orleans late in February 1864. Tried by Judge Atocha, the charges were raiding and stealing on the plantation of Mr. Darden in the Lafourche parish. The trial ended March 2, 1864, with ten men convicted.

The guilty committed outrages act so barbaric as to rival the most beastly practices of soldiery in the dark ages. After robbing the plantation to an unlimited extent of chickens, geese, ducks, and everything else to which they took a fancy, they entered a Negro hut and before the eyes of the mother and father, each and all of them violated a Negro girl. After they had satisfied their brutal lust, they stripped the poor girl of such little articles of jewelry

---

62 Ibid., p. 6.

as she had upon her person, tore off her clothes and left. The oldest of the group, thirty-three-year-old Private Bryan Hopkins of Brooklyn, was found to be the ringleader and sent to Tortugas[63] for life at hard labor with a ball and chain. Jordan N. Lee, an eighteen-year-old private from Poughkeepsie, was another participant; he stood at the girl's head with a drawn bayonet to keep her from struggling: he was sent to Tortugas for ten years. The rest of the brutes were privates between the ages of sixteen and twenty years of age: all sentenced to three years.

After the trial, Colonel Molineux issued the following order on March 5, 1864:

"the disgraceful conduct, so revolting in its character, having been perpetrated by soldiers, is a stain upon our uniform and flag. It is inconceivable how men wearing the United States uniform can so far forget their manhood and descend to such vile and loathsome acts. The Colonel Commanding hereby gives notice, absolute and positive, that any and all acts committed on peaceable citizens and harmless negroes, of such character as to reflect dishonor and disgrace upon us as soldiers, will be met with severe and summary punishment."[64]

---

[63] Fort Jefferson on Dry Tortugas Island: In September 1861, the first prisoner soldiers appeared, those sentenced by Courts-Martial to confinement and hard labor for acts such as mutinous conduct. President Lincoln then substituted imprisonment on the Dry Tortugas and the ratio of soldier to prisoners was about four to one. In June 1864, the ratio was almost equal, with 653 soldiers and 753 convicts. In November 1864, only 583 soldiers guarded 882 prisoners and eight were able to escape. On 24 July 1865, four special civilian prisoners arrived. These were Samuel Mudd, Edmund Spangler, Samuel Arnold and Michael O'Laughlen, who had been convicted of conspiracy in the assassination of President Abraham Lincoln.

[64] G.O. Molineux to regiment, *acts committed on peaceable citizens and harmless negroes, March 5, 1864*, From the collection of E. L. Molineux NYS Military Museum & Veteran Research Center.

Chapter Six

"IGNORANCE & SUPERSTITION HAS BEEN THE WHOLE CAUSE OF THE WAR & SLAVERY HAS CAUSED IGNORANCE."

THE RED RIVER CAMPAIGN

On March 7, 1864, a brigade of cavalry commanded by Colonel N.A.M. Dudley rode through camp. The brigade's movement brought with it an eerie sense of foreboding for the men of the 159th NY. They now believed their peaceful stay in Thibodaux would not last much longer.[1] The cavalry unit was en route toward the next military movement—the Red River Campaign. The campaign was set in motion after seven months of meaningless adventures and countless preparations by the Union army performed with no serious military objective.

General Banks, however, was planning his next move: the capture of Mobile, Alabama, an idea known, and agreed upon by General Grant. But Banks's idea was postponed to instead pursue a foothold in Texas. The Union recently gained positions along the Texas coast, renewing Major General Halleck's desire for a combined naval and military operation on the Red River. The idea behind his plan was threefold: cut off the Confederates' supply from Texas to their forces in the east, split the Confederate line, surround their military force, and gain an outlet for northern Louisiana's cotton and sugar.

His plan included the great fleet of ironclads commanded by Admiral Porter, along with three available armies. Orders were for General Banks's army to move northward by the Atchafalaya River, General Steele's troops to advance from his position in Arkansas, and from Vicksburg, General Sherman's force. General Grant, Banks, Sherman, Steele, and Admiral Porter received corresponding instructions and the Red River expedition was commenced.[2]

---

1 Tiemann, William F. *The 159th Regiment Infantry, New York State Volunteers*, In the War of the Rebellion 1862 -1865, (Brooklyn New York 1891), pp. 66-67.
2 Irwin, Richard B. *History of the 19th Army Corps*. G. P. Putnam's Sons New York, (The Knickerbocker Press 1892), ch. 23, par. 1.

General Banks, knowing the expectations of Halleck, began making the necessary arrangements for the enormous undertaking. Texas being the objective, Banks would lead his army up the Red River: the shortest and best route.

Banks committed a large cavalry force to operate on the plains that lie beyond Mansfield, Louisiana, instrumental in defeating the Confederate mounted forces in the region. Success required a large force of infantry and artillery. Before the Union army passed Shreveport, the combined armies of the Confederacy in the region had to be met and beaten.

Food and provisions were scarce in the Red River country, and once Shreveport was occupied by Union troops, the large army would continue to march an additional 300 miles across a desert. An immense provision train was critical for the troops' survival. Admiral Porter's cooperation with his Navy was necessary to keep open the long line of supplies to be transferred up the Red River. The highest water levels in the upper Red River, necessary for Porter's fleet, would predetermine the date of the Union army's movement.

General Sherman traveled from Vicksburg to New Orleans and met with General Banks concerning the aid expected from the Army of the Tennessee, reaching an agreement March 1, 1864. Admiral Porter and Banks previously arranged for his large fleet of gunboats to be at the mouth of the Red River in time for the rising of the river waters. Sherman's agreement with Banks was to send 10,000 picked men of his army, with Porter's fleet, to arrive in Alexandria, Louisiana on March 17, 1864. General Banks agreed his troops would march north by the Teche, meeting them there. General Steele's army in Little Rock would move out at the same time, meeting the combined forces and the fleet at the Red River.

The plan was a massive undertaking: three Union armies and a naval fleet, hundreds of miles apart, concentrating at a remote point far inside the enemy's lines, a location on a river difficult to navigate, currently obstructed by low waters.

Admiral Porter's Flotilla, Harper's Weekly [3]

Admiral Porter reached the mouth of the Red River in early March. He assembled a great fleet of nineteen ironclads, fifteen of the heavier class and four of the lighter: in all carrying 162 guns.[4] General A. J. Smith boarded his men at Vicksburg on a fleet of nineteen river transports, joining Admiral Porter at the mouth of the Red River the following day.

On March 10, 1864, Colonel Molineux ordered regimental and company commanders to have all stragglers and convalescents fit for duty returned to their companies without delay. Six days remained until the time General Banks agreed to be at Alexandria with his army.

On the morning of March 12, 1864, the combined fleet entered the Red River at the head of the Atchafalaya River. Admiral Porter took nine gunboats and the army transports descending the stream until reaching Simmesport. Lieutenant Commander Phelps took the remainder of the fleet to ascend the Red River.[5]

That same day, Major Waltermire was relieved of command of the

---

[3] *Admiral Porter's Flotilla*, Illustration from Harper's Weekly, Library of Congress # 2006691852.

[4] Irwin, Richard B. *History of the 19th Army Corps*. G. P. Putnam's Sons New York, (The Knickerbocker Press 1892), ch. 23, par. 6.

[5] Ibid., ch. 23, par. 7.

159th NY by Lieutenant-Colonel Gaul returning from Albany, NY, with a new company of ninety-two enlisted men, mostly veterans of the 14th New York Volunteers including three officers. They joined the regiment and were assigned to Co. G, many of the existing men from the "old G" were transferred to other companies within the regiment.[6] John was among the forty-eight men from the "old G" transferred to other companies within the 159th NY. John was assigned to Co. B, originally organized as a Brooklyn Company. Other men from Co. G who transferred into Co. B were 1st Lieutenant George W. Hussey from Brooklyn, Private James Drewett of Stockport, Private John Gault of Claverack, Private Charles Snyder of Ghent and Privates Silas W. Richmond, John Hennessy, Clarence Chaplin, and John Dailey, all from Kinderhook.

On March 13, 1864, General Grover arrived in Thibodaux to reorganize and take over command of the division, constituted as the 2nd and the brigade also the 2nd. This included the regiments from the 159th New York, 131st New York (Metropolitan NY regiment), 19th New York and 1st Louisiana. The 159th NY was thereafter part of the 2nd Brigade, 2nd Division, 19th Army Corps commanded by Brigadier-General H. W. Birge.[7]

On the afternoon of March 14, 1864, Brigadier General Mower arrived at the Confederate-held Fort DeRussy, located south of Alexandria. Just before nightfall, Mower ordered two brigades to sweep the walls of the fort, forcing the enemy to surrender. Commander Phelps with his fleet burst through a Confederate blockade nine miles below the fort. Arriving in the ironclad gunboat *Eastport* he fired one shot from his 100-pound parrott and saw the white flag displayed. Mower's losses were three men killed and thirty-five wounded. The Union captured twenty-five officers and 292 men along with ten guns. When the news reached Admiral Porter, he at once ordered his fastest boat to Alexandria.

The first of the Navy fleet arrived in Alexandria on March 15, 1864, just as the last Confederate boats were making good their escape above the falls on the north side of town. General Kilby Smith and his division followed in the remaining transports with the fleet of gunboats. Landing in Alexandria during the afternoon of March 16, 1864, they relieved the naval detachment

---

6 Tiemann, William F. *The 159th Regiment Infantry, New York State Volunteers*, In the War of the Rebellion 1862 -1865, (Brooklyn New York 1891), p. 67.

7 Ibid., p. 68.

sent ashore earlier to occupy the town.[8] Marching orders were received by the 159th NY. The men prepared for the move and four days later broke camp, marching to Terrebonne station where they took rail cars to Algiers, arriving March 21, 1864.[9]

That evening, General Mower with two divisions (1st and 3rd of the 16th Corps) along with the 1st Cavalry brigade commanded by Colonel Lucas, surprised the Confederate encampment of Colonel William Vincent, capturing his entire regiment along with four guns from their battery during a heavy rain and hailstorm. A few of Vincent's men managed to escape in the confusion of darkness, but 250 Confederate soldiers were caught along with 200 horses. This was a heavy blow to Confederate commander Taylor; it deprived him of the only cavalry he had available for scouting until the arrival of General Green from Texas.[10]

On March 17, 1864, General Banks departed New Orleans. Seven days later, he arrived at Alexandria on the transport Black Hawk, a sidewheel steamer. He took command of the combined forces of Franklin and A. J. Smith. General Grover received orders to return to Thibodaux and change to river transportation.

On March 24, 1864, the 159th NY embarked on the transport, *James Battel*, a shallow draft steamer sailing up the Mississippi and Red River, arriving in Alexandria three days later. The regiment went into camp at Bayou Rapides off the banks of the Red River.[11] The 3rd brigade had already landed at Alexandria the day before, and the 2nd brigade arrived on the 28th. By March 30, 1864, Banks's entire force was in Alexandria: 31,000 Union officers and soldiers ready for duty with an additional 4,600 cavalry and ninety guns.

The Red River was slowly rising but still too low for the gunboats to pass over the rapids. The *Eastport*, an ironclad ram, tried to pass, only to get hung on the rocks in imminent peril for nearly three days. It was finally forced off by a whole brigade pulling on ropes to the rhythm of field music. *Eastport* was the first gunboat to pass over the rapids. This was important to the Admiral. He was unwilling to expose the boats of lighter draught and armament to the

---

8 Irwin, Richard B. *History of the 19th Army Corps*. G. P. Putnam's Sons New York, (The Knickerbocker Press 1892), ch .23, par. 10.

9 Tiemann (n2), p. 68.

10 Irwin (n1), ch. 23, par. 15.

11 Tiemann, William F. *The 159th Regiment Infantry, New York State Volunteers*, In the War of the Rebellion 1862 -1865, (Brooklyn New York 1891), p. 68.

risk of capture if sent up alone. The hospital boat *Woodford*, a sidewheel steamer, followed the ironclad and was wrecked in the attempt. It took three days for the next five boats to pass over the low water. On April 3, 1864, the last of the twelve gunboats and thirty transports selected to accompany the expedition to Shreveport floated safely above the rapids obstructions. The rapids' bare rocks in the riverbed presented an obstacle, creating a division of the fleet. Supplies needed to be landed at Alexandria, loaded into wagons, hauled around the rapids by road, then reloaded for shipping up the Red River. This logistical issue made it necessary to establish depots in the town and above the falls.[12]

Banks's army, in the advance of Shreveport, crossing Cane River [13]

The main army started their move toward Shreveport leaving Grover's Division (including the 159th NY) to garrison and guard Alexandria and the military stores. The town became the depot for supplies needed while the campaign was engaged. In addition to the infantry regiments, there was a squadron of cavalry and battery of artillery with orders to protect the stores located there.

The major threat for Molineux were "guerillas," the country was infested

---

12 Irwin, Richard B. *History of the 19th Army Corps*. G. P. Putnam's Sons New York, (The Knickerbocker Press 1892), ch .23, par. 17.

13 *Banks Army Crossing Cane River*, Illustration, artist C.E.H. Bonwill, Library of Congress # 94507514.

with them, a group of freebooting rebels having no allegiance, not even to the Confederate army. Liable to attacks, Molineux ordered the men to move into town and throw up breastworks. If contending with these lawless men roaming the country wasn't enough, there was also the Confederate cavalry to deal with. Because of the enemy's cavalry threat, General Grover received orders from the Assistant Adjutant-General to construct additional defenses obstructing the roads leading into Alexandria. Barricades of cotton bales, barrels filled with earth, or any other suitable material were built to support the construction of earthworks already created.[14] The mounted cavalry and State troops were posted to cover all approaches into town and give warning of any appearance of enemy. Two gunboats were positioned to assist in protecting the town and its approaches.

The men of the 159th NY were comfortably quartered at their post in Alexandria and in excellent spirits. Initially there was not much to keep the men occupied other than doing provost duty, but as the campaign pushed forward their post became bustling with activity.

On April 4, 1864, Brigadier-General Birge arrived in Alexandria and assumed command of the brigade, relieving Colonel Molineux from his duty. Molineux was temporarily detached from the brigade to organize the "State Troops" an independent organization of "loyal men of Louisiana," natives of the state who were enlisting to support the cause of the Union.[15] Lieutenant-Colonel Gaul was assigned commander of the 159th NY and President of Court Martial.

In John's next letter to his family, he expressed his belief that out of ignorance, the people of the South initially feared Union troops occupying their towns. But by John's recounting, they learned, "the yankees are not so bad after all" and in fear of the rebels returning, wanted the Union force to stay.

Letter #26 - April 4, 1864

Alexandria, La

I thought I would write to you this morning to let you know that I am

---

14 O.R. *Assistant Adjutant-General to General Grover, construct additional defenses*, Series I – Vol. XXVI – Part 1, p. 19.

15 Tiemann, William F. *The 159th Regiment Infantry, New York State Volunteers*, In the War of the Rebellion 1862 -1865, (Brooklyn New York 1891), p. 69.

well. I feel pretty good although I have had a fever and aged some and am not quite so fleshy as I was. I received your letter No. 14; I was glad to hear from you and thankful that you enjoy good health and like your new home. Alexandria is somewhere about 350 miles North West from New Orleans if you look over the map you will find it. I think we are in a healthier climate than in Thibodaux. The weather is quite cool and feels healthy, there seems to be some prospects of our staying here through the summer, still we can't tell. The Rebel's held this place for about three weeks and our folks drove them out. We captured 500 prisoners and one battery of 6 cannons; we have a large force ahead of us of our own men. The morning we came here there left 30,000 men on the way for Shreveport near the Texan line where the Reb's are gathered, and the next morning there left they tell me 40,000 more, and I can't tell how many more are here but the whole Country is covered with camp, we have an awful force here. I like this place very well; Alexandria is a place like Thibodaux on the Red River. When we came here there was 23 Iron Clad Gunboats instead of 14 as I wrote in my last, and some 30 or 40 steamboats it looked like a seaport on dry land. They have mostly gone on to Shreveport, the headquarters of our army is established here, and I think we shall stay here to keep it (i.e.) that is our division Gen Grover, Division 159,131, 90, 91 NY 1st La. Constitute our Brigade. I am happy to hear that Eb[16] and Permelia has hired out so well and suppose they are boarded also. I hope they may do well. I think it is the best thing they could have done. Let me know their Post Office address is. You said you was disappointed in not getting a letter, perhaps that letter was to Hudson post office, it doesn't seem right to direct my letters to Hudson in care of Ed Roraback but if you get them all right, but they may lay some time before you get them. You better see the Post Master and see they are sent to Ed Roraback and so you get them. I sent $30 the 22nd of March to you; you will find it in the express office in Hudson. I think we will be paid about the middle of May again if we stay here. I have felt good since yesterday as if I was going to hear good news. I don't know as I can write to Martha, for some time my postage stamps I put in the box were all packed up, I have but one with me which I send

---

16 Ebenezar Duntz: Age 27, he is a carpenter from Gallatin, NY. Courting Permilia Mambert.

today to you. You wrote how Jobe & Walt was turned around, I always was turned around there too, see that they go to school. Tell John M. Welch that I would like to be discharged from the service and if he can affect it, I will do as I have written to you. My impression is that this war will end this summer if we have any luck at all for the reb's are fleeing before us like fleeing from the wrath to come. This place is lonesome for the citizens are all gone, homes all empty except a few, but they begin to come back and in a few weeks they will find that the yankees are not so bad after all. When we left Thibodaux, a great many cried, they thought we were going away and that the reb's would come back again the more they became acquainted with the yanks the better they liked them. Ignorance & superstition has been the whole cause of the war & slavery has caused ignorance. But that is dead. I could write home laughable instances, but I can't today unless I can get some more paper. The boys in the 128th are all well & doing well. G. Potts[17] I saw yesterday that he is well. I must close, write often, give my love to all and take good care of your health. I expect to get home, trust in God.

To my family from J W Mambert [18]

Levee at Alexandria, La. during the Civil War [19]

---

17 George W. Potts, Private 128th Regiment Co. I: Age 34, promoted corporal July 1,1861; captured in action on October 19, 1864, at Cedar Creek, VA, and paroled. Farm laborer from Churchtown, NY. Married with five children ranging in age from 4 to 11 years.

18 *Letter #26, April 4, 1864, Alexandria, La.* John W. Mambert letters from the Civil War.

19 Levee at Alexandria, La [Photographed between 1861 and 1865, printed between 1880 and 1889] Retrieved from the Library of Congress # 2013647481.

Because Alexandria was situated near the center of Louisiana on the Red River with a level plain of longleaf pine forests, its location was considered a natural site for settlement. The area was originally obtained through a land grant from Spain in 1785 to Alexander Fulton who established an outpost there. Fulton, originally a Pennsylvania trader, purchased additional land from the Choctaw, Tensas, Apalachee and Pascagoula tribes through his store. He laid out his plan for Alexandria in 1805. The settlement grew rapidly because of its strategic geographic location and soon became a center of transportation for trading and agriculture. The river above Alexandria was not passable by boat half the year, giving the town control of navigation on the Red River. During the time the river's rapids were visible, all cargoes had to be transported by land around this impediment and reshipped. In certain seasons, the upper river was so shallow, boats could not make the trip southeast to Alexandria. As a result, merchants constructed warehouses to store goods moving from south Louisiana to western Louisiana and Texas while waiting for the next rise in the river.[20]

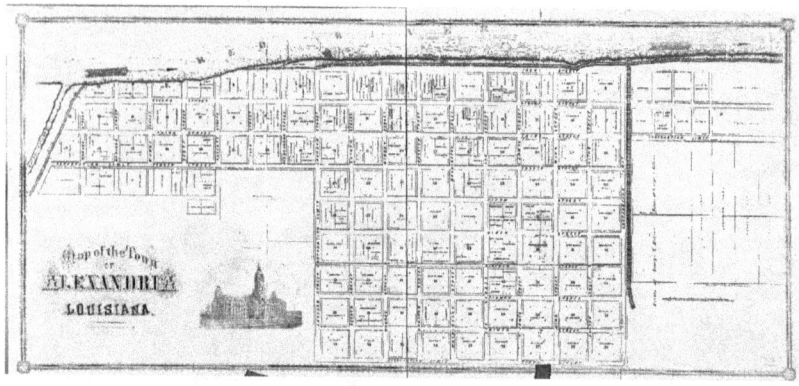

Map of the Town of Alexandria, Louisiana [21]

General Grover oversaw the logistics of supplying materials and more importantly, food rations to the troops marching northwest from the stores in Alexandria. River transport was the only option available to move vital

---

20 The History of Alexandria, *The early years*: http://www.alexandria-louisiana.com/alexandria-louisiana-history.htm

21 Bringhurst R. W, and Lith Braden & Burford. *Map of the town of Alexandria, Louisiana*. [Indianapolis, Ind.: Braden & Burford, 1872] Map. Retrieved from the Library of Congress # 2004629015.

supplies and communication from the front to Alexandria and New Orleans.

As the huge force advanced further, increasing the distance from the stores, there were issues with getting supplies to the troops, and communication with the front. The problem was obvious: no boat at Grover's disposal was able to maneuver over the shallows surrounding Alexandria. The men responsible for shipping supplies fell behind (three days late) getting 200,000 rations to General Smith's troops. Grover was pleading with headquarters upriver to send a transport down from Grand Encore. The following day, they manage to get the *Polar Star*, an armed side-wheel steamer over the rapids, resolving the communication and transport issues.[22]

Grover received instructions from Banks to have the quartermaster make Alexandria a clearing house for all cotton in the region, shipping it to New Orleans and coordinating the transactions. While the large mass of troops above were fighting the enemy in pursuit of Shreveport, in Alexandria, they were collecting, tagging, and shipping cotton to New Orleans. All sugar and molasses in the region was seized and transported to Alexandria for use by the army and the commissary.

Later, Major General D. Hunter wrote to Grant about the corruption he witnessed during the campaign:

"The Department of the Gulf is one great mass of corruption. Cotton and politics, instead of the war, appear to have engrossed the army. The vital interests of the contest are laid aside, and we are amused with sham State governments, which are a complete laughing stock to the people, and the lives of our men are sacrificed in the interest of cotton speculators." "The vicious trade regulations, or the vicious administration of them, have filled the enemy's country with all kinds of goods except military supplies, and these they have been smart enough to capture. If this course is continued, we cannot look for a speedy termination of the war." [23]

John next writes to family about the enemy troops in Louisiana deserting the Confederate army and joining the Union in great numbers:

"The Reb's are coming into our lines fast and profess to be good union men"

Records indicate 5,000 Louisiana men joined the Union army (along

---

22 O.R. *Grover was pleading with headquarters*, Series I – Vol. XXXIV – Part 3, p. 71.
23 O.R. *Hunter to Grant May 2, 1864*, Series I – Vol. XXXIV – Part 3, p. 390.

with 29,000 Black men), but that number could have been much higher including Jayhawkers.

Letter #27 - April 12th, 1864
Answer to letter No. 15

Alexandria La.

To Mrs. John W. Mambert and family,

Last night I read a letter from you and one from Martha, and one from Ram's. Your letter No. 15 and a book by Nix Nax. I was happy to hear from you that you were all well and doing well and you will be gladdened to hear that I am well and getting along well, although I can't say that I enjoy good health. I have lost considerable flesh, and so on 1st day I had an attack of the Ague[24] and fever, and that keeps me just about, so I am not really sick, now enjoying good health. The climate is against me. The boys of our Regiment are all very well and seem to enjoy themselves. In my past letter I wrote to you about our journey here. On the morning when we arrived here and the next morning there left from here, 70,000 troops besides those which remained, how many I can't tell but you have no idea what a war is prevailing in our Nation. I hope the people of the North won't look upon the Rebellion with the least degree of allowance but to unite and help put it down. The people of Louisiana have put down slavery of themselves and say by a large majority too, that Louisiana shall forever be a free state. They have seen the fruits of the free labor syndrome and am very pleased with its results. Yesterday a large portion of our troops left here and went up the river. Our Regt. And a part of the 90th N.Y. is expected to do provo duty in this place. If it should be so we may stay here all summer. We expect to move into the village today and quarter in houses, if so, it will be nice, we have 3 or 4 batteries here and troops enough to support them. The 128th lays off ¼ mile from

---

24 Ague: A sharp fever, most usually malaria. Chills and fever, heavy sweating. Also called "aksis." Likely Malaria, both an acute and chronic disease caused by protozoa are small (but not simple) organisms transmitted to humans by female mosquitoes.

us, they are all well, tell Jobe, Lou Brooks[25], George Potts and the rest. We expect an attack here. Our headquarters are established here, and we feed all the army in this section and those who went from us, I mean those who are gone on up the river to Shreveport. The Reb's are coming into our lines fast and profess to be good union men on whom the Reb's have committed degradation and conscripted them contrary to their wishes. They take the oath of obedience and beg to be armed and help fight, so General Banks expected a Company of them. They said they wanted to fight on their own hooks and wanted to wear their citizens dress, so they mounted and armed to the teeth and out they went and was gone a short time and in they come with a rebel Captain of the Company which they had left, but a madder man you never see. He said he would not have cared if he had been taken by the Yankees but to be taken by the Jay Hawkers as they are called, he did not like it.

So, you know what Jay Hawkers are and if the Reb's catches them, you may have known what the result will be. Since then, the Reb's have been caught and bought here every day, and they scour the country and catch them, and find cotton, you have no idea what cotton in bales we are shipping from here to New Orleans. Last week the Reb's got down the river, and some 5 or 6 that is all fired in our direction, they attacked a transport, but it was a gun boat, they call them mules, and our troops poured the grape and canister[26] into them, but they could not hit them very well. The river being low and the banks high, so the Reb's told them to hold on and not shoot, so they stopped, and the Reb's told them to come and take them aboard, they surrendered as prisoners of war. They could have run back in the woods, but they wanted to surrender so our boats took them on board and a jollier set you never see, to think they was with the Yankees. They asked them why they shot, they said they had to fight or else it was not right, they said they were all right now. So, you can see how much these people want to

---

25 Lewis C. Brooks, 128th Regiment Co. G: Age 24, mustered in as corporal August 25,1862; promoted Sergeant Nov 1, 1863. A farm laborer from Livingston, NY.

26 Canister: A projectile, shot from a cannon, filled with about 35 iron balls the size of marbles that scattered like the pellets of a shotgun.

fight. I could write you a good many instances of the same kind. You don't know how the poor jayhawkers suffer; they are hunted, shot, and hung. Men have laid in the woods for months waiting for the Union army to come to their relief, and wherever our army goes the people seem to be in a state of starvation. They will come in the lines for something to eat, and some who come in a carriage with 2 horses and a Negro to drive, and a mean woman, and beg for something to eat always looking sheepish with a vail over their faces. Alexandria is a place about the size of Kinderhook, but the buildings are poor, well one half of the people north would not live in them, but there are some nice houses. Red River is a stream a little larger than Black River but very deep and the banks of such color as the red hill, only it is clay and sand. The country is level and cover with some white and yellow pine, with some black and white oaks, and some bitter nuts and locust. The climate is I think much healthier than in Thibodaux. The bank of the river is steep and falls down which causes it to look like a mud hole. On our way up here, I saw a high hill or bluff as they called it. I felt so glad to see a hill I took off my hat and gave three cheers for this is the first hill I saw in Louisiana, but there is no such thing as a stone and in our camp each Company dug a well and you get water in but 4 feet, most anywhere. I think that this summer will wind up the war, I think General Grant and Old Abe Lincoln are honest and both firm in opinion. And I hope too that the people will elect Old Abe for another term; if so, it will bring this Rebellion to a close much sooner than any other way for he has been firm to his word and if he is elected this Rebellion will never break out again. But if another is elected and a compromise is made with the aristocracy of the south in less than 20 years, we will have another war for the north will have to do as the south says or else, they will say we will decide. The majority of the people of the south are for the Union, but it is ruled by the aristocracy as England is ruled by the Lords. The majority have no education, there is no free speech, no schools nor anything to enlighten the people, so you see ignorance is the mother of crime. I will here stop on this point for the present. You must tell Jobe and Walt that I am glad to hear that they have done so well in the fur business, but I don't want Jobe or Walt to hire out by the month at all, for one day is long enough for them, besides Jobe is ruptured and I want him to be careful of himself and above all I want him and Walt to go to school and learn what they can. If Jobe wants to work let him do so, but don't hire him out by the month to wear him out in his young days. I received his lock of hair, but it doesn't look so red as it did, and he doesn't improve much in writing. I got a letter from Martha and Rams, and I will write to them today if we don't move into town. I am glad to hear that you will attend to the business at home in settling up, I hope you will. I don't want you to think that and worry about home, although often think of home but I don't worry about it for I have only 18 months more to stay here if fate my full term. I expect I will soon hear how it is settled; I don't understand how you got $25 instead of $14 on that note. I don't want you or Mother to take or use for if you do you can't collect the principle. I have never given such authority, I hope you have not done so, I took it that it was for rent instead of interest of the Braswell place. I feel as if I shall return home again and see you all, I must close.

P.S. the trees are all leaved out and it is full of flies as it can stick, this tells you how the weather is. You write as usual and direct as usual put plain 159 Regt. N.Y. Vol. I will write on the receipt of the next letter.

Remember me. J. W. Mambert, Musician [27]

NO WORK FOR POOR WHITES IN THE SOUTH

Life was dismal for poor Whites in the South. They were extremely poor, not owning property and tenants to the wealthy. The planters had no desire to employ poor Whites while they were able to enslave people to do their work for free. Poor Whites faced severe unemployment, unable to compete with slave labor. Struggling to make ends meet, some poor Whites turned to a life of crime, living on the lam, while others found occasional odd jobs and seasonal agricultural work. This type of labor kept many poor White men on the move in search of jobs. Separated from their families, the dire

---

[27] *Letter #27, April 12th, 1864, Answer to letter No. 15, Alexandria La.* John W. Mambert letters from the Civil War.

situation fractured households, leaving women as family heads for much of the year. In this time and place, there was no chance of a better life for poor Whites. At the same time, many Irish and Germans immigrated to the US creating even more competition for jobs.

The poor white population of the South," made particularly inviting targets for a southern legal system dominated by slaveholders, who generally incarcerated them for behavioral, nonviolent "crimes" such as trading, drinking and other social interactions with slaves and free blacks. On the eve of secession, slaveholders were still jailing poor whites for small amounts of debt, publicly whipping thieves, and auctioning off debtors and criminals (for their labor) to the highest bidder. In addition to the region's sophisticated legal system, the Old South also had an extremely effective extra-legal system to keep the lower-class Whites in their places. From vigilance committees to minutemen groups, these organizations helped maintain both slavery and the southern social hierarchy, ultimately forcing a divided region to wage an unwanted war."[28]

JAYHAWKERS

Jayhawkers were Unionists who would rob, burn out and murder rebels in arms against the federal government. Large numbers of these men took to the woods and swamps to evade Confederate enrolling officers, declaring they had no intention to fight for either Federal or Confederate Governments. However, when Banks's army reached Alexandria, hundreds of Unionists emerged from the woods and swamps. Armed primarily with double barreled shotguns and revolvers, leading members of these groups formed an alliance with the Union sharing all information they had on the enemy.[29] Taking the oath of allegiance to the United States, General Banks organized the jayhawkers into a regiment known as the First Louisiana Scouts. This was the group of men Colonel Molineux organized as state troops when he was detached from the 159th NY in Alexandria.

In a letter to his sister, Molineux wrote:

---

28 Merritt, Keri Leigh, *Masterless Men, Poor Whites and Slavery in the Antebellum South*, (Cambridge University Press 2017), pp. 5-6.
29 [HOUSTON] TRI-WEEKLY TELEGRAPH, August 29, 1863, p. 2, c. 3 The Louisiana Jayhawkers. Camp Stonewall Jackson, near Washington, La.} August 17th, 1863.} Editor Telegraph.

"General Birge had arrived and relieved me of the command of the Brigade and that I had been placed on duty "organizing the state troops." This is no more or less than organizing and commanding the Union refugees, scouts, partisans, deserters and jay hawkers who flock to our lines and wish to enter our service as independent companies of scouts. They are brave hardy men, many sincere for the cause, others fighting for plunder and pay and to address private wrongs and injuries; some probably stay in the confederate service. The majority, I am satisfied with and we can use all to good advantage." [30]

At the close of the Red River campaign most of this unconventional group of men disbanded from the Union army remaining in the Alexandria region. The swamps in LaSalle and Catahoula parishes were crawling with rebel deserters and Jayhawkers who remained plundering secessionist, mostly planters, taking whatever they needed then burning the homes and occasionally killing the residents.

The Jayhawker situation eventually struck a breaking point with Confederate General Alfred Mouton who denounced them as outlaws and traitors. This triggered him to issue "secret" special GENERAL ORDERS on June 12, 1863.

"Information has been received that there are bands of outlaws, deserters, conscripts, and stragglers from a point above Hineston, on the Calcasieu River, in the parish of Rapides, down to the lower parishes, extending into the parishes of Calcasieu, through to the Bayou Teche, which are committing depredations, robberies, and incendiarism, and who are openly violating the Confederate laws, with arms in their hands."

The order continued stating:

"These bands, beyond the pale of society, must be exterminated, especially the leaders; and every man found with arms for the purpose of resisting the operations of the Confederate laws, or against whom satisfactory evidence may be given, must be executed on the spot. No prisoners should be taken. Such as are not sufficiently guilty to deserve immediate execution must be liberated, and, if conscripts, ordered to report forthwith." By order of Brigadier-General ALFRED MOUTON:[31]

---

30 James I Robertson Jr Civil War Sesquicentennial Legacy Collection at The Library of Virginia, *Letter written Apr. 10, 1864.* – Edward L. Molineux collection.

31 O.R. *Secret Special Order by General Mouton,* June 12, 1863.

BATTLE AT SABINE CROSSROADS

While the 159th NY was guarding Alexandria, moving supplies to the battle front, and loading cotton on transports, General Banks's forces were marching toward Shreveport and advancing on Pleasant Hill. Leading the army was General Albert Lee, commander of the cavalry forces, supported by Colonel Harai Robinson commanding the 3rd brigade and Colonel Thomas Lucas who commanded the 1st brigade. Robinson's men were pushing the advance guard of the Confederate cavalry until about 2:00 p.m. in the afternoon when at Wilson's farm, three miles beyond Pleasant Hill, he came upon the main body of the Confederate force under the command of General Green.

The enemy was on rising ground posted on the skirt of the woods. Robinson dismounted his men and engaged the rebel force. Faced with such firm resistance, Lucas was ordered in to support Robinson and save him from being driven off the field by the unwavering charge of the enemy. Lucas dismounted his men. The two brigades charged together on foot, driving the Confederates from their position, pursuing them seven miles beyond Pleasant Hill to Carroll's sawmill. In this engagement, General Lee suffered a loss of eleven men killed, forty-two wounded and nine missing.[32]

Brigadier General Thomas Ransom, commander of the 13th Corps, started marching his troops at 5:30 a.m. in the morning. By two o'clock in the afternoon, the head of his column arrived at Pleasant Hill and went into camp. Brigadier General William Emory, commanding the 19th Corps, closely followed Ransom, and arrived late the following morning. Brigadier General A. J. Smith, commander of the 13th Corps, was a good day's march behind Ransom and Emory.

When Lee found himself so obstinately opposed, he sent a message to Major General Franklin, asking that a brigade of infantry be sent forward to his assistance. Franklin declined his request, thinking that after a hard day's march, the infantry would be too worn out to advance at night. Even so, Banks joined Franklin an hour or two before midnight, ordering Franklin to send a brigade to Lee, reporting at dawn.

Meanwhile, the whole of Green's cavalry corps joined General Taylor's

---

[32] Irwin, Richard B. *History of the 19th Army Corps*. G. P. Putnam's Sons New York, (The Knickerbocker Press 1892), ch. 23, par. 30.

while two divisions of the Confederate army commanded by Major-General Sterling Price arrived from Arkansas taking a post within supporting distance of Taylor. Taylor now had about 16,000 Confederate soldiers ready to give a good fight against their Union opponent.

Banks, on the other hand, had his army stretched out the length of a long day's march on a single narrow road. With no elbow room and numbers so great, the men were restricted in their ability to move if under attack. Banks was even more burdened by twelve miles of wagons transporting all his ammunition and stores. This weakened his army tremendously due to the number of men necessary to guard them through the heart of the enemy's country.[33]

Banks's column was closing up in readiness to meet the enemy with its full strength. Riding ahead, Ransom himself arrived on the field at 1:30 p.m. in the afternoon. At this time, on Lee's orders, Colonel William Landram, commander of the 4th division, pushed forward his 19th Kentucky regiment, deployed as skirmishers, supported strongly with the rest of Emerson's brigade. They drove Green's troopers across open ground, and well into the woods beyond the crest of a hill. Here, Lee was joined by Nims, bringing his guns into the battery across the road.

On the left of Nims, two howitzers were placed, detached from the 6th Missouri cavalry. On the right and left of the horse artillery, Emerson formed his troops. Colonel Joseph Vance with the 2nd brigade took position on Emerson's right. Colonel Lucas dismounted, extending the line of battle to the right, with him two sections of battery. Covering the flanks in the forest, Colonel Nathan Dudley deployed the 8th New Hampshire on the right, and on the left the 3rd and 31st Massachusetts as skirmishers.

Directly in front of Banks stood Taylor in order of battle, covering the crossing of the ways that lead to Pleasant Hill, to Shreveport, to Bayou Pierre, and to the Sabine. On his right was the cavalry of Brigadier General Hamilton P. Bee, then Major General John George Walker's three brigades of infantry across the main road, and on Walker's left, Brigadier General Mouton's infantry division supported on his left by the cavalry brigades of Brigadier General's James Patrick Major and Arthur P. Bagby. Neither side could move forward without bringing on a battle.

General Banks rode to the extreme front line, arriving to view the enemy

---

33 Ibid., ch. 24, par .5.

before him covering the crossing leading to Pleasant Hill. Taylor's army was ready for battle. Banks at once ordered Lee to hold his ground, sending orders to Major General Franklin to bring the Union column forward quickly.

Late in the afternoon, after the two opposing lines had stared at each other for two hours or more, Taylor ordered his attack with a vigorous charge by Mouton's division. The Union infantry on the field numbered 2,400 soldiers, Banks's whole force available at the time was 5,000 strong. In opposition, Taylor was advancing with 10,000 men, his line widely overlapping and flanking those of Banks when the two forces convened on the field.[34]

At 4 p.m., Banks ordered his full infantry to the front. He then rode to the hill where the battery of Captain Ormand Nims's guns were positioned. Banks's troops were defiant, desperately struggling to hold their ground. The situation appeared hopeless. Enemy troops were gaining on the Union flanks. Brigadier General Cameron, commander of the 3rd brigade, along with the 13th Corps, had four battalions deployed, fighting hard against the rebel force.

The Union line began to crumble, and the rear fell apart. Once the Union retreat began and the battle line was broken, it was difficult to maintain any sense of organization. Soldiers had to move back and find enough open space to reestablish a strong battle line formation. A mile behind the battlefield, the retreating Union column came upon a cluster of wagons tangled and stuck fast in a marsh; it was the cavalry train. A failed attempt to move the wagons to safety led to disaster.

On the front line, Nims's battery was still holding firm. He had already lost more than half of his horses and upon retreat, was compelled to abandon three of his guns. Now, the remainder of his battery was absorbed in the wreck of the wagons, losing fourteen more guns to make matters worse. Two additional batteries lost a section of their guns, while two other batteries lost all their guns.

In all, twenty guns were lost, three on the field and seventeen in the jam, along with 175 wagons, eleven ambulances and over 1,000 draught animals. Attempts to get past the obstruction forced the infantry to turn wide for a long distance and push their way through the woods. It soon became a state of confusion: men, wagons and horses crowding to the rear, and this was only the beginning of what was to come.[35]

---

34 Ibid., ch. 24, par. 12.
35 Ibid., ch. 24, par. 16.

Emory was met by an aide-de-camp with orders from Banks to: "Move your infantry immediately to the front, leaving one regiment as guard to your batteries and train." Further down the road, Emory exchanged a few words with a wounded officer and ordered his division to march double-quick. A little further down the road they met stragglers, a swollen stream of fugitives crying that the day was lost. From Emory down to the smallest drummer boy, every man knew that the hour had come to show what the 1st division was made of. The leading regiments and flankers fixed their bayonets, the staff officers drew their swords; not a man fell out of step as a steady even pace quickened.

Emory's men forced their way through the confused mass in a zealous attempt to reach a position where the enemy could be held in check. This was not an easy task. It was not until after they passed the retreating soldiers and stray bullets that they knew the enemy was close. Although Emory found an opening to deploy his men on the battlegrounds, the advantage was small in comparison to the task before him.[36]

The spot was a small clearing near a fenced farm and narrow road. The enemy was coming at a fast pace. Emory was determined to sacrifice his leading regiment to gain time and room for the division to form a line. Emory sent orders to his three brigade commanders: General Dwight was to hold the center position at any cost, General McMillan to the right and Colonel Benedict to Dwight's left. Emory rode up and led forward the 161st New York (organized from Chemung, Steuben, Chenango, and Broome counties). He deployed the regiment widely, as skirmishers, across the whole front of the division in the very teeth of the Confederate line of battle. Rapidly they advanced with wild yells, firing heavily. Throughout the charge, Emory's men stood their ground bravely long enough for the entire 1st division to deploy a line of battle. The enemy charged toward Emory's men. The 1st brigade's full line was kneeling, waiting and ready, opening fierce fire at point-blank range. They threw off the attack with a heavy loss to the Confederate army. The brunt of the attack was endured by the 29th Maine holding the center and by-road. Although it only lasted twenty minutes, the fight was severe.[37]

---

36 Ibid., ch. 24, par. 18.
37 Ibid., ch. 24, par. 19.

General Banks's losses in the battle of Sabine Cross-Roads were as follows:

DIVISION	KILLED	WOUNDED	MISSING	TOTAL
Cavalry	39	250	144	433
Cameron's	24	99	195	318
Landram's	28	148	909	1,085
Emory's	24	148	175	347
Staff of Nineteenth Corps	0	3	0	3
Total	**115**	**648**	**1,423**	**2,186**

At 1:00 a.m., Banks's army retreated back to Pleasant Hill. In a report from Rear Admiral Porter to General Sherman, he stated:

> The fugitives arriving at Grand Encore reported that the army was cut to pieces, and I hear that when the general and staff arrived at Pleasant Hill, he had lost all command of himself. I do not wonder at that. An uneducated soldier may be cool and pleasant enough in the hour of victory, but the best general is known for his hour of defeat. General Banks lost all his prestige and the men talked so openly about him that our officers had to check them and threaten to have them punished.[38]

The events of that day deeply affected Banks. Whatever hope he had of taking Shreveport would be ending soon. General A. J. Smith was ordered by General Grant to return to Vicksburg within the next two days. This meant that Banks would have to be on the Mississippi River, ready to move with his full force against Mobile by May 1, 1864. So Banks made up his mind to retreat his remaining force to Grand Encore.

In his decision to withdraw, he chose the second-best route, the decision that his enemy, General Taylor was hoping Banks would make. It would have been safer for his men if Banks had chosen Blair's Landing as the destination, receiving support from his gunboats and General Kilby Smith's provisional division. Taylor sent a detachment of cavalry to Blair's Landing, gathering intelligence just in case, poised to seize the crossing at Bayou Pierre.

---

38 O.R., report, *Rear Admiral Porter to General Sherman*, Series I – Vol. XXXIV – Part 3, p. 171.

Banks gave the order for the wagons to move out to Grand Encore early in the morning on April 9, 1864. The wagon train was escorted by General Lee commanding the 19th Corp cavalry and Colonel William Dickey's colored brigade. Banks positioned his army at Pleasant Hill to cover the withdrawal.[39]

BATTLE AT PLEASANT HILL

Pleasant Hill was a good location to fight a battle, but not for any length of time. It was too remote to be a supply base and there was no water except for the cisterns in the village. There was barely enough water to supply the villagers, much less so now that two armies were drawing from it. There was never enough water to supply the needs of the men and the animals for a single day. The lack of water made it impossible to stay any longer than necessary to cover a safe withdrawal of the army's wagon train.

At first light, Taylor noted that his adversary had already departed Sabine crossroads. He quickly advanced his entire force to search for the Union army. He discovered Banks in position at Pleasant Hill.

By noon, the last of Taylor's infantry arrived on the field, exhausted from their long march that started the previous morning. Taylor gave his soldiers two hours rest before any thoughts of battle against the Union army. Two miles south, the Union army stood with Emory firmly holding the right line; his position was a short distance north of the village in front of a fork in the roads leading to Mansfield and Logansport. Brigadier General Dwight's 1st brigade formed an extreme right flank, Colonel Shaw's 2nd brigade was posted at an open approach. Colonel Benedicts 3rd brigade was positioned in a ditch crossing a slight hollow. The right of his line rested tiered behind the left line of Shaw. Benedict's front was hidden by a light growth of reeds and willows. An artillery battery under the command of Lieutenant Southworth of the 1st division held the hill between Dwight and Shaw. Captain Henry W. Closson's reserve artillery positioned themselves to fire over the heads of Benedict's men. Brigadier General James W. McMillan's 2nd brigade was behind Dwight and Shaw.

About 400 yards beyond Benedict's position, slightly overlapping his left, the line was extended by the 16th Corps with General A. J. Smith's

---

39 Irwin, Richard B. *History of the 19th Army Corps*. G. P. Putnam's Sons New York, (The Knickerbocker Press 1892), ch. 25, par. 1-11.

1st and 3rd divisions strongly posted in the woods. They were covering the crossing of the roads where Emory was positioned. On the right of the 3rd division was Captain Hebard's 1st division battery. Some 500 hand-picked men from Colonel Lucas 1st brigade cavalry mounted, along with a similar number from Colonel Gooding's 5th brigade cavalry. They were positioned along with the 16th Corps. Lucas watched with rapt attention all possible approaches for any sign of Confederate troop movement.

General Taylor formed his Confederate line of battle. On the left of the Mansfield Road, he positioned General Hamilton Bee with two brigades of cavalry supported by General Camille Polignac. On the right side of the road was General John Walker and General Thomas Churchill along with three regiments of cavalry covering his right flank: they were moving behind the tree lines out of the Union sight, following the upper road to Sabine.[40]

Night was approaching. Banks was about to give up on the notion of any attack by Taylor's army, when the battle began.

It was just after 3 p.m. when the Confederates began their move; Churchill's men began their line of march through the woods with Parsons leading. Churchill's column extended to the lower Sabine Road which enters Pleasant Hill from the southwest—he mistook it for Fort Jesup Road which approaches the village from the south. In his error, he changed the lines: front to left. With this, the two lines of Parsons and General James Tappan made a diagonal charge down the left flank on the front of Benedict, when they were supposed to have fallen on the flank and rear of Mower's 3rd division. Benedict was now in a bad position and his indirect order of attack exposed his weakness. His skirmishers, now driven back through the woods, uncovered his right and center lines which until then had been hidden. Churchill's men descended the slope firing as they came upon the Union line and for one brief moment the fight was hand to hand. Benedict's men were driven out of the ditch, drifting to the cover of the woods where Mower's soldiers lay in wait.

When Green heard Churchill's musketry, he launched his cavalry along with Bee's two brigades in a reckless charge against the left of Colonel Shaw's line. This wild sweep by the enemy was quickly halted by the close range of muskets belonging to the 14th Iowa and the 24th Missouri regiments. Many fell from rebel

---

40 Ibid., ch. 25, par. 5-11.

saddles, among the victims were Bee and his two brigade commanders Buchel, and Debray. The beaten remnants of the cavalry fled in confusion.

Eighteen Confederate cannons, some of which belonged to the Union the previous day, were now concentrating their fire on the six guns of Southworth's 25th NY battery. By sheer mass they overcame him. Shaw's front was vigorously attacked by the enemy, also menacing his flank. Shaw continued putting up a strong fight even while in great danger of being cut off from the rest of the army. Not a moment too soon, A. J. Smith ordered a division to assist with withdrawing Shaw's troops. As Shaw retreated, General Dwight found himself attacked in front by one of Major General John Walker's brigades and on his flank and rear by Brigadier General James Major's cavalry. Under severe attack by the enemy, Dwight quickly repositioned the 29th Maine and the 161st NY, holding firm under fire coming all at once from all sides.[41]

For a moment, the advancing enemy appeared to be the masters. Along the Union front, no one appeared in position to help save Dwight and his men, far off on the right, standing alone, gray uniforms surging on them from all sides.

But far away, out of sight from the plain, an event was occurring that would cost the Confederates the battle. Brigadier General Mosby Parsons, following after the defeated Benedict, ordered his right flank to attack Colonel William F. Lynch's brigade positioned at the skirt of the woods left of Mower. Lynch struck hard and began doubling up the Missourians' soldiers. Observing Lynch's move, A. J. Smith ordered his line forward. Shaw now positioned his brigade to the right of Mower. Emory emerged, leading McMillan from reserves to improve Dwight's situation, restoring the line on Dwight's left.

For the Confederates, victory appeared imminent...until it wasn't. The long lines of A. J. Smith, Mower riding at the head, emerged from the woods—a sweeping unbroken front. To the right, McMillan joined his brigade, all of them steadily closing in on the enemy.

At this sudden turn of events, the Confederate forces of Parsons, Tappan and Walker began their retreat: The Confederate army was in a state of confusion. Their disordered ranks were pushed back about a mile with a loss of five guns. After nightfall, Taylor's infantry and part of his cavalry fell back six miles.

---

41 Ibid., ch. 25, par. 13-17.

The battle came to an end with the Union controlling the field. Three of the guns lost at Sabine Crossroads were now back in Union hands.⁴²

Reinforced Union Troops Repel Further Attacks at Pleasant Hill ⁴³

The losses to the Union army at Pleasant Hill were 152 killed, 859 wounded, 495 missing: in all, 1,506. Of these, nearly half were in Emory's division which reported eight officers and forty-seven men killed, nineteen officers and 275 men wounded, four officers and 374 missing: in all 725. The Confederate losses were estimated by Taylor at 1,500.⁴⁴

Unfortunately for General Banks, General Hallock's plans for the Union army to enter Texas through Shreveport did not happen. The Red River campaign was a complete failure. The campaign was considered a major blunder by the majority of Union officers who participated; the Confederates held the Red River for the rest of the war.

The Opelousas Courier, a Confederate newspaper, printed the following version of the two field battles between Taylor and Banks:

> After a forced suspension of two months we today resume our regular issue. Since our last number, great and glorious events have transpired in our poor Louisiana, and the invaders of our soils have paid dearly their pretentions of subjecting our state.

---

42 Ibid., ch. 25, par. 19.

43 "*Battle of Pleasant Hill, Louisiana, April 9, 1864* artist impression," House Divided: The Civil War Research Engine at Dickinson College, https://hd.housedivided.dickinson.edu/node/41848

44 Irwin, (n2), ch. 25, par. 26.

We give elsewhere the details of the success of our arms in the Northern portion of Louisiana. Believing that we are now delivered forever of the presence of Yankees, we will endeavor to give our readers the most reliable news we can get, in attendant better times.

We had every reason to hope that we would be enabled to get details of the battles of Mansfield and Pleasant Hill, but we have been deceived once more. It is therefore impossible for us to give our readers what we have not.

The two battles are two Confederate victories. Our troops fought like veterans and whipped the Yankees most deservedly. The federal loss is estimated at 2,000 killed and 3,000 wounded; 2,500 prisoners, 21 pieces of artillery, 300 wagons and a large quantity of small arms. Besides the above guns and wagons and small arms which we have in our possession, the Yankees have destroyed a large number of wagons, stores, & c.

Our loss is also very severe, considering the many able officers who have been killed; 1,000 killed and wounded, and a few hundred prisoners.

The Yankees are now in Alexandria, where they are invested and dare not adventure out. For the last 18 to 20 days there have been many skirmishes around Alexandria, in which we have invariably repulsed them back into their quarters and taken some prisoners. It is currently reported that the federal army is, and has been since the battles, on half rations. Their communications with New Orleans is intercepted by our batteries, which have already destroyed two gunboats and five transports, several of the latter loaded with troops and provisions from New Orleans. It is very probable that ere this army reaches the city it will have lost over 15,000 men.[45]

In most situations, reporters traveled with the troops writing their stories by recounting events, but this was not always the case. Sometimes, the newspapers had to generate their stories using information provided through gossip and rumors. Editors loved to sensationalize information, and of course show favoritism to their troops—accuracy was not a requirement.

In a letter to his sister, Colonel Molineux expresses his thoughts about the

---

45 *The Opelousas Courier*, May 14, 1864, Library of Congress, Natl. Endowment for the Humanities, Chronicling. America.

Red River campaign:

"I suspect this Red River Expedition was a cotton and political expedition and a great military blunder and so received by our best leaders. I now know it to be as I suspected and our defeat in front proved the military error. Alas, that blood and treasure should be wanted to enrich a few and give power to a party! As good soldiers we do our duty, but we feel at the same time that our sufferings do not do any good to the ending of the war. For a few bales of cotton a few more votes for the next elections. We are now here, the river falling, supplies scarce, gun boats and transports aground and must remain so in hopes that the June rise (which frequently fails here) will free our boats. If (God forbid!) the river does not rise we must either support the fleet with a large army until next March or blow them up (Porter's fleet). As a military man, of course, I look at it seriously and yet I do not do so in public nor must you let this letter go out of the family."[46]

RETREAT TO ALEXANDRIA

On April 23, 1864, the Union army fell back to Cane River where the rebels were strong in force. The soldiers reached the enemy's location, and an attack was made by two Union divisions who assaulted the enemy position. They drove the rebels off, securing safe passage for the Union soldiers to retreat back to Alexandria.[47]

Major-General D. Hunter arrived in Alexandria on April 27, 1864, to meet with General Banks and review the situation. He also delivered communications from Lieutenant-General Grant. Grant's orders were clear: Banks was to immediately return to New Orleans and prepare to carry out his previous instructions the moment his troops returned. He was also ordered to leave his troops in the field under the command of the senior officer.[48]

The next day, Hunter reported to Grant the state of affairs in Alexandria:

"I regret very much to find affairs here in a very complicated, perplexing, and precarious situation. You have, of course, had the particulars of the fights.

---

46 James I Robertson Jr Civil War Sesquicentennial Legacy Collection at The Library of Virginia, *Letter written Apr. 17, 1864.* – Edward L. Molineux collection.

47 Tiemann, William F. *The 159th Regiment Infantry, New York State Volunteers*, In the War of the Rebellion 1862 -1865, (Brooklyn New York 1891), p. 71.

48 O.R., *Halleck to Banks, April 27, 1864*, Series I – Vol. XXXIV – Part 3, pp. 306-307.

The situation at present is this: We have some six, eight, or ten gunboats, among them two monitors, above the rapids, with no possibility of getting them out. The whole question is, then, reduced to this: Shall we destroy the gun-boats or lose the services at this critical period of the war of the 20,000 men necessary to take care of them? My opinion is, of course, to destroy the boats. Why this expedition was ordered I cannot imagine. General Banks assures me it was undertaken against his opinion and in earnest protest."[49]

Banks made it clear from the beginning he wanted nothing to do with the expedition, but Halleck did, escaping blame for the disaster, successfully pinning the great blunder on Banks who would pay the price.

THE GREAT DAM

A great undertaking to be remembered as Bailey's Dam, occurred out of necessity in the month that passed since Porter's fleet ascended the rapids above Alexandria. The paused fleet, now anchored, was about to face its next challenge. The water level in the river had fallen more than six feet, exposing river rocks for almost a mile. The large fleet needed to descend over them. At a water level of seven feet the heavy gunboats could barely float; in that section of the river, the water level had declined to thirteen feet.

Admiral Bailey proposed the construction of a dam. The challenge was to raise the water above the dam seven feet, backing it up to float the gunboats over the upper rapids where the river was at its widest. Bailey designed his plan. Troops were carefully selected to work on building the dam which started on April 30, 1864.[50]

General Birge was ordered to Baton Rouge leaving Colonel Molineux in command. His brigade consisted of the 159th NY, 13th Connecticut, 1st Louisiana along with some cavalry and artillery. Molineux and his men crossed the river by pontoon bridge[51] from Alexandria to the opposite side

---

49 O.R., *Hunter to Grant April 28, 1864*, Series I – Vol. XXXIV – Part 3, p. 308.
50 Irwin, Richard B. *History of the 19th Army Corps*. G. P. Putnam's Sons New York, (The Knickerbocker Press 1892), ch .28, par. 2-3.
51 Pontoon Bridge:(pronounced pawn-TOON) A floating bridge, which was constructed by anchoring a series of large, flat-bottomed boats across a waterway and then laying wooden planks across them. The planks (the "chess") were anchored by side rails and then covered with a layer of soil to protect it and to dampen sounds. Pontoon bridges were extremely important to the outcome of several battles, including Fredericksburg.

(where Fort Randolph and Buhlow State Historic Site are located today in Pineville). From there, the brigade had a clear view of the area and could give adequate support covering the troops constructing the dam. They set up camp, remaining there until the dam was completed.

The first evening, flames were spotted about one mile from town. It was determined the fire was started by the enemy as they passed behind Alexandria. This move revealed to Molineux the enemy's intentions of cutting off supplies and hampering the Union retreat downriver.[52]

During the construction of the dam, Molineux's soldiers held additional positions to secure the safety of the navy. The Navy was constantly being engaged with the modest, but active forces of Taylor.

On May 4, 1864, the sounds of cannons firing could be heard coming from the riverbanks above town. The next day, heavy musketry was heard firing downriver. It was later discovered that a large force of rebels with several batteries established themselves downriver. They attacked the Union transport John Warner along with two small gunboats, disabling them.[53]

Destruction of the U.S. transport John Warner by confederate batteries on Red River [54]

The dam's construction began on the north bank where a wing-dam

---

[52] Tiemann, William F. *The 159th Regiment Infantry, New York State Volunteers*, In the War of the Rebellion 1862 -1865, (Brooklyn New York 1891), p. 71.

[53] Ibid., p. 72.

[54] *Transport John Warner*, Frank Leslie's Illustrated Newspaper 1864, C.E.H Bonwell, Library of Congress # 9750156.

was constructed of large trees. The tree butts were tied to cross logs, the tops laid toward the current. To keep them in place, they were weighted with stone and brick obtained by tearing down buildings in the Alexandria neighborhood. On the south bank, a crib was made of logs and timbers then filled with stone, bricks and heavy pieces of machinery taken from the neighboring sugarhouses and cotton gins. When completed, there remained an open space of about 150 feet between the two wings where the rising water rushed through. This gap was closed by sinking four of the large Mississippi coal barges belonging to the navy.[55]

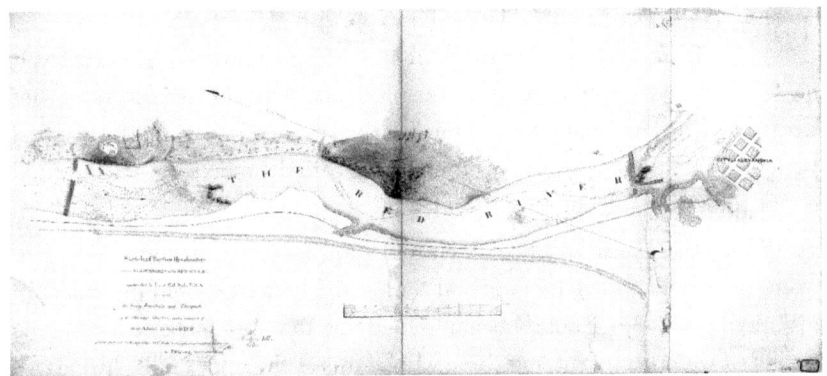

Sketch of the two dams to extricate the heavy Ironclads and Transports [56]

On May 8, 1864, Molineux sent a request to headquarters for a small cavalry force of 100 men for scouting on his side of the river; they were needed to replace the Louisiana scouts withdrawn from his command.[57]

That same day, the dam was completed. The water rose over five feet at the upper fall, creating a depth of eight feet. Three light-draught gunboats had already steamed up and took prompt advantage of the rise, passing the upper fall and moving safely to the pool formed by the dam.

During the early morning hours of the following day, tremendous pressure of pent-up waters surged against the dam. Rushing in a roaring torrent, the river pushed two of the four sunken barges out of position, swinging them

---

55 Irwin, Richard B. *History of the 19th Army Corps*. G. P. Putnam's Sons New York, (The Knickerbocker Press 1892), ch. 28, par. 3.

56 *Sketch of the two breakwaters above Alexandria in the Red River*, Gerdes, F.H. & Lewis G De Russy. [1864] Map. Library of Congress # 99446418

57 O.R., *Molineux to Capt. Hibbert, May 8, 1864*, Series I – Vol. XXXIV – Part 3, p. 508.

furiously against the rocks below. This created a gap sixty feet wide in the dam. The other boats of the fleet were not quite ready to go when they were unexpectedly exposed to great danger through the newly formed gap. The Admiral immediately mounted a horse and galloped to the upper falls. He called out to the Lexington, a sidewheel steamer to run the rapids. Instantly the vessel was under way with a full head of steam. It made the plunge— every man holding his breath in suspense.

For a moment it seemed lost as it swayed and almost disappeared in the foam and surge, only to be greeted with a mighty cheer when it was seen riding to safety below. Three more boats promptly followed down the chute.

Six gunboats and two tugs were still imprisoned above as the water levels quickly reduced. Uncertainty took the place of celebration for the army that worked so hard to create the volume of water needed to release the trapped vessels. Now the river level was diminishing, and the men knew there would not be another natural rise that year.[58]

While others were skeptical that the remaining vessels would ever be free from the waters above, for Admiral Bailey, the break exposed the defect in his initial plan. He promptly began remedying the situation by dividing the weight of water, constructing three wing dams at the upper falls. Just above the exposed rocks, a stone crib was built on the south side directly opposite a previously constructed tree dam. Just below the rocks projecting diagonally from the north bank, a bracket dam was constructed. This forced the whole current into one very narrow channel, creating a new rise in the river, gaining six feet. The work was completed in three days and three nights, allowing the remainder of the fleet to pass the rapids free of any danger. (For his part in the design and execution of this great undertaking, Admiral Bailey received the thanks of Congress on June 11, 1864, and was made a brigadier-general by President Lincoln.)

---

58 Irwin, Richard B. *History of the 19th Army Corps*. G. P. Putnam's Sons New York, (The Knickerbocker Press 1892), ch. 28, par. 4.

Bailey's Dam [59]

BACKSTORY: HALLECK BLAMES BANKS

Before the decision to march the Union army through Shreveport was finalized, Chief of Staff Henry Halleck promoted the idea of the Red River campaign to General Banks a year earlier. At the time, he implied to Banks that other major commanders, namely Sherman and Steele, favored an expedition. Halleck said they agreed that the Red River was the shortest and best defense for Louisiana and Arkansas, also suitable as a base of operations against Texas.[60] General Grant however, considered the campaign a strategic distraction. He planned for Banks to push east and capture Mobile and Atlanta as part of a coordinated series of offensives during the spring of 1864. Banks also disagreed with Halleck on attacking Shreveport. Banks was hoping to mount sea expeditions to capture the Galveston and Mobile areas.

While preparations for Halleck's Red River expedition were in their infancy, Banks sent a detailed report to him listing all concerns regarding the campaign. Halleck responded by withdrawing any previous suggestions regarding the Red River campaign. There was no evidence that Halleck provided Banks's report to Grant, who had been asking for updates on the overall operations at the time. Halleck then began referring to the Red River expedition as Banks's plan, establishing a pattern of deflecting any responsibility for its possible failure—based on Banks's report.

---

59 *Bailey's Dam*, Frank Leslie's Scenes and Portraits of the Civil War, Page 402, Wikimedia Commons

60 O.R., *Halleck to Banks*, Series I – Vol. XXXIV – Part 2, p. 15.

By mid-April, Halleck was referring to the Red River campaign as Banks's disaster, calling for his removal as commander of New Orleans. However, President Lincoln was not having it. He had too much respect for Banks.[61]

By the beginning of May 1864, Halleck was instructing Grant to persuade Lincoln to remove Banks from general command as a military necessity and not a suggestion. As time progressed and the results of the campaign continued to deteriorate, negative information was provided by other commanders deeply involved in the campaign. Sherman stated that the Red River campaign was "One damn blunder from beginning to end." General Grant eventually decided the removal of Banks was essential, sending him back to New Orleans.[62]

General orders from President Lincoln assigned Major General E.R.S. Canby[63] to take Banks's command. On January 15, 1863. Officials promoted Canby to major general of volunteers on May 7, 1864. Also on that day, the War Department placed the Department of the Gulf under the command of the newly created Military Division of West Mississippi, to be commanded by Henry Halleck's old classmate, General Canby. On May 11, 1864, Canby arrived in Louisiana and assumed command.

THE BURNING OF ALEXANDRIA

On May 11, 1864, after loading their baggage onboard the transport boats, the remainder of Banks's army proceeded to cross the river on the pontoon bridge.[64] His army continued on in retreat from the failed expedition marching out of Alexandria on the road running parallel to the Red River. As the soldiers departed the city, it was set afire.

Flames were set to nearly every part of the city beginning with Front Street. Strong winds spread the fire rapidly. The courthouse was the only building on

---

61 O.R., *Banks report to Halleck*, Series I – Vol. XXXIV – Part 3, p. 192.
62 O.R., *Halleck to Grant*, Series I – Vol. XXXIV – Part 3, p. 409.
63 Edward Richard Sprigg Canby was from Piatt's Landing, Kentucky. He attended the United States Military Academy at West Point in 1835, among his classmates was Henry Halleck. Like many Civil War generals, Canby's first combat experience came during the Mexican American War. When the Civil War erupted, Canby was in command of Fort Defiance, in the New Mexico Territory. When the War Department began making wartime assignments, Canby received a promotion to the rank of colonel with the 19th U.S. Infantry.
64 Tiemann, William F. *The 159th Regiment Infantry, New York State Volunteers*, In the War of the Rebellion 1862 -1865, (Brooklyn New York 1891), p. 74.

the city square not damaged, but later it was set afire from the inside and every record of the Parish was consumed. The Episcopal and Methodist churches were burned, but the Catholic church stood unharmed. Most of the houses were wooden structures quickly devoured by the raging flames.

Everything in Alexandria lying north of the railroad was gone within a few hours. Nothing was left. Nine-tenths of the town was consumed by the fire including the business district and most of the fine residences belonging to the 4,500 inhabitants now suffering from the ravages of war. The burning of the city left the citizens fleeing. There was no safety in the streets where the heat was intense and the smoke suffocating. Residents rushed to the river's edge by the high bank of the levy to escape the burning heat. The federal torch destroyed their dwellings, household goods and the last morsel of provisions. The people were left destitute.

As expected, the citizens who took the oath of allegiance to the Union desired to leave with the Federal army since their husbands who joined the Union army, were off fighting for them. The people begged General Banks to be allowed to board the transports. The officers wanted to let them on, but the peremptory order from Banks was "not allow any of the white citizens to board them."

Ironically, when Banks first arrived in Alexandria, he informed the residents that his occupation of the country was to be permanent, and he intended to protect all who would come forward and take the oath; those who did not were threatened with banishment and confiscation of their property. Following the oath, an election was held in Alexandria and delegates were sent to the Constitutional Convention, then in session in New Orleans. There was also a recruiting officer appointed in town and over 1,000 male residents were mustered into the United States service (thus the reason husbands were gone). There were quite a number of permanent citizens who were promised protection by Banks, but now their homes and property were reduced to ashes and the residents were being turned out into the world with absolutely nothing except the amnesty oath.[65]

Judge Thomas C. Manning was one of the seven commissioners, Confederate Governor Henry Watkins Allen appointed to document the atrocities committed by Federal troops. Manning represented Rapides

---

65 Cowardin, James A., and John D. Hammersley *"Burning of Alexandria, La."* The Daily Dispatch [Richmond]: August 11, 1864. Richmond Dispatch. Microfilm. Ann Arbor, Mi: Proquest. 1 microfilm reel; 35mm.

Parish[66] where he collected sworn testimony from five residences affected by the Union soldiers, two of whom witnessed the burning of Alexandria.

Jacob Walker, a merchant whose store was on Front Street testified on June 27:

"I have resided in this town (Alexandria) twenty-four years, and am a native of Germany — am fifty years old. This town was fired on the morning of Friday, May 13th, between 8 and 9 o'clock, A.M. Several Yankee soldiers broke into the store on Front Street next to mine and pilfered the tobacco, sugar, and lard, which were the sole contents. While the party were below [in the building] another set went into the second story, and immediately afterwards the house commenced burning."

Another resident named Lewis Texada testified:

"On the morning of the day the town was burned I heard Capt. Francis, whom I understood to be on Gen. Banks' staff say to the daughters of Dr. Davidson, one of the citizens, that Gen. A, J. Smith gave verbal orders to his troops to burn and destroy, and that he would be court-martialed for it."

He went on stating:

"free passage to New Orleans, as he said by orders from Gen. Banks. He denied that Gen, Banks approved or countenanced the burning that had] been accomplished, and was, as I understand, repelling the natural suspicion of the citizens that his chief, who was the commander of the army, was the cause of the disaster."[67]

Every public record of Alexandria was burned in that fire. The state of Louisiana granted the town a new Charter in an Act dated September 29, 1868.

As the Union army marched out of Alexandria, General Taylor found himself not only footloose in the region but also able to use the upper Red River. He removed Bailey's obstruction in the river behind Banks's withdrawal of his army, relieving the fear of future invasion by the Union.[68] The Confederates held the Red River through the rest of the war.

---

66 Young Sanders Center, Franklin LA: *Official Report relative to the Conduct of Federal Troops in Western Louisiana during the Invasion of 1863 and 1864.* Compiled from Sworn Testimony, Under Direction of Governor Henry W. Allen, Shreveport, (April 1865), p. 4.

67 Ibid., p. 79.

68 Irwin, Richard B. *History of the 19th Army Corps.* G. P. Putnam's Sons New York, (The Knickerbocker Press 1892), ch. 28, par. 10.

THE JOHN WARNER

On May 14, 1864, as the 159th NY was marching away from Alexandria, they came upon the location where the enemy previously attacked the Union transport John Warner. The ground was covered with letters from the mail captured on the transport. Countless articles had been brought ashore by the Confederates and scattered everywhere along the banks. Among the items were remnants of a large mail the regiment had sent only a few days prior. One Captain found his ordinance returns and several soldiers recovered letters they had written home. A journalist, following the Union army, searched with no luck for a report he sent to a Connecticut newspaper; it was assumed to be now providing information to the enemy.

The enemy constructed rifle pits and breastworks, fresh campfires where the "Johnnies" roasted their meat and baked their cornbread. [69] The campfires indicated the enemy was close. The 159th NY detailed as pickets were stationed near the river.

MANSURA PLAINS

The following afternoon, after marching all day, the regiment was clearing a heavily wooded area and came upon a large flatland known as Mansura Plains. Here, the 159th NY along with their division found the enemy in a line of battle, opposing their further progress. The Union force was at once advanced, ordering a brigade in front to support the cavalry who had been skirmishing with the rebels since early morning. The enemy made an impressive charge, then retreated in a hurry.

The regiment was suffering under unbearable heat and a scarcity of water. They resumed their march for a short time, then halted for the night. Just as the men began cooking their meals, Colonel Molineux received orders to move the second brigade immediately to the front—the enemy was driving the Union cavalry back.[70] Molineux asked the General if he should bring the first brigade, since it was closer to the front.

The General answered, "No; the first brigade is too small, too damn

---

69 Sprague, Homer B. (Homer Baxter), b.1829., *History of the 13th Infantry Regiment of Connecticut Volunteers, during the Great Rebellion,* (Hartford Conn. 1867), pp. 207-208.

70 Tiemann, William F. *The 159th Regiment Infantry, New York State Volunteers,* In the War of the Rebellion 1862 -1865, (Brooklyn New York 1891), p. 75.

small! I want the second brigade."

"Fall in! Fall in!" echoed instantly along the brigade line. Leaving their coffee steaming, the men were off. "By the right flank at a double-quick" were the orders for the 159th NY and 13th Connecticut regiments in front. They were on a run through Marksville traversing the length of the main street more than two miles through choking dust.

"File right!" was ordered. Into the open plain the soldiers ran, each with a gun on his right shoulder. Winding through the plain, they moved like monstrous serpents with glittering scales and banners. Rushing closer, they advanced parallel to their cavalry line just ahead and actively engaged. The enemy was shelling the regiment.

"Halt! Front! Fix bayonets!" rang out the loud powerful voice of Colonel Molineux just as General Grover came up and announced, " Boys," he said, "very likely the cavalry will charge you, if they do take it cool and fire by rank, remember the rear rank fires first."

The soldiers responded with a tremendous, 'Hurrah!' which the cavalry repeated. The enemy replied with a yell like the screech of a thousand wildcats but with no charge. Nightfall was approaching.

The Union cavalry, knowing they were supported by the infantry, used their Spencer carbines with surprising strength. The batteries on both sides went to work. The vivid spark of cannons firing unceasingly, and hostile shouting continued as darkness fell. At 10 p.m., the battle ceased.[71]

A drove of hogs crossed the front line of battle, suffering severely. After the engagement, fires were made as the men refreshed themselves with coffee and roast pork.[72] The men lay on their muskets in line that night, exhausted by the long march followed by three miles at double-quick. Their feet were raw and blistered, throats and nostrils dry with dust. Exhausted, the men sank into a deep sleep.[73]

The following morning, the remaining Union army caught up with the 159th NY and their brigade positioned in the advance. They formed a line of

---

[71] Sprague, Homer B. (Homer Baxter), b.1829, *History of the 13th Infantry Regiment of Connecticut Volunteers, during the Great Rebellion,* (Hartford Conn. 1867), p. 210.

[72] Tiemann, William F. *The 159th Regiment Infantry, New York State Volunteers,* In the War of the Rebellion 1862 -1865, (Brooklyn New York 1891), pp. 75-76.

[73] Sprague, (n2), p. 211.

battle with the enemy who returned with a stronger force in their front. The Union army began their march: over fences, through cane and corn fields, pushing the enemy for a few miles until they came upon Mansura Plains, halfway between Alexandria and Morganza on the Mississippi River.

The enemy made another stand. As the 159th NY approached with their brigade, the enemy opened on them with artillery.

They faced each other on the battlefield: a prairie that went on for miles. The Union army was drawn up in rank by brigades. While the 159th NY was advancing, the brigades of the 13th and 16th Army corps formed to their right and rear with the cavalry in front. Positioned on both flanks were their batteries with cannons hidden by the smoke of their rapid discharge.

A more glorious spectacle could not be imagined: As far as the eye could see, troops were marching and taking position, cannons booming, flags flying; rifles glistening in the bright sun. So impressive were these lines that they would not be quickly forgotten by the men who witnessed such magnificence.[74]

General Grover came dashing from the rear of the line to the front with a section of artillery; the horses at full gallop quickly wheeled the cannons into the position Grover was indicating as he looked through his field glasses toward the enemy. Cannons readied, the order was given, and they opened fire on the Confederate forces. After an artillery duel of nearly three hours, the Confederate force decided not to wait for the remaining Union division to be ordered into the battle. Breaking from the line, the enemy made a rapid retreat. Union forces pursued them for miles, finally giving up the chase after pushing the enemy far enough away.

---

74 Tiemann (n1), p. 76.

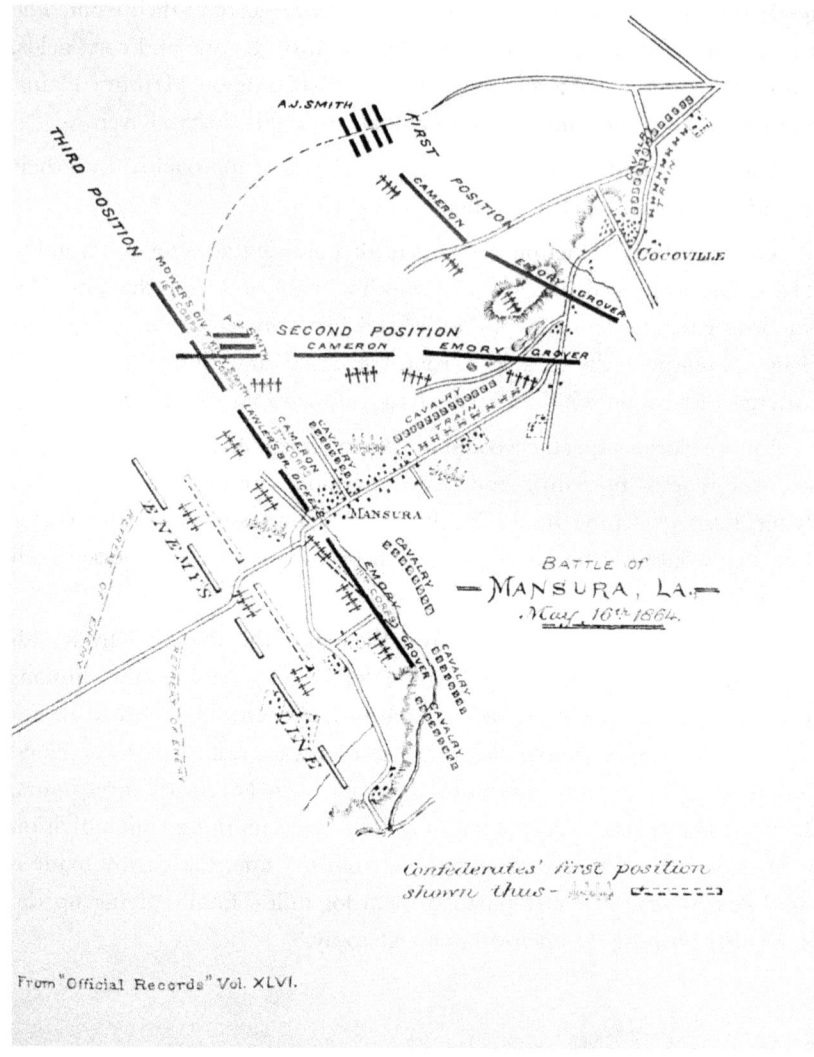

Battle of Mansura Plains [75]

THE LONG FATIGUING JOURNEY

Early in the morning of May 17, 1864, the 159th NY commenced marching, arriving in Simmesport that afternoon (they were there a year prior to the day).

---

75 Illustration, *Battle of Mansura Plains*. Tiemann, William F. *The 159th Regiment Infantry, New York State Volunteers*, In the War of the Rebellion 1862 -1865 p. 77

During the entire march, the enemy was constantly harassing the men from across the river and bayous, which accomplished little. The following day, the men were ferried over the Atchafalaya River on the steamer Cumberland where they went into camp at Semmes plantation (the same location they occupied a year earlier). Two days later, the 159th NY marched back to the ferry crossing to give support to Major-General A. J. Smith, who was engaged with the enemy the prior day. They arrived only to learn that their assistance was not needed since Smith's men successfully defeated the enemy. The regiment then proceeded to march six miles along the river where they camped for the evening.

The next morning, the men started their march completing eleven miles before halting at midnight, exhausted. Four hours later, they were awakened with orders to move, but didn't start marching until 8 a.m. They marched twenty miles in extreme heat. Several of the men experienced heatstroke. The dust on the road was ankle deep, rising up in choking clouds at every step. The road itself was surrounded by high bushes preventing air circulation and making it almost impossible to breathe.

Finally, on May 22, 1864 at 4 a.m., the men marched the last four miles of this long fatiguing journey. The 159th NY arrived at Morgan's Bend where they established their camp between the levee and the river. As the men came into sight of the river, tired and footsore as they were, they broke out in loud cheers. Once camp was made, they enjoyed a bath in the river's cool, refreshing waters. Shortly after the regiment was settled, several new recruits reported to the regiment.

The mail arrived, making the rest more pleasurable, especially since the men had not received mail in many days. John himself received a couple of letters with good news on the family's affairs. [76]

Letter #28 - May 23, 1864

Camp along the bank of the Mississippi at the mouth of Red River

> I must say that I am well only I have been tired out in marching. I know that you will be glad to hear from me for, our communication has been cut off for some time as I had no chance to send letters to

---

76 Tiemann, William F. *The 159th Regiment Infantry, New York State Volunteers*, In the War of the Rebellion 1862 -1865, (Brooklyn New York 1891), pp. 77-78.

you. I received your letters 19 and 20 yesterday. I was glad to hear that all our business was so well settled that makes me feel good, I was afraid you might have some trouble but before May, I sent $30 to you the 22 of March I think you have not written whether you have it or not, you must call at the express office for it. I ordered it to be left there until called for. We have had a very hard fight here, or I might say 2 or 3 fights. I can't write to you all about it, but I was through one fight on the plains of Mansura, it was a regular artillery fight or duel. I was with the Reg all through, there was no infantry fight, but one of our men was lightly wounded and that I was one that helped carry the stretchers. The cannon balls flew, our regiment laid down behind a ridge thrown up for a fence and I went and sat on the stretcher so I could see and dodge the balls, the ball would strike the ground behind and before me and the shells could burst in front of the regiment and the iron would squeal and the dust would fly when a ball would strike the ground as if a cannon went off only the smoke was red instead of white. There I saw one of the prettiest sights I ever saw such a vast army moving all in times of battle as far back as you could see, and about 3 or 4 miles long on a plain, a little rolling, it was the pretty sight I ever saw, and then such firing of cannons along the whole of the front line and we was in front. One shell struck some 10 paces from under the doctors horse between his fore leg and hind leg, I saw the ball come and he was coming straight for me and I was fixing to dodge and I see soon he would not reach me and that he would miss the doctor about 2 feet, but such a dust I could but just see the doctor. The one that was wounded was behind me, the ball came right down out of the air and struck the ground and bounced and hit his leg at a glance and knocked him 3 or 4 times over. The balls were about 22 inches long, they were shells but a few bursts nearer me. How I wish that Uncle Allen Mambert could have been with me through that battle. It was serious but after all, I could not help but laugh sometimes to see how some would dodge and how the smoke would fly. I can't write much. We fought nearly a week at different times and fought our way out. There was two of our men captured by the Reb's and 3 boats sunk. I thought your letters were gone but by the numbers they are all right, 19 and 20 are the last, 17 and 18, I wrote an answer to, and I don't know if you received it, or not, if not it was captured, but I think you got it. We have had hard times marching, we have halted here since the night before last, I don't know where next, it's so wonderful, not that the sun burns like the shining of a hot fire, it is so dry and dusty. I will close by saying that you must take good care of everything and also remember me.

J.W. Mambert [77]

Colonel Molineux wrote in his diary about the picturesque battle, particularly the cannon ball strikes (sounding very similar to John's encounter):

"A rebel battery has complete range on us, and as the men require cheering up, I exert myself and ride up and down their front. While doing so a shell flew over my head and exploded not far to the rear. This was followed by another which exploded directly in front of me, throwing splinters and a cloud of dust around. The firing looks beautiful with the flash of guns in the dusk of the evening. A spent shot came right at a group of drummer boys, with a hop, skip and jump they scattered; but one of them in trying to get out of the way, ran directly into it, tripping himself and rolling over and over."[78]

THE SICKLY ENCAMPMENT

The encampment location in Morganza was believed to be of advantage with good water for use in all purposes—so they thought, but soon discovered differently. The truth was, the location was perhaps one of the sickliest in which the 19th Corps had ever encamped. The heat was oppressive and growing more unbearable every day. Crude shelters were constructed of bushes and leaves cut fresh from the neighboring thicket, often renewed; it gave little protection from the intensity of the sun. The levee and dense undergrowth kept any breeze from reaching camp, and the soil in which the wagons were constantly driven raised clouds of suffocating dust covering everything. The doctors believed many men were taken sick from inhaling so much of it.[79] If it rained, it became a sea of fat

---

77 *Letter #28, May 23, 1864, Camp along the bank of the Mississippi at the mouth of Red River.* John W. Mambert letters from the Civil War.

78 James I Robertson Jr Civil War Sesquicentennial Legacy Collection at The Library of Virginia – Edward L. Molineux collection. *Diary entry, no date.*

79 W.F. Tiemann Collection, *letter June 2,1864 Morganza.* N.Y.S. Military Museum and Veterans Research Center Archives.

black mud when the mosquitoes were at their worst. And the sickly season was close at hand; the field and general hospitals were soon filled with the unwell, followed by many deaths from disease.[80]

Colonel Molineux, who was quite ill, was sent to New Orleans on June 1, 1864 to recuperate. In a letter home he wrote:

"I expect you have all wondered at my last letters. It is because I have been sick and am still unwell. I left Alexandria on the retreat weak and unwell although my back was all right. Through the march I forced myself by my will to keep well, but when we reached Morganza I was obliged to give up, that is to say, keep to my tent, although still in command of the brigade. Intense, old-fashioned headache had been and were the chief of my troubles. The place is a horrid spider- vermin- snake infested sand plain. No shade, no air, nothing but dust and sun, sun, sun." further along, "the doctor applied for me to go to Baton Rouge, or New Orleans or somewhere in the shade but that old brute General Emory refused it, adding some remarks about officers wishing to get to New Orleans to dissipate. So as Franklin and Grover were absent, I had to continue to bake five days longer. When Grover came back, he gave me leave at once and I am here." [81]

CANBY'S MARINE PATROL

General Canby's Military Division of West Mississippi intended to use the 19th Corps as somewhat of a marine patrol, or coastguard, keeping the main body of the infantry ready to embark on a fleet of transports specifically assigned for the service. Canby's concept was a quick dispatch of troops having the availability to move up or down the Mississippi or any of the adjacent waters which could be menaced or attacked by the enemy. The orders for the organization and the Corps equipment to be used in this manner was a well-thought-out plan, advantageous for the military, but for some reason, never put into practice.[82]

In a letter home, Captain Tiemann shared his thoughts on Canby's plan:

---

80 Irwin, Richard B. *History of the 19th Army Corps*. G. P. Putnam's Sons New York, (The Knickerbocker Press 1892), ch. 29, par. 13.

81 James I Robertson Jr Civil War Sesquicentennial Legacy Collection at The Library of Virginia – Edward L. Molineux collection, *Letter written June 3, 1864*.

82 Irwin (n1), ch. 29, par. 8,14

"there is no news of any kind from this quarter. The guerillas still infest the river banks & trouble our gun boats & the river boats occasionally. They disabled one of our tin clads the other day about 2.5 miles above here. If we are to patrol the river & keep it free from these scoundrels, I do not see why they do not set us to work. We are still lying quietly in the woods."[83]

For eleven days, the 159th NY, 13th Connecticut and 1st Louisiana occupied the dreadful conditions before moving their camp two miles to a more suitable location. It was a charming spot: shady and cool, between the levee and river in a grove of young cottonwood trees. The men began constructing their new camp by thinning the trees. They left five-foot-high stumps along the company streets where tents were erected. The men put up "dog tents" as the canvas shelter under which two men could barely crawl. Inside, the men made bunks on shorter stumps about two feet from the ground. This proved to be a good arrangement, especially since they endured heavy showers with the most terrific lightning and thunder for the next ten days. Regular rations of whiskey and quinine were issued by order of the surgeon to keep the men healthy.[84]

The Camp at Morganza Bend Louisiana [85]

In the following letter home, John expresses his frustration with the war

---

83 W.F. Tiemann Collection, *letter June 7, 1864, Morganza*. N.Y.S. Military Museum and Veterans Research Center Archives.

84 Tiemann, William F. *The 159th Regiment Infantry, New York State Volunteers*, In the War of the Rebellion 1862 -1865, (Brooklyn New York 1891), p. 79.

85 *Camp at Morganza Bend Louisiana. United States Louisiana*, None [Photographed between 1861 and 1865, printed between 1880 and 1889] [Photograph] Library of Congress # 2013651623.

and the affliction it has caused the country. He again conveys his anger toward the aristocrats he believes are the culprits for all the suffering. Reading his Northern newspapers, John knows there are sympathizers (Copperheads) who support the South, having no true idea of the miserable life the people live. John writes:

"anyone who sympathizes with the Southerners planters who have caused this war is not fit to live under the United States government."

Letter #29 - June 1, 1864
Answer to letters 21 & 22, Mississippi River

Camp on banks of Mississippi, mouth of Red River

To Mrs. J. W. and Family

I am thankful to God that I am permitted this privilege of writing to you. I received your letters, all according to numbers, and yesterday I received No. 22, it was late at night when it was brought to my tent with the star. I put it in my pocket and thought I would not open it until morning, I would dream over it, and sure enough I had a very sweet and pleasant dream about home and how glad I was. But one thing I was surprised at which I will not mention. I must say to you that I have not been very well since our march, it has been much too much for me. I have had diarrhea and have been reduced in flesh considerably. This climate doesn't agree with me or rather the food doesn't. I think it is the food. I think I am getting better, but I am weak, I can hardly walk. Our Regt. is over half sick with diarrhea. I am out of money, and I have to do the best I can, but I expect to be paid in a short time. I sent you $30 about the 22nd of March from Algiers. I have never heard whether you received it or not. Please let me know. I ordered it to be kept at the Express Office in Hudson until called for and soon after I sent the receipt, Lambert Dingman[86] lives in Valatie, he owes me $3, try and get that. I wrote once before about it.

---

86 Lambert Dingman, Sergeant 159th Co. F: Age 26, promoted to 2nd L.T. C.O. I. Discharged. A farm laborer in Kinderhook, NY. Lived with mother and father, no siblings.

The war news you hear as much about it as I do, if Grant is successful and captures Richmond that will end the Rebellion. I am happy to think and know that you have paid Mag and that you had good luck in getting all our money in. One thing I must say is that I am glad that you write every week if it is only 2 words in it satisfies my thoughts about home. I can't write as often, and I have no chance to send a letter for we were hemmed in for three weeks and no mail could go through. The Reb's destroyed the mail between the mouth of Red River and Alexandria one up and one down, but as luck would have it, none of yours were destroyed. I can't tell whether one of mine was destroyed or not, but I think not. I get the paper regularly as the mail comes. I see that Permelia and Eb are back. I am sorry they could not stay one year, however, don't work where you don't like it and don't hire out for a year one month is long enough at a time. I have been hired out for three years and I don't like it, but I will stay unless I can get off honestly. 17 months more and out unless sooner discharged. I must say that I have laid and slept on the ground ever since the 19th of March on the rubber blanket I got, I take no cold on the march, whenever night overtakes I spread out the rubber, lay right down, spread a shelter over it and go to sleep. The weather is very hot, I wish you could have one such a day. There are hot breezes like the blaze of a fire here. I am glad that Parmelia has written me a letter and I hope you will all write. You must write to Martha. One word to John's Martha if I am rightfully informed, you have found Christ is the sinner's friend of your savior. I hope that you have given him your whole heart and that you may prove to be a blessing to all by whom you are surrounded, trust in him and do his will and he will bless you.

La. June 4th, 1864, while writing this letter of June 1st I was informed that we would be paid and so I did not finish until today. Last night we were paid and today or tomorrow I will send you $35 or $40. I want some myself for I have suffered for want of something that I could eat. My appetite was gone, and I was quite feeble, but I am on the gain. This morning I went out and bought 2 lbs of cheese and 20 cents worth of crackers, sweet ones, it cost me $1.20 so you see how far money goes, I got 12 crackers for 20 cents. I was paid $48, I owed 50 cents; then I pay to send the money by express and that cost $1. I mean to keep a little on hand in case of necessity. They sell potatoes here for 14 cents per lb. and onions the same; butter 60 cents; cheese 50 cents. Tobacco, I don't use now; have not for nearly 2 months, it sells for $2 a plug here, large plugs. Lemons for 4 bits or 50 cents. Tell Jobe and Walt they must be good boys and go to school and learn well and then they will soon be men. Tell them to write.

I suppose you live well sometimes on eels that you catch in the mud creek. I suppose Job and Walt set hooks and catch eels. I'm glad that John has plenty of work and at good prices and now is the time to make some extra ahead. I hope that this war will soon be over at least by fall so that I may return home, and I shall be a happy man. I have seen enough of war and sufferings, but I want to see rebellion crushed and our Country restored to peace and anyone who sympathizes with the Southern planters who have caused this war is not fit to live under the United States government, and if they were not ignorant of certain facts they would soon know how to decide.

Our Regiment is now encamped one mile below where they were 3 days ago. The 128th is encamped about one half mile from here. Our army here reaches nearly 4 miles along the banks of the river like a city. You have no idea what war is going on, but I think the end is in sight. I forgot to say that our Regt. is encamped in a grove of cottonwood trees so thick you can scarcely walk through. They are as large as a man's leg, all of one size.

I forgot to say that I am sorry to hear of so much sickness and so many deaths and at least of old Uncle Coontie Silvernail[87], but if I ever live to return again, which I trust in God I may. I think I shall find a great change and you will see a great change in me for I am 10 years older than I was before I left home. I want you to give my love to all our friends and relatives. Tell grandpap

---

87 Conrad "Uncle Coontie" Silvernail: Age 82, a farmer in Livingston, NY. Died April 18, 1864, buried in Churchtown, NY.

Sheldon and granny that I am doing well, for me in this climate. Tell him to write and tell grandpap Mambert the same. Tell them both that I should have liked to had them with me one half day in the battle on the Plains of Mansura. I must close, write and Trust in God for my safety and welfare.

Yours John Mambert [88]

FAITH: A DAILY ESSENTIAL

In John's letters, it is evident that his faith was a daily essential; the assurance of things hoped for were important to him. Throughout his letters, he refers to his faith in the Lord and how he trusts God's guidance. John and his family were members of the Dutch Reformed Church in Claverack, NY, where the Reverend Sluyter held sermons. This was also where John's parents and grandparents attended church services.

John nurtured his family with the same belief he was taught: Maintain a holy life in relation to God and each other and extend God's kingdom into the society in which you live. John's daughter in-law Martha, recently accepted Christ, becoming a believer in the word.

John wrote: "trust in him and do his will and he will bless you" as he does daily, trusting that the Lord will protect him and return him safely to his family–the quicker the better.

A new regiment, the 14th New Hampshire, 750 strong, arrived June 8, 1864, attached to the 2nd brigade. A fine-looking body of men in their clean new clothes presented a sharp contrast to the 159th NY in their old, well-worn, and rather dilapidated uniforms.

In the afternoon of June 11, 1864, the 19th Corps was reviewed by Major-

---

88 *Letter #29, June 1, 1864, Answer to letters 21 & 22, Mississippi River.* John W. Mambert letters from the Civil War.

General William H. Emory.[89] The weather was clear when the regiment left camp heading for the parade grounds. But soon the rain began descending in a tropical torrent, drenching every soldier, and reducing the field music to discord. Still, the soaking rain did not interrupt the ceremony, although the inharmonious music played on.

On June 19, 1864, the 159th NY received orders to prepare rations for ten days. That evening, the regiment along with the 13th Connecticut and a detachment of cavalry, embarked on the transport Ohio Belle, a sidewheel steamer. After a lengthy delay getting the remainder of the division on additional transport, the fleet cast off from the docks the following afternoon. The force numbered 5,000 infantry, 500 cavalry and four pieces of artillery, all under the command of Brigadier-General Grover.

There were several Union transports and gunboats now steaming up the Mississippi heading 60 miles for Tunica Bend above Morganza. The fleet was being shot at by numerous guerillas as they proceeded along the river. The infantry disembarked on occasion, managing to capture a few prisoners. They continued navigating through the night to Fort Adams arriving at 6 a.m. the next morning. Scouts were sent in different directions, but the enemy had already fled.

No trace of them could be observed from the summit of a hill on the slope where the old fort stands.[90] The officers who climbed to the top gained a captivating and extensive view of the surrounding country. At 5 p.m., the cavalry and officers boarded the steamer and began a hurried return to the camp near Morganza, arriving at 10 p.m. The men were thoroughly baked by the heat of a crowded Mississippi steamer during the summer season: it

---

89 Major-General William Hemsley Emory: (September 7, 1811 – December 1, 1887) he was a prominent American surveyor and civil engineer in the 19th century. Served as a brigade commander in the Army of the Potomac in 1862 and was transferred to the Western Theater. He was promoted to brigadier general of volunteers on March 17, 1862. He later commanded a division in the Port Hudson campaign. He subsequently returned to the East as the commander of the Nineteenth Corps, serving in all the major battles in the Shenandoah Valley Campaign of 1864, especially at the Battle of Cedar Creek, where Emory's actions helped save the Union army from a devastating defeat until Major General Phillip Sheridan's arrival.

90 Tiemann, William F. *The 159th Regiment Infantry, New York State Volunteers*, In the War of the Rebellion 1862 -1865, (Brooklyn New York 1891), pp. 80-81.

was like an oven.[91]

The trip up to Fort Adams Mississippi was the one and only time the 159th NY regiment traveled outside of Louisiana while attached to the gulf expedition. (Fort Adams was one of the country's first military fortifications constructed in 1799. Originally built as the port of entry from Spanish Louisiana, it was later garrisoned for the War of 1812.)

John wrote about his appreciation for the state of Mississippi in comparison to Louisiana. He also updates the family on the regiment's health, which is not good. There are high numbers of men reporting sick, most of them suffering from diarrhea.

Letter #30 - June 22,1864

Camp Mud – Mouth of Red River on the banks of the Mississippi, Morganza, 159, Reg 1 – La

To: Mrs. John Mambert & Family

I am quite well and enjoying good health thanks be to God. I hope you are enjoying the same blessing. I received your letter which was not numbered, should have been numbered 24 and I was glad to hear that you were all well. I have just returned from an expedition up the Mississippi River. We started on the evening of the 19th with our division loading 7 transports with troops and 4- or 5-gun boats, 2 Regt's on a boat, 300 Calvary and one Battery. Our object was to drive a squad of Reb's from the bank of the Mississippi between here and Natchez which had fired into our boats while passing. Next morning we found ourselves at the place and we landed unmolested, and our Calvary went out, but found no Reb's, they returned at sunset when we all went on board and went up to Fort Adams in the state of Miss. Arrived there at daylight, went ashore, and encamped. Sent out our Calvary and they returned at about 6 pm and saw no Reb's, so we went aboard and this morning and we found ourselves back again, it was more like a trip of pleasure than anything else, and now we are looking for a mail to hear from you. I thought I would

---

91 Sprague, Homer B. (Homer Baxter), b.1829, *History of the 13th Infantry Regiment of Connecticut Volunteers, during the Great Rebellion*, (Hartford Conn. 1867), p. 217.

wait to write until we received it. We expect to hear good news from Grant and Sherman. If they are successful, then together with the fall's election will bring this Rebellion to a close. We expect to lay here perhaps all summer, and maybe winter here, but we can't tell. We will patrol this river to keep the Reb's from crossing. I like it in Mississippi much better than here in La. The old Fort Adams is an old work built over 200 years ago and all you can see is the place where it stood, the country on the East side of the river is hilly and fruit grows there, the corn is nearly fit to cook. Young potatoes as large as your fist. Our camp is in a very sickly place. Jim Races[92] son is dead. He died the 18th of this month. His mother is a sister of Mart Cooper's wife. He was sick with the diarrhea but was about camp on the morning of the 18th Ed Rote[93] slept in his tent. Ed got up for Roll Call which is pretty early. Robert Race[94] did not get up. Ed supposed he was asleep so Ed thought he would not disturb him. At breakfast time he thought he would wake him and to his great surprise he found he was dead. So unexpectedly they die here in this climate. He was buried with military honors and with a good coffin, as they call it in this country. Let his mother know our Regt. is sickly. I have seen 130 men reported sick in our Regt. one morning, nine out of ten with diarrhea. There are not more than three out of ten that can live on government rations, and if it wasn't for the pay the soldiers get, there would be seven out of ten dying in this climate. You stated you had not received a letter from me, and I have written. I don't worry about me; trust in God, he will deliver me. My impression is that I shall be home by next winter. The 128th Reg. was not cut up as you wrote there was one killed and a few wounded. The one killed was a Texan just enlisted. Job Kells[95] and

---

92 Johnathan Jim Race: Age 48, a farm laborer in Greenport, NY. Father to Robert.

93 Edward Rote, Private 159th Regiment Co. E: Age 19, a laborer in Greenport, NY. Lived with his brother and family.

94 Robert Race, 159th Regiment Co. E: Age 18, died of disease in Morganza, L.A. June 18, 1864. From Greenport NY. Left behind his parents and six siblings ranging in age from 5 to 14 years old. Reviewing census records, it appears Robert's true age was 16.

95 Job Kells, Private 128th Regiment Co. G: Age 23, wounded in action at Port Hudson, La. A farm laborer in Churchtown, NY. Lived and worked for a relative, Fitz Tipple and family.

Lew Brooks and Bob Brush[96] are doing well and George Roots[97] and Lt. Boils[98] also. The weather is very warm here and showery, heavy thunder. I must give you our market at the sutlers; potatoes old 15 cents per lb.; onions the same, dried Cod 20 cents, butter 60 cents, cheese 50 cents, a can of peaches holding one qt. $1.25 a can of the same kind filled with chicken, beef mutton or turkey 3/$1, condensed milk weighing 1 lb. 75 cents; lobsters the same as oysters, a tumbler full of raspberry, current or fruit 50 cents in small bottles that is jelling; whiskey I have known them to pay $7 for a canteen full ice water, lemonade or beer 10 cents a glass and at the above prices it takes three men in one shanty to hand it out and make change. They can't begin to supply half the soldiers needs, a full stock only lasts 3 or 4 days and many times less. I must close.

John W. Mambert.[99]

TROPICAL DYSENTERY

Chronic diarrhea and dysentery caused more combined deaths than any other diseases during the war. Of the soldiers suffering from chronic diarrhea, 162 of every 1,000 men died. The disease was distinguished by prolonged severe daily diarrhea with frequent liquid or unformed stool. These symptoms were often accompanied by joint pain, weight loss and bodily weakness. Though chronic diarrhea was attributed to scurvy, physicians and surgeons also thought chronic diarrhea was possibly the result of lingering diseased vapor from the nearby stagnant waters, particularly since relatively healthy soldiers tended to develop chronic diarrhea in groups rather than individually.

---

96 Robert "Bob" Brush, 128th Regiment Co G: Age 33, a farm laborer in Claverack, NY. Lived with parents and seven siblings ranging in age from 7 to 26 years old.

97 Charles "George" Root, Private 159th Regiment Co. G: Age 37, enrolled as Corporal, was reduced in rank October 1, 1864. He was a seaman from Claverack, NY. Lived with his parents.

98 Stephen Boyle, Private 128th Regiment Co. I: Age 45, a farmer from Taghkanic, NY. Married with five children ranging in age from 7 to 13.

99 *Letter #30, June 22,1864, Camp Mud, Morganza. To Mrs. John Mambert & Family.* John W. Mambert letters from the Civil War.

John mentioned that 130 men in the 159th NY were sick with diarrhea. At the time, less than 600 men remained in the regiment; almost a quarter of them were on the sick list for diarrhea, four times the average. Most cases of chronic diarrhea were caused by intestinal bacteria such as E. coli or Salmonella. Amoebiasis, caused by ingestion of a freshwater amoeba species common only in tropical or subtropical regions, could explain some cases that occurred in Louisiana, Mississippi, and Alabama frequently distinguished as "tropical dysentery."

A soldier's diet consisted mostly of pickled beef heavily salted and in need of soaking prior to cooking, also coffee and hardtack were staples with the large biscuit dipped in coffee to make it more palatable. The regiment's commander and surgeon must have realized the waters nearby were causing sickness. Due to this later realized insight, Major Commander Waltermire issued the following order on June 25, 1864:

"until further orders the enlisted men of this command will be allowed to bathe only before the hours of 8 A.M. and after P.M. and will not swim beyond fifty feet from shore nor enter the waters within ½ hour after meals. The further use of spring water for drinking is strictly prohibited"[100]

The location of the regiment's sickly encampment in Morganza was ranked second highest in deaths by disease only beaten by St. Louis, MO.

CANBY TAKES CONTROL

Toward the end of June, the 19th Army Corps underwent the last of its many reorganizations, this time at the hands of General Canby. The 1st and 2nd divisions were left as they had been during the campaign that was coming to an end. The 13th Corps was broken up, seventeen of the 13th Corps best regiments were selected to form the 19th Corp new 3rd Division commanded by General Michael Lawler.

General Emory, suffering from the effects of the Louisiana climate and personal hardships of the campaign, applied for a leave of absence assuming there would be no movement during the summer. Canby granted Emory his leave and assigned General John F. Reynolds to command the 19th

---

100 G.O. *Waltermire to Regiment June 25, 1864.* Regimental Company Books for the 159th N.Y.S.V., National Archives Washington DC.

Corp. He also gave Emory's 1st division to General Benjamin S. Roberts, a total stranger to the officers and soldiers. Emory, learning of the changes in command to be made, gave up his leave of absence. Reynolds noted the deep and widespread disappointment among the officers and men of the corps when they heard Emory was departing. He persuaded Canby to leave the command of the 19th Army Corps to Emory.

General Grover kept command of the 2nd division along with General Birge, leaving Molineux and Sharpe as brigade commanders. A fourth brigade was added, commanded by Colonel David Shunk of the 8th Indiana, composed of his regiment, the 24th and 28th Iowa, the 18th Indiana, and regiments from the disbanded 13th Corps. Later, the 1st Louisiana was taken from Molineux's brigade and remained in the Gulf; in its place was added the 11th Indiana and the 22nd Iowa.

General Grant's orders to Canby were the same as those he gave to Banks: "go and take Mobile." This order to Canby was important, but secondary to the overall comprehensive plan adopted by Grant for the spring campaign. If Canby marched on Mobile distracting the enemy, Grant might be able to withdraw a large part of the force from General Sherman's campaign, or better yet, get Mobile to surrender without a struggle. This gain in control of the Alabama River would give General Sherman a secure base for supplies and a safe line of retreat if the need arises. The occupation of a line from Atlanta to Mobile would, as Grant had previously remarked, "once more split the Confederacy in two."[101]

---

101 Irwin, Richard B. *History of the 19th Army Corps*. G. P. Putnam's Sons New York, (The Knickerbocker Press 1892), ch. 29, par. 7,17,18.

Chapter Seven

"My advice would be for you not to enlist,"

John junior must have shared in a letter to his father his desire to join the military. In the next few letters John writes home, he tries to convince him not to enlist. John makes his case stating many examples of why it would be a bad decision. In his final plea to his son, John wrote:

"I think it is right to enlist, but knowing what I now know, I would not enlist for all the money you could pile between here and New York 10 feet high."

A CHANGE OF SCENERY FOR THE 159TH NY

While the military division of West Mississippi maintained control of the Southern waterways, General Grant ordered General David Hunter, commander in the Shenandoah Valley, to march to Charlottesville, Virginia, by way of Lynchburg. He also sent General Phil Sheridan, along with his cavalry, to Charlottesville where he was to convene with Hunter. General Lee, commander of the Confederate army, learning of the movement, sent General Jubal Early[1] to intercept the Union forces. Moving with speed and promptness, Early got to Lynchburg preceding Hunter.

Hunter, now believing his position precarious, decided to make his escape across the mountains into West Virginia instead of retreating through the valley. His retreat left the gates of the great Shenandoah Valley wide open for Early, who immediately took advantage of the situation, marching north through the unguarded valley.

The valley was now wide open to the enemy. No military action was happening in Louisiana, so General Grant informed General Canby to

---

[1] Confederate General Jubal Early: A graduate of the Military Academy at West Point in 1837. Commissioned a 2nd Lieutenant in Company E of the 3rd United States Artillery and sent to Fortress Monroe. He resigned his commission the following year returning to Virginia to study law. Obtaining his law license, Early was elected to the Virginia Legislature the following year. He voted against secession during the Virginia Convention in April 1861, but, once his state seceded, he remained loyal to Virginia. *Britannica.com*

postpone plans on Mobile, Alabama, ordering the 19th Corps immediately to Hampton Roads, Virginia. Canby's orders were to send the 1st and 2nd divisions, putting General Emory in command of the detachment.[2]

On July 2, 1864, the two divisions began their move down river to Algiers. The 159th NY broke camp and embarked on the sloop transport *Lancaster No. 4*. Sailing down the river, they made stops at Port Hudson and Baton Rouge.

After a hot and tedious two-day voyage, the screw steamer, crowded with troops and baggage, arrived in Algiers in the evening. The troops were quartered at the Belvidere Iron Foundry and surrounding property.

The next morning, Lieutenant-Colonel Gaul resigned due to contracting an illness during the campaign. Major Waltermire was promoted to Lieutenant-Colonel replacing Gaul.

REGIMENT COLORS

Colonel Molineux, along with the men of the 159th NY, enjoyed a pleasant event on July 8, 1864, when they were presented by the 23rd Regiment National Guard of Brooklyn with regimental flags, including the State and National flags. And they bestowed the colonel with a magnificent sword and equipment. The flag was inscribed, "Irish Bend, La., April 14, 1863" and "Port Hudson, La., May 25, June 14, 1863" with two guidons, the figures "159" all made completely of silk with heavy gold fringe. The colors were carried by the 159th NY through all subsequent campaigns and battles. The colors were returned to Colonel Molineux when the regiment was mustered out of service.[3]

---

2 Irwin, Richard B. *History of the 19th Army Corps*. G. P. Putnam's Sons New York, (The Knickerbocker Press 1892), ch. 29, par. 21.

3 Tiemann, William F. *The 159th Regiment Infantry, New York State Volunteers*, In the War of the Rebellion 1862 -1865, (Brooklyn New York 1891), pp. 82, 85.

Flag presented to Colonel Molineux by the 23rd Regiment National Guard of Brooklyn [4]

SWELTERING & STUFFY SITUATIONS

The weather in Algiers was intensely hot; the regiment was dealing with the brutal heat as best they could. Their encampment was across the river from New Orleans. The city was so close, the men had an endless supply of rum. This did not help the sweltering situation, on the contrary, trips by boat to obtain more spirits were numerous, adding to an already unhealthy situation. Life in camp was getting disorderly—one man had a bayonet speared through his arm by a comrade during an inebriated stunt. Assistant-surgeon Briggs distinguished himself by knocking out the biggest and heaviest man in the regiment: who just happened to be drunk and disorderly. After the regiment "stewed" for two weeks, they marched to the levee embarking on the side-wheel steamer *Cahawba* with all regimental property.[5]

---

4 *159th Regiment NY Volunteer Infantry Regimental Color*, Civil War. New York State Military Museum & Veterans Research Center.

5 Tiemann, William F. *The 159th Regiment Infantry, New York State Volunteers*, In the War of the Rebellion 1862 -1865, (Brooklyn New York 1891), pp. 85-86.

US Transport "Cahawba", artist Alfred Rudolph [6]

The steamer was used as brigade headquarters. Colonel Molineux and his staff were already on board along with the 131st New York and 22nd Iowa. Again, the troops endured overcrowded transportation, stuffing the steamer with 1,350 men. To help the men tolerate the packed circumstances, everything possible was done to keep them content.

They left port at midnight under sealed orders not to be opened for twenty-four hours after departure. When the orders were opened, the brigade was instructed to proceed to Fortress Monroe and report. The transport passed the mouth of the Mississippi river on the evening of July 18, 1864, passing Tortugas and Key West two days later.

Not one soldier onboard that sidewheeler was disappointed by their departure from the Louisiana lowlands, with its poisonous swamps, filthy water, overpowering heat, and intolerable mosquitoes. Yet, there was not one man on that steamer who did not feel a tightness of the throat when thinking about all he had seen and suffered, especially when remembering many of his less fortunate comrades who succumbed to the dangers and trials of which he himself was now turning his back on—hopefully for the last time.

The 159th NY regiment arrived off the coast of Fortress Monroe 1,771 miles from New Orleans at 12 a.m. on July 24, 1864. The passage was quick,

---

6 Illustration, *US Transport "Cahawba"*, artist Alfred Rudolph. Gift of J.P. Morgan, Library of Congress # 2004660624.

only six and a half days; this time, with pleasant weather throughout the voyage. Colonel Molineux was ferried to Fortress Monroe and received orders for the troops. When he returned to the steamer, they proceeded up the James River arriving the next day at Bermuda Hundred, Virginia, with orders for the soldiers to disembark immediately.[7]

SIEGE OF PETERSBURG

On July 25, 1864, the men unloaded the regimental stores from the steamer and proceeded to march six miles to the entrenchments on the frontline of the siege at Petersburg. The regiment set up camp near the headquarters of the 10th Corps commander Major-General Birney.

The 159th NY position had been attacked by the enemy twenty-seven times before their arrival. The breastworks were constructed of dirt seven feet high, on top of which were sandbags to protect the sharpshooters. At intervals of 250 yards, batteries were positioned, each mounting ten to twelve cannons. The area containing batteries number five and six was nicknamed "Fort Slaughter" because it proved to be extremely fatal for the enemy on each assault attempt.

The Confederate and Union pickets spoke regularly with each other and exchanged papers. But a general order forbidding communication with the enemy ended that four days before the 159th NY arrived.[8] The enemy, a strong force in their front, provided nonstop firing while in the trenches or on picket duty; it appeared to them they were in the trenches at Port Hudson all over again.[9]

On July 27, 1864 at 3 a.m., they struck their tents and six hours later marched half a mile to the front establishing camp near the bomb-proofs.[10] The men proceeded back into the trenches at 8 p.m., exiting two days later, moving their camp behind the breastworks. On July 31, 1864 at 2 a.m., they moved from the front, marched back to the river, and embarked on the

---

7 Tiemann, William F. The 159th Regiment Infantry, New York State Volunteers, In the War of the Rebellion 1862-1865, (Brooklyn New York 1891), pp. 86-87.

8 W.F. Tiemann Collection at The N.Y.S. Military Museum, *Letter July 28, 1864, In the Field.*

9 Tiemann (n1), pp. 86-87.

10 Bomb-Proofs - Secure quarters constructed for the soldiers near the front lines of battle or constructed within a fort.

gunboat steamer *Winona*. They sailed in the afternoon, stopped at Fortress Monroe for a short time, then navigated up Chesapeake Bay. On August 1, 1864 at 6 p.m., the regiment arrived in Washington, DC. They disembarked the steamer and marched up Pennsylvania Avenue arriving at "Soldiers Rest" near the railroad depot.

SOLDIERS REST & BRUIN

Soldiers Rest was one of the largest military facilities erected in Washington during the Civil War. It was on the north side of Capitol Hill, at North Capitol Street and Delaware Avenue NW next to the B&O Railroad. The Rest provided lodging and hot meals to new recruits from the North who were on their way to join the armies of the Potomac and Shenandoah Valley. It also accommodated soldiers enroute back to the battlefront and assisted those recently paroled from Confederate prison camps. When it closed in March of 1866, Soldiers Rest had provided services to 974,000 men.

Similar stations were supported by the United States Sanitary Commission, a relief agency approved by the War Department on June 18, 1861, with the purpose of aiding sick, wounded and traveling Union soldiers. The agency depended on thousands of women volunteers to perform numerous duties, from collecting donations to working as nurses in hospitals, and assisting at rest stations and refreshment saloons. They also sponsored Sanitary Fairs in Northern cities, raising millions of dollars used to send food, clothing, and medicine to Union soldiers.[11]

The 159th NY received a new recruit in the early spring of 1864, and he became a very important comrade. His name was Bruin. At the time of his enlistment, he was very young, and yes: he was a bear cub. During the regiment's campaign in Alexandria, Lieutenant Governor James Wells gave him to Colonel Molineux as a gift. The colonel then presented Bruin to the regiment where he was immediately placed in the command of the drum corps. Bruin soon became a favorite, leading the regiment while they marched. He attracted constant attention as the main attraction, the reason why the 159th NY became known as the "bear regiment" or the "Molineux

---

11 Magnus, Charles. *Soldiers Rest, Washington, D.C.*, Smithsonian, National Museum of American History (CC0).

bears" wherever they went.[12]

The men had great fun marching on Pennsylvania Avenue with the citizens stopping to watch Bruin at the head of the column led by the drum major. On that day, some 500 people came to see Bruin as the regiment marched. Once in a while, as the people naturally pressed too close to the bear, the drum major would lengthen Bruin's chain, allowing him to head for the crowd, scattering them in all directions.[13]

Colonel Molineux mentioned Bruin in a letter home:

"By the way, the new Lieutenant Governor Mills of this state presented me with a fine male bear cub. He is such a playful good-natured fellow that I thought of giving him to Child to adopt, but on second thought preferred the benefits of a liberal southern education for his growing mind and gave him to the regiment. The drum corps have him in charge and he is a grand favorite and attracts universal attention. On grand occasions and parades he marches majestically at the head of the regiment with the whole drum corps beating the march behind him."[14]

While researching additional documentation on Bruin, illustrations and photos have not surfaced. However, it has been documented that the 12th Wisconsin Regiment Co. E also had a bear mascot named Bruin. Wisconsin's Bruin was the pet of a young soldier, Harlan Squires. He and his father enlisted in the regiment together in November of 1861; Bruin became the company mascot also marching at the head of the regiment.[15]

While at the "Rest," the regiment was provided with a bountiful supper followed by a good night's sleep. Their baggage was loaded on railcars and reported going to Chambersburg, Pennsylvania. Chambersburg was burned three days earlier by order of Confederate General Jubal Early. The devastating fire destroyed over 550 structures and left over 2,000 people homeless—similar to the Union burning Alexandria a few months earlier.

---

12 Tiemann, William F. *The 159th Regiment Infantry, New York State Volunteers,* In the War of the Rebellion 1862 -1865, (Brooklyn New York 1891), pp. 88-89.

13 W.F. Tiemann Collection, *Letter Aug 4, 1864, Tenleytown D.C.* The N.Y.S. Military Museum and Veterans Research Center.

14 James I Robertson Jr Civil War Sesquicentennial Legacy Collection at The Library of Virginia – Edward L. Molineux collection, *Letter written June 6, 1864.*

15 Telzrow, Michael, Russell Horton, Kevin Hampton. *Wisconsin in the Civil War,* Wisconsin Veterans Museum/Wisconsin Department of Veterans Affairs, p. 143.

HEADING FOR THE SHENANDOAH VALLEY

On July 11, 1864, a large force of Confederate soldiers commanded by General Early crossed the Potomac River heading for Washington, D.C., determined to capture the Union Capitol. The battle lasted for two days at Fort Stevens on the north side of the Capitol. Early found it too well-defended for a successful attack and retreated, instead doing damage in the states of Pennsylvania and Maryland. In defense of another possible attack, the 159th NY (attached to the 19th corps, along with many other units throughout the Union) was transported to Washington, D.C. for protection.

The 159th NY was ordered to move out on August 2, 1864. Starting at 5 a.m., they marched in columns by company up Pennsylvania Avenue, passing the Capitol, Treasury and White House, continuing on to an encampment six miles northwest of the city. They settled into camp at Tenleytown, high on a hill near Fort Gaines, not far from the chain bridge crossing the Potomac. The weather was intensely hot, 98 degrees in the shade. The regiment had to endure the intense "Louisiana-like heat" for the next three days.[16]

On August 12, 1864, the 159th NY once again received orders to move. They were to unite with additional forces under the command of Major General Philip Sheridan in the Shenandoah Valley, where Sheridan was currently dealing with the rebel commander Early.

General Sheridan was a career military officer, receiving his appointment to the United States Military Academy in 1848. Graduating in 1853, he started his service commissioned as a brevet 2nd lieutenant assigned to the 1st U.S. Infantry Regiment at Fort Duncan, Texas. At the height of his military career, President Grover Cleveland appointed him to the rank of General of the Army in 1888.

Sheridan was born in Albany, New York, but grew up in Somerset, Ohio. He was of small stature, 5 feet 5 inches tall. Nicknamed, "Little Phil," Abraham Lincoln described him accordingly:

"A brown, chunky little chap, with a long body, short legs, not enough neck to hang him, and such long arms that if his ankles itch he can scratch them without stooping."

Two days after the 159th NY received orders to join Sheridan, they were

---

16 Tiemann, William F. *The 159th Regiment Infantry, New York State Volunteers*, In the War of the Rebellion 1862 -1865, (Brooklyn New York 1891), pp. 89-90.

roused at 1 a.m. After eating their breakfast at 4 a.m., they started marching over the Leesburg Pike. For the next two days, the regiment marched 36 miles over rough terrain. At 12 a.m. they arrived in Leesburg, Virginia, where they proceeded to set up camp a quarter mile from town.

On August 17, 1864 at 4 a.m., the regiment started marching; they continued until 10:30 a.m. when they were halted, going into camp four miles east of Snickers Gap, Virginia. The men were just getting comfortable, resting their tired feet, when at 5 p.m., they were ordered to strike the camp. The regiment pressed forward rapidly; the enemy was attempting to cut them off before uniting with the 6th and 8th Corps. At 9 p.m., they reached the Shenandoah River at Castleman's. The men stripped off their shoes, stockings, and trousers to keep them dry while attempting to cross the river, looking like a group of highlanders wading in the waters. Adding to their discomfort, rain began to fall.

The regiment arrived in Berryville, Virginia, 69 miles from Washington after marching thirty-three miles in thirteen and a half hours. Exhausted, the men threw themselves on the wet, muddy ground, no blanket or covering, overcome by sleep under a relentless rain.

Soldiers fording the river, artist Edwin Forbes [17]

On August 18, 1864, at 5:30 a.m., the 159th NY started marching without breakfast. Shortly after they began, they were halted to allow the 6th and 8th Corps to pass. The regiment took advantage of the short break to get a mouthful of food. The commissary was empty. The men had to survive on foraged corn and apples, with coffee. While stopped, they saw Major-General Sheridan (a short, stout, quick-moving man who appeared to have enough determination for

---

17 Illustration, *Soldiers fording the river*, artist Edwin Forbes. Gift of J.P. Morgan. Library of Congress # 2004661504

a dozen men) passing the long line of soldiers en route for the valley. The regiment continued to march four more miles until stopping to set up camp in a wooded area one mile from the road. They remained in camp throughout the next day.[18] Colonel Molineux, feeling the strain of the difficult journey they were enduring, shared his views about their latest crusade in a letter home:

"This campaign is one of suffering to men and officers. Short rations, constant labor, and little sleep, but I hope we will accomplish something for the cause."[19]

The change from the fatiguing climate of Louisiana to brisk air, crystal waters, rolling wheat fields and beautiful blue mountains of the Shenandoah was like a tonic for the men of the 159th NY. Daily, their spirits grew. The number of men reporting for duty increased as more men came off the sick list. The excellent road conditions and open country allowed the soldiers to achieve long marches with ease and comfort.[20]

Colonel Molineux wrote:

"you may well imagine how I love and live in this cool, fresh mountain air. Most of the officers and men complain of it being too cold (our southern blood having become thin), but to me it is only fresh and delightful, although I do find a camp fire uncomfortable when we stand at arms at 3:30 a.m. The scenery of this valley is perfectly magnificent,"

Sheridan's Wagon Trains in the Valley, artist Alfred Rudolph [21]

---

18 Tiemann, William F. *The 159th Regiment Infantry, New York State Volunteers*, In the War of the Rebellion 1862 -1865, (Brooklyn, New York 1891), pp. 91-92.

19 James I Robertson Jr Civil War Sesquicentennial Legacy Collection at The Library of Virginia – Edward L. Molineux collection, *Letter written Aug. 4, 1864*.

20 Irwin, Richard B. *History of the 19th Army Corps*. G. P. Putnam's Sons New York, (The Knickerbocker Press 1892), ch. 30, par. 26.

21 Illustration, *Sheridan's Wagon Train in the Valley*, artist Alfred Rudolph. Gift of J.P. Morgan. Library of Congress # 2004660442.

THE CHESSBOARD

General Grant's forces were reduced by Early's prior military engagements. In the valley, the current conflict between the two armies was deadlocked. Grant was willing to spare the 6th and the 19th Corps from their quiet campaigns elsewhere and bring them to the hostilities in the Shenandoah Valley. The two added corps along with two divisions of cavalry strengthened the Army of the Potomac.

General Lee, commanding the Confederate army, not only decided against withdrawing Early from the valley, but he also sent reinforcements: General Richard Anderson and General Joseph Kershaw's divisions of infantry. He added Fitzhugh Lee with one division of cavalry and Captain Wilfred Cutshaw's battalion of artillery. Lee believed this move would strengthen Early, enabling him to hold his ground.

While the generals were in the throes of strategic planning, it was harvesting time in the valley, a season of urgency to reap what was sown or go hungry. Generals on both sides of the war had goals related to the harvest. Lee's goal was to control the valley during the harvest because they needed the crops to feed the troops. Grant's message to Sheridan was to destroy the crops to keep the Confederate army weak and hungry.

Initially, Sheridan was positioned at Cedar Creek; he found the spot to be "a very bad one." Withdrawing his force, he led them to the farthest edge of the valley where the Potomac River and the Shenandoah Valley meet. While retreating from Cedar Creek, Sheridan enforced Grant's new policy of making the valley useless to the Confederate armies, burning all the grain and carrying off all the animals above Winchester:

"I have destroyed everything eatable," were Sheridan's words.[22]

General Sheridan assessed the surroundings and saw only one supportable position for defense—Halltown, Virginia. The Confederate defense rested on Fisher's Hill, Virginia. Between these two positions was a wide plain: like a chessboard between players. The two forces began a series of moves. Each side watched and waited for their adversary to weaken, make a mistake, or overlook an advantage.

---

22 Irwin, Richard B. *History of the 19th Army Corps*. G. P. Putnam's Sons New York, (The Knickerbocker Press 1892), ch. 31, par. 10,13.

Once Sheridan positioned his force in Halltown, a series of seesaw movements began where Sheridan would return to Halltown, and Early would advance to Bunker Hill. Early found as his objective the Baltimore and Ohio railway, but Sheridan's objective was Early: Whenever he found Early at Bunker Hill wreaking havoc on the railroad or canal, he took a step forward to the Clifton-Berryville line and pushed Early's troops to Winchester. The alternating play between the two armies could have continued until the end of the war. But other events intervened.[23]

At 9 a.m. on August 20, 1864, the 159th NY marched into position left of the road leading to Harper's Ferry, three miles south of Charles Town, West Virginia. They joined the rest of their brigade now composed of the 159th NY, the 131st NY, 3rd Massachusetts cavalry (dismounted), 11th Indiana and 22nd Iowa, with Colonel Molineux in command. They remained the 2nd Brigade, 2nd Division and a detachment of the 19th Corps.[24]

General Sheridan's army remained at rest, while Early occupied Bunker Hill and General Anderson in Winchester. Then, on August 21, 1864, the two enemy forces moved simultaneously to attack the Union army. Heavy firing was heard in the direction of Summit Point, south of the 159th NY's position.

The Union infantry took position in line. As their cavalry fell back, the enemy moved forward. The 159th NY was posted on the crest of a hill and put to work throwing up breastworks. At 9:30 p.m. they received orders to move, marching toward Harpers Ferry holding the extreme rear while covering the movement of their brigade.

BATTLE AT HALLTOWN

At 3 a.m., August 22, 1864, they reached Halltown, six miles south of Harpers Ferry where they halted in camp; the enemy followed close behind. At 4 a.m., skirmishing began. One hour later, the men marched out of camp and into an open field to the left of the road. The 159th NY was deployed as skirmishers. Advancing one quarter mile, they constructed breastworks and became pickets for their division.

---

23 Ibid., ch.31, par. 10.
24 Tiemann, William F. *The 159th Regiment Infantry, New York State Volunteers*, In the War of the Rebellion 1862 -1865, (Brooklyn, New York 1891), p. 92.

The following morning, the regiment was relieved and returned to their camp in Halltown. On August 24, 1864, the regiment moved their camp behind the breastworks. That afternoon, under the command of Colonel McCauley, they joined the 22nd Iowa and 11th Indiana advancing on the skirmish line for reconnaissance.

The 159th NY moved forward across an open field in front of a wood line, engaged almost immediately with a large enemy picket line strongly reinforced. The regiment pressed forward bravely against the continuous roar of the musketry, pushing the enemy through the woods. Holding their position on the edge of the woods, heavy firing exposed the presence of a much larger enemy force; orders were given for the 159th NY to fall back.

The retreat was difficult. A large number of rebels were in hot pursuit. Even so, the retreat was executed in good order. Three companies held in reserve were deployed to strengthen the pickets, but they were halted by the enemy's advance, so the first line reserves were withdrawn.

Officers witnessing the engagement spoke highly of the 159th NY regiment's conduct, praising the soldiers in the official report. For the next three days, the regiment was awakened at 4 a.m., standing under arms each day.

During the battle at Halltown, ten men of the 159th NY were wounded, two fatally.

COMPANY B, LOSSES AND WOUNDED
Killed
Private James J. Lenfesty

Wounded
Private William Sherman, prisoner
Corporal Martin Smalix

COMPANY C, LOSSES AND WOUNDED
Wounded
Private Christian Schnack

COMPANY G, LOSSES AND WOUNDED
Killed
Sergeant Frank W. Kurtz

Wounded
Sergeant Egbert S. Covey
Corporal Nicholas R. Shultis
Private Russel Van Deusen

COMPANY H, LOSSES AND WOUNDED
Wounded
Captain Wells O. Pettit

COMPANY I, LOSSES AND WOUNDED
Wounded
Corporal Joseph O. Reed [25]

Molineux wrote home:

"The people of the North seem to imagine that the rebel force in this vicinity has retired. This is a mistake in my opinion and I cannot help imagining that they are trying to either draw on us to some favorite position or to outflank us. However General Sheridan is, in my opinion, a shrewd safe soldier and will not, I trust, be outwitted. As we advance, we do so cautiously and entrenched ourselves. Indeed, it is a campaign of the shovel and pick".[26]

In the letter below, John begins to reason with Junior why he should not enlist. Most of his argument is based on his personal feelings about the war, giving many examples of the horrible things that could go wrong and change, or end his life.

But John did state one logical reason why he believed his son could not join: "the enrollment act," passed the previous year. John assumed it included an exemption that could relieve Junior from enlisting...it did not. In any case, John attempted to dissuade his son from joining.

Letter #31 - August 27, 1864

Camp near Halltown, West WV.

I sent a letter yesterday with Mr. Bergie[27]. After he left yesterday, we had quite a hard fight, but such firing of cannon and shells and musketry, it was awful. I don't know how many men we lost but quite a number and today the Reb's have left our front and are gone, we drove

---

25 Tiemann, William F. *The 159th Regiment Infantry, New York State Volunteers*, In the War of the Rebellion 1862 -1865, (Brooklyn, New York 1891), pp. 93-94.

26 James I Robertson Jr Civil War Sesquicentennial Legacy Collection at The Library of Virginia – Edward L. Molineux collection, *Letter written Aug. 30, 1864*.

27 Thomas Berridge, Sergeant 159th Regiment Co. A: Age 24, promoted to Commissary Sergeant. A carpenter from Hollowville in Claverack, NY. Born in England, married with no children.

them back yesterday. I think we will leave this place before tomorrow at this time, it is near 7 o'clock. I received your letter 34 dated Sept. 21st and Permelia's letter of July 14th in answer to your letter about enlisting. My advice would be for you not to enlist, you may enlist for heavy artillery as the rest has and after you are here then you will be turned into infantry. That is the way they do it, I have seen 2 or 3 regiments serve so. I think you could not stand the hardships, and one year would kill you of the first engagement, you might lose an arm or a leg or your head if you could be in one battle you would never want to be in another one. I don't think they can take you for I am in the service, and you have the whole of the family to see to and if you go away who will take care of them. My advice would be not to enlist. You think this war may end if they get a new president, it may, but if it doesn't you will see one of the worst wars that ever has been recorded on the pages of history. It will either end or be 10 times worse. I don't care who they elect for president. If the Southern people are bound to independence or extermination, then it will be a war of extermination. If I was at home, I would not enlist again not knowing what the future would be for $50,000 money would not alter just me, if you enlist you deprive yourself of all your liberty and everything else, you are a complete ass. If you can any way get clear do so, for if you go you take your life in your hands. If you had served one or two years you could tell better. As soon as we get to winter quarters, I will try to come home, I can't before. We expect to start some time tonight, but where we don't know, we are as dumb as an ass. I received Permelia's letter with 10 cents in to send a letter back. I will send one when I can, the 10 cents come very good for I got some cheese which tasted good. Butter, I have not tasted it since June, I hardly know how it looks. The weather is pleasant, and all is quiet tonight. It is dark and I can't see.

John Mambert
Write, give my love to all.

I like this country well. When I come back, I am going to the Western states, Iowa I think, I will buy you all small farms. We have the 22nd Iowa Regt. in our brigade. I am now writing by candlelight. They are all beating drums and it roars. They fought yesterday about ¼ of a mile from my tent and when they commenced, I laid asleep and woke up and of all the roaring of cannon and musketry the earth trembled. I thought hell was to pay. I got up quick and rubbed my eyes and by and by I saw our drum major come running on a keen sum they had thrown a shell close to him and his eye was white as two eggs, he got down and puffed and blowed he got scared from the shell. He did not say a word for almost half an hour. I could not think at first what was the matter, he is an awful coward. Now I will close by saying don't enlist, get clear if you can for you know but little about a soldier's life. If you enlist you will have to carry a musket no matter if it is in the heavy artillery. That is the way they have to do it here. A fifer has it the easiest of anybody, but if you don't enlist as a fifer musician, they will make you hold a gun. The papers must set forth musician and the word private struck out or they will hold a person as a private. I have seen all these things done. William Moon was served so he thought he enlisted as a musician and they made him take a gun, you see I must close.

J.W. Mambert [28]

On August 28, 1864, the 159th NY struck tents and began marching at 7:30 a.m. They marched twelve miles from Harpers Ferry to a position behind breastworks south of Charles Town. Each morning at 3 a.m. they formed their line, standing under arms until 5:30 a.m. During the day, the men kept busy with daily drills, details for pickets, and maintaining the breastworks.

There was little to eat, mostly green corn which the men roasted, and bitter green apples. The northern climate, at first appreciated by the men of the 159th NY, was now insufferable. It was a drastic difference in temperature compared to Louisiana.[29] They felt the cold weather terribly, especially at night and in the early morning when the temperatures were coldest.

BATTLE OF BERRYVILLE

General Sheridan ordered a massive advance on September 3, 1864. He instructed his infantry to assemble near Berryville. The 159th NY was

---

28 *Letter #31, August 27, 1864, Camp near Halltown, West WV.* John W. Mambert letters from the Civil War.
29 Tiemann, William F. *The 159th Regiment Infantry, New York State Volunteers,* In the War of the Rebellion 1862 -1865, (Brooklyn, New York 1891), p. 95.

awakened at 3 a.m., struck their tents and prepared four days rations for their haversacks. At 6 a.m. they marched twelve miles southeast, eventually halting in a large field. They moved into position along the 19th Corps line, one mile west of Arcadia Farm where they encamped. The 6th Corps was located to the right and the 8th Corps on the left towards Berryville.

The regiment had been in camp only a short time when heavy firing was heard in their front. The men were ordered to form a line across the pike. With the firing growing quicker and heavier, the 159th NY along with the 22nd Iowa and 13th Connecticut were ordered to the scene of action forming a line slightly to the right of the 8th Corps. The men marched forward under heavy artillery fire, at short musket range of the enemy occupying the crest of a hill.

As it was growing dark, the firing began to diminish. The men were halted and ordered to take what rest they could in line, finding it impossible to sleep under heavy rain and bitter cold temperatures. During the night, steady fire from muskets and artillery continued as both sides targeted their opponent's muzzle flashes. The rebels fired their artillery on the Union ambulances working behind the regiments, ceasing only when the rolling of the ambulances stopped. Darkness, rain, and low ammunition finally ended the battle.

Captain William McKinley[30] of the 23rd Ohio regiment wrote:

"the heavens were fairly illuminated by the flashes of our own and the enemies' guns."

McKinley enlisted as a private with the 23rd Ohio Regiment in June 1861. McKinley and his regiment were sent to the Shenandoah Valley to pursue the enemy. He was promoted to captain and transferred to General Crook's staff. During the Berryville battle, McKinley had his horse shot out from under him.

The next morning, Kershaw's division skirmished with the 8th Corps while Early led two divisions north to strike the Union right flank. Additional Union troops were drawn in, leaving the 159th NY on the line as pickets to receive the enemy's shelling. One man was killed from Co. G, Henry

---

30 William McKinley became the 25th president of the United States thirty-two years after the Civil War. He served from 1897 until his assassination on September 6, 1901; shot by Leon Czolgosz, an anarchist.

Karcher.³¹ The regiment was relieved at 7 p.m., marching back a half mile.³² General Early, marching up on the 19th Corps, found out the 6th Corps were only one mile further north; he chose not to attack.

The following day, the regiment constructed three lines of rude breastworks in double quick time—bayonets, tin cups, old shoes, everything was brought into requisition to accomplish the work. It was completed under heavy rain which persisted all that day and the next. Three days later, Colonel Molineux's brigade left camp at 10 a.m., moving parallel with the breastworks heading west in the direction of Winchester Pike.

After marching six miles up the valley, they spotted rebel-mounted sentries from a Confederate cavalry. One of the rebels was wounded by a sharpshooter but managed to escape. Molineux learned the enemy was in strong force a short distance in their front; the decision was made to return to camp without engaging. As heavy rain and thunderstorms continued for the next three days, the surgeon ordered whiskey rations for the men to help keep them healthy.

On September 11, 1864, President Lincoln appointed that day one of thanksgiving and praise for triumphs at Mobile, Alabama, and Atlanta, Georgia. Both happened to be victories General Banks wished for, instead of the humiliating loss pursuing Shreveport—costing him militarily. Admiral Farragut seized Mobile Bay on August 5, 1864, followed by General Sherman's victory in Atlanta, September 1, 1864.³³

MISERABLE CAMP LIFE

For seven days it rained steadily. On Sunday, religious services at brigade headquarters were stopped abruptly by a thunderstorm. The men felt weary and miserable; there was no escape from the constant soaking.

One day, a soldier was drummed through the camp to "The Rogue's March," escorted by seven men with muskets and bayonets. The bottom of a hardtack box was broken through and placed on his head, resting on his shoulders. On one side of the box was written, "THIEF" on the other,

---

31 Henry Karcher, Private 159ᵗʰ Regiment Co. G. Age 28, died in action near Berryville, Va. Enlisted in Hudson, NY.
32 Tiemann, William F. *The 159ᵗʰ Regiment Infantry, New York State Volunteers*, In the War of the Rebellion 1862. -1865, (Brooklyn, New York 1891), p96.
33 Ibid., p. 97.

"STOLE PROPERTY WHILE GUARDING IT."[34]

On September 13, 1864, the 159th NY supplied a detail for picket duty, their assigned cavalry captured the 8th South Carolina Infantry with its officers and battle flags. Extra precautions were taken; there were rumors of impending enemy retaliation. The night passed quietly with no sign of the enemy. The next day, heavy rain fell…again.

THIRD BATTLE AT WINCHESTER

On September 16, 1864, the 159th NY participated in a brigade drill, the first held in months. This created a stir among the soldiers; it was assumed some type of military action was being contemplated. The next day, General Grant visited General Sheridan, and orders were received for the troops to move without wagons or baggage.

On September 19, 1864 at 2 a.m., Sheridan put his force in motion (three hours earlier than he planned with Grant.) The army began marching up the pike toward Winchester, Virginia.

The enemy was rumored to be in strong force in Winchester. As morning light shone brighter, they heard heavy firing ahead of them.[35] Slowly threading his large army through the long gorge leading to Winchester, Sheridan may have underestimated the difficult crossing. To make matters worse, General Horatio Wright's wagon train, loaded with artillery, was en route to the battlefield. Although the troops began crossing at 6 a.m., the head of the column could only move at a snail's pace, taking more than four hours to get across.[36]

Pressing on, the men passed hospital tents alongside the road. Surgeons were busy caring for the wounded brought to them from the front. At 9 a.m., the 159th NY marched along the crest of a hill crossing Opequon Creek, filling their canteens as they passed. The creek flows at the foot of a broad and thickly wooded gorge with high steep banks. The ravine rises to the level of a rolling plain nearly three miles long with Winchester standing on the

---

34 Sprague, Homer B. (Homer Baxter), b.1829, *History of the 13th Infantry Regiment of Connecticut Volunteers, during the Great Rebellion,* (Hartford Conn. 1867), p. 224.

35 Tiemann, William F. *The 159th Regiment Infantry, New York State Volunteers,* In the War of the Rebellion 1862 -1865, (Brooklyn New York 1891), pp. 97-98.

36 Irwin, Richard B. *History of the 19th Army Corps.* G. P. Putnam's Sons New York, (The Knickerbocker Press 1892), ch. 32, par. 3.

western edge. Here and there, the high ground was covered with large oaks, pines, and undergrowth intersected by brooks called runs, the largest being the Red Bud Run which flowed into Opequon Creek.

Shortly after crossing the creek, a line of battle was formed with General Grover's division in the front. The 4th Division with Molineux's brigade was positioned in the rear of the first line, at supporting distance. The 159th NY then moved forward to the top of a hill where they could observe the battlefield.[37] On the right, General Emory extended the line of battle to Red Bud Run by posting Colonel Jacob Sharpe and General Henry Birge's brigades from Grover's division, along with Molineux and Colonel David Shunk in the second line. The 9th Connecticut troops were deployed as skirmishers to cover the right flank of Birge. General William Dwight's two brigades formed on the right and rear of Grover in a tiered formation parallel to the regiments already on the line. This formation supported Grover's line and covered the flank against any turning movement by the Confederates.[38]

The 19th Army Corp, Grover's Division at 3rd Battle in Winchester, artist Alfred Rudolph [39]

At 11:45 a.m., the Union troops heard the sound of Sheridan's bugle, this was then repeated from the corps, divisions and finally the brigade headquarters. The whole army began moving forward with great spirit, all while under heavy fire from the enemy.[40] The remaining brigades in the Union front moved to the right as they advanced, and the brigade of the 6th Corps was on the left. Unintentionally the 6th moved too far left creating a gap in the front line;

---

37 Tiemann, William F. *The 159th Regiment Infantry, New York State Volunteers*, In the War of the Rebellion 1862 -1865, Brooklyn New York 1891, p. 98.
38 Irwin (n1), ch. 32, par. 5.
39 Illustration, The 19th Army Corp, Grover's Division at 3rd Battle in Winchester, artist Alfred Rudolph.
40 Ibid., ch. 32, par. 8.

Molineux's brigade was then ordered to move and fill the void.

Deployed in parade order, the 159th NY and the 22nd Iowa were positioned directly in front, attracting the attention of the enemy's battery. The regiment, exposed to a deadly flank fire, was then assaulted under a heavy fire of cannon and musketry from the enemy, who was shelling them as fast as their cannons could be reloaded and fired.[41]

3rd Battle at Winchester, artist Alfred Rudolph. View from behind the Union line facing the Confederate army with Winchester in their background. The 6th Corps on the left, 19th Corps in the middle and the 8th Corps on the right.[42]

The 159th NY received the order: "Commence firing!" They returned fire on the advancing column of Confederate soldiers. The enemy's fire from artillery and musketry was overwhelming, forcing the whole brigade to be halted, but never stopping the brave men of the 159th NY from maintaining a steady fire on the enemy.

When Molineux's brigade was ordered to "fill the gap," it expanded nearly 400 yards wide. Confederate commander Jubal Early saw this opportunity and marched a fresh brigade toward the gap causing the Union flanks to separate. Immediately from the Union line came orders from Lieutenant Handy, "Cease firing!" followed by, "Retreat! Retreat!!" Molineux, looking to his right, saw the line rapidly falling back, followed closely by the large enemy army close on their heels.[43]

General Early ordered his Confederate line to sweep forward. With renewed force, the enemy broke in on the right of the Union brigade. They surrounded both flanks of Molineux's forces, leaving his men exposed. The 159th NY and the 13th Connecticut were holding steady under partial cover of a gully before being

---

41 Ibid., ch. 32, par. 11.

42 Illustration, *3rd Battle at Winchester, artist Alfred Rudolph*.

43 Tiemann, William F. *The 159th Regiment Infantry, New York State Volunteers*, In the War of the Rebellion 1862-1865, (Brooklyn New York 1891), p. 99.

swept back by the enemy. The men of the brigade positioned on the right, paid a heavy penalty for their earlier action of creating a gap in the line. The brigade on the left was less exposed to the enemy and able to stand firm. Molineux's brigade had a difficult time restoring order; reforming their line, they were ordered forward with the rest of the brigade.

The enemy charged upon a section of the 1st Maine battery posted in the center of the division. Fortunately, the 131st NY saw the enemy's activity. Taking advantage of a small, wooded ravine for cover, they changed direction and held their fire until they could see the shoulders of the enemy; once in their sights, they poured a heavy fire on the rebs. More support came from the 11th Indiana, part of the 3rd Massachusetts cavalry (dismounted) the 131st NY and 176th NY. They quickly rallied, opening a heavy well-sustained fire upon the advancing enemy, moving them back in chaos.[44] Unfortunately, in their retreat the Confederates swept across Molineux's remaining men cut off from the rest of the brigade. Prisoners were taken by the enemy including Captain William F. Tiemann and nineteen men; all of them bravely holding their posts through the battle.[45]

The countercharge, driving the whole Confederate line back, went beyond the positions from which they originally advanced their attack. The Union army held their ground, waiting for either enemy movements or general orders. By 1 a.m., the fight was over.

---

44 O.R., Colonel Edward L. Molineux, *Battle Reports on Winchester.* Serial 090, p. 330.
45 Irwin, Richard B. *History of the 19th Army Corps.* G. P. Putnam's Sons New York, (The Knickerbocker Press 1892), ch. 32. par. 11.

Battle at Winchester, Charge upon the Rebels at the Stonewall, artist Alfred Rudolph [46]

The Confederate army was greatly outnumbered from the start of the battle, and after the engagement, with the number of losses sustained, they were in no condition for the offensive. From the defensive, they had no option other than escape. Once the disorderly retreat of Early's men commenced, there was no stopping them. The Union troops pursuing the enemy were halted at Kernstown. Sheridan, mindful that his men were on their feet since early morning, fighting a long, hard battle, decided against sending his infantry in pursuit of the fleeing enemy.

During the engagement, the men saw every musket, cannon, and saber put to use under the command of their young and vigorous commander, General Sheridan. They looked upon a decisive victory, ending with the retreat of their enemy. Sheridan successfully drew out the enemy forcing Early to become ousted by the uniting movements of the Union brigades, sending him South. Sheridan himself openly rejoiced, and the men, catching the enthusiasm of their great leader, went wild with excitement. Sheridan, accompanied by his corps commanders, Wright, Emory, and General George Crook, rode their horses down the front lines as mighty cheers were heard; giving new life to the wounded, and consoling the last moments of the dying.[47]

---

46 Illustration, *Charge upon the Rebels at the Stonewall*, artist Alfred Rudolph. Gift of J.P. Morgan, Library of Congress # 2004660452.

47 Irwin, Richard B. History of the 19th Army Corps. G. P. Putnam's Sons New York, (The Knickerbocker Press 1892), ch. 32, par. 22.

Losses to the Army of the Shenandoah compiled by the War Department were 5,018; this included 697 killed, 3,983 wounded, 338 missing. Of the three infantry corps engaged, the 19th suffered the heaviest loss: 2,074 men. Of the brigades, Birge's suffered the most: 107 killed, 349 wounded, sixty-nine missing, a total of 525 men. Molineux's brigade was next: fifty-eight killed, 362 wounded and eighty-seven missing, a total of 507 men.

The total numbers for the 159th NY lost in the battle were five men killed, four officers and forty-nine men wounded, one officer and nineteen men prisoners, a total of 78 men.[48]

COMPANY A, LOSSES, WOUNDED AND MISSING:
Killed
Corporal John D. Tator   Private – Delbert Van Deusen

Wounded
1st Lieutenant Edward Duffy   Corporal Henry A. Osborne
Private Daniel Jennings

Missing
Captain William F. Tiemann   Private Thomas Ward
Corporal; Silas W. Perry   (died POW Salisbury, NC)

COMPANY B, LOSSES, WOUNDED AND MISSING:
Wounded
Private Fredrick Lawrence   Private Stephen Loughlin
Private Silas W. Richmond

Missing
Corporal George W. Hatfield

COMPANY C, LOSSES, WOUNDED AND MISSING:
Killed
Sergeant Augustus Wendt

Wounded
Sergeant James Fitzgerald   Private John Schermerhorn
Sergeant Jonas A Kellerhouse   Private Christian Schnack
Private Eugene A. Edwards   Private Freeland Wheeler
Private Myron Staats   Private William Shufelt
Private Austin Gailor

Missing

---
48 Ibid., ch. 32, par. 24.

Private Grovsvenor Smith   Private Door DeWitt
(died POW in Richmond, VA)

COMPANY D, LOSSES, WOUNDED AND MISSING:

Killed
Private Patrick Fitzgerald

Wounded
Private Frank W. Kisters   Private Isaac Morris
Private Conrad Smith

Missing
Joseph Treitlein (died POW Salisbury, NC)

COMPANY E, LOSSES, WOUNDED AND MISSING:

Wounded
Private James decker   Private Robert Proper
Private John W. Almstead   Private William H. Coyle
Private George A. Benzie

Missing
Sergeant DeWitt McNeill

COMPANY F, LOSSES, WOUNDED AND MISSING:

Killed
Private Floyd C. Nichols   Private William Colgan

Wounded
Corporal William Callaghan   Corporal Terence Mackey

Missing
Private Bartholomew Doser   Private Charles W. Mott
(died POW Salisbury, NC)

COMPANY G, LOSSES, WOUNDED AND MISSING:

Wounded
Corporal David Post   Private Dennis Sheehan
Corporal Jacob Hallenbeck   Private Anthony M. Michael
Private James H. Coe   Private John Toomey

Missing
Private Caleb Brady
Private Jeremiah Meagher
Private Henry Sherman
  (died POW Salisbury, NC)
Private Sabor S. Spaulding

COMPANY H, LOSSES, WOUNDED AND MISSING:

Killed
Private Henry E. Lander

Private Washington Adams

Wounded
Sergeant Martin Traver
Private Augustus G. Jenkins
Private Benjamin Leonard

Private William H. Frier
Private Charles Powell
Private Robert Hurley

Missing
Private John L. Lougea
   (died POW Salisbury, NC)

Private George Millott

COMPANY I, LOSSES, WOUNDED AND MISSING:

Wounded
Private Eugene A. Corey
Private William H. Wagoner

Private Walter C. Houck

Missing
Private Thomas Scott

COMPANY K, LOSSES, WOUNDED AND MISSING:

Wounded
Private Edward Brophy
Private John Kane
Private John Coughlan
Private John M. Kewan

Private George A. Hoffman
Private William D. Tanner
Private Henry Hahn

Missing
Private Bernard Doolin
   (died in Salisbury, NC
Private Timothy Dolan
Private Andrew Goshia
   (died POW Salisbury, NC)[49]

---

49 Tiemann, William F. *The 159th Regiment Infantry, New York State Volunteers,* In the War of the Rebellion 1862-1865, (Brooklyn New York 1891), pp. 100-103.

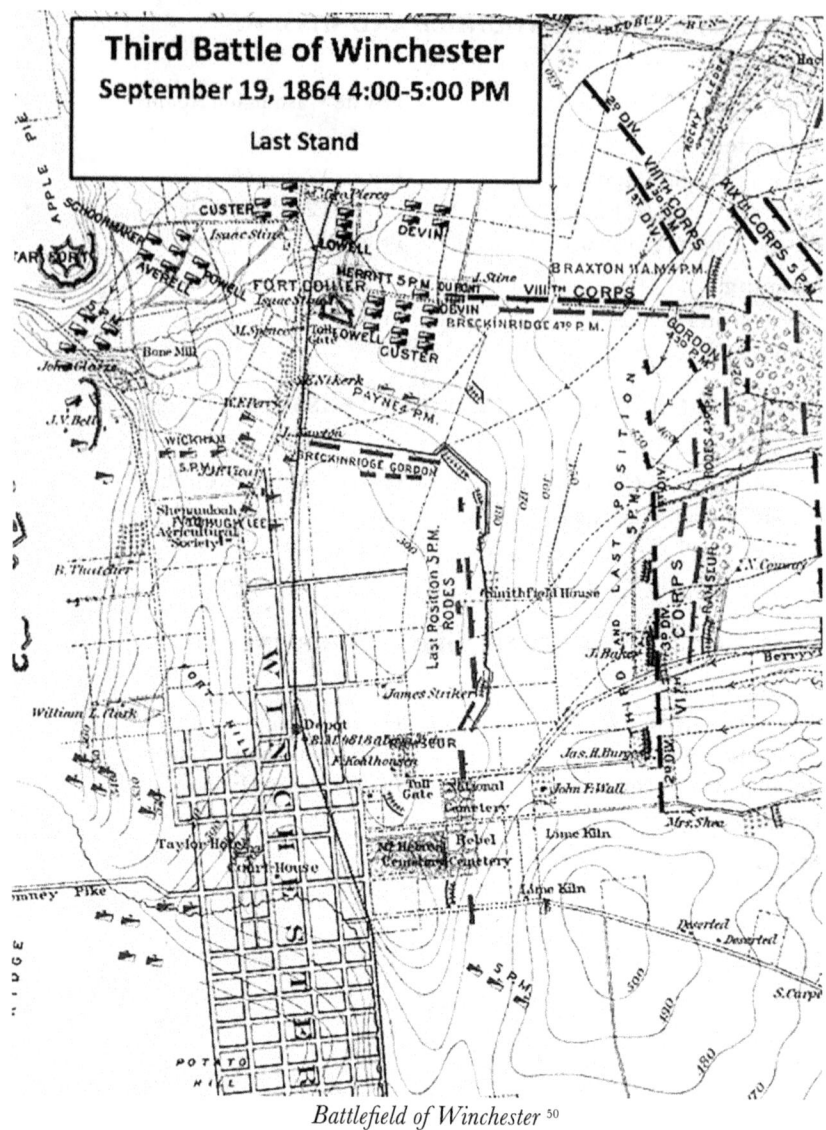

Battlefield of Winchester [50]

---

50 Author, Bvt. Lt. Col. G. L. Gillespie, Major of Engineers, U.S.A. under the authority of the U.S. Secretary of War. Portion of an 1873 map titled "Battlefield of Winchester, Va. (Opequon) [September 19, 1864]" prepared by an order of Lt. Gen. P.H. Sheridan under the authority of the Secretary of War.

ESTABLISHING MEDICAL FACILITIES

As the 159th NY moved forward toward the battle crossing Opequon Creek, the regiment medical personnel's priority was to establish the primary medical station and field hospital. Every regiment in the Union Army had a surgeon and an assistant surgeon commissioned by the state enlisting the troops. The surgeons were listed as officers on the muster rolls and were permanently attached to their regiment; they were never detached unless an urgent situation arose. Originally the 159th NY was organized with two surgeons: Charles A. Robertson from Albany and William Y. Provost from Brooklyn, plus an assistant surgeon Caleb C. Briggs from Columbia County. Caleb was a country doctor with a medical practice in the hamlet of Martindale, not far from John's home.

Because this was a planned engagement, John, along with the other field musicians, were assigned by Provost and Briggs to prepare splinters and bandages for the next day. Instructions to the men were specific and needed to be strictly followed. Once the hospitals were established, duties assigned to individual musicians could change depending on the circumstances.

The Surgeon Provost needed to get the field hospital set up quickly after the regiment crossed the creek marching toward the front. The 6th and 8th Corps established their hospital at the residence of Daniel T. Wood near the creek's crossing. The 159th NY continued marching about a mile and a half following the Red Bud Run. As they got close to where the 19th Corps entered the battlefield, the regiment passed the first hospital established for their corps at the residence of Charles Wood; his house and mill on the Red Bud Run. A quarter mile farther, the 159th NY came upon an abandoned factory building just behind the line. That's where Provost established the regiment's field hospital.

Assistant Surgeon Briggs, accompanied by his orderly carrying the hospital knapsack filled with emergency supplies, established their primary station just outside musketry range of the battlefield. As the wounded fell, Briggs ordered them to be picked up quickly and brought to his location. To speed up the process, stretcher bearers were assigned (with the drum corps and other musicians, including John) to retrieve men injured on the battlefield.

The wounded soldiers were initially walked or carried to Briggs's field station where they were triaged by severity of injury. Men with wounds so serious they could not be helped, were given sedatives to make them as comfortable as possible. Those requiring immediate attention were treated at the field station by Briggs with the help of his orderly. Wounded soldiers were given liquor to counteract shock before receiving first aid. Briggs's equipment consisted of sponges, basins, lint, and bandages. The seriously wounded were placed in an ambulance and taken to the regiment field hospital, a short trip behind his location. Those with an injury that did not require emergency treatment were also sent to the field hospital out of battle range.

Dr. Briggs supervised the stretcher bearers to the best of his ability. He needed to make sure they were swift and attentive in taking the wounded off the battle line, allowing healthy infantry soldiers to keep their eyes on the battle. Almost immediately, wounded men were streaming in from the front, at first slowly, but soon the numbers quickly increased.

Wounded soldiers being carried off the field, artist Alfred Rudolph [51]

U. S. Army Medical Director, Surgeon James T. Ghiselin described the scene that day:

"field hospitals were established during the day on or near Opequon Creek, and their locations were well protected from the shot of the enemy by wooded hills. They were all in the immediate vicinity of good roads. The wounded, as a general thing, received good care, and had nourishing food promptly administered to them. During the latter part of the day, however, it was impossible to collect all the wounded, as the army pushed on so rapidly, thereby increasing

---

[51] Illustration, *Wounded Being Carried Away*, artist Alfred Rudolph. Gift of J.P. Morgan, Library of Congress, # 2004660079.

the distance for the ambulances. Quite a large number, therefore, remained on the field that night, many being concealed in the thick woods where they had fallen. At 9 p.m. the same night the general commanding ordered me to have all the wounded taken to Winchester, and the field hospitals broke up as rapidly as possible. For this purpose, a detail of medical officers to remain was made from each corps, a certain proportion of ambulances was ordered to be left, and the chief quartermaster placed at my disposal all the empty army wagons. Asst. Surg. H. A. Du Bois, U. S. Army, assistant medical director, was ordered to remain and take general direction of the removal of the wounded and of the establishment and organization of hospitals in Winchester."[52]

John sent a message off to his family quickly letting them know he survived the great battle. Knowing it would be in print back home, he did not want them to worry about his well-being.

Letter #32 - September 21, 1864

Winchester

To Mrs. J.W. M. & Family,

Our battle was fought on the 19 and 20 and an awful battle it was, the victory is ours. I passed through the battle safely and am all right. I am at the hospital taking care of the wounded. Our loss is heavy, but the Reb's are three to our one. Our Regiment is all cut up, I was in the midst of the battle, it was awful such cannonading you never heard and such shelling. Our forces are following up the Reb's and are as far as Fisher's hill. We have captured an awful lot of Reb prisoners you will see in the papers. I don't feel very well from passing through such a battle and not very well when we commenced. You will see in the papers our loss of our Regt., we lost heavily. I have only a few moments to write if this I do to let you know that I am all right, so you need not worry thus far.

Yours, John W. Mambert

Trust in God for my welfare
Write soon and direct as your last.[53]

---

52 O.R., *No. 8. Report of Surg. James T. Ghiselin, U. S. Army, Medical Director*, Middle, Military Division, of operations August 27-December 31. Serial 090, p. 141.
53 *Letter #32, September 21, 1864, Winchester, To Mrs. J.W. M. & Family.* John W. Mambert letters from the Civil War.

ORGANIZING THE WOUNDED

Several weeks before the Winchester battle, Ghiselin felt a major event was stirring in the valley, so he ordered the post quartermaster at Harper's Ferry to keep 300 hospital tents on hand for such an emergency. After the engagement, he requested twenty additional medical officers be sent to Winchester along with supplies for 5,000 wounded. He also asked for an experienced surgeon to take charge of the medical facilities. Surgeon John Brinton was given urgent orders to supervise the construction of a large tent hospital at Winchester; he arrived the next day.

On September 21, 1864, ambulances transported all the wounded soldiers from the battleground to hospitals established in Winchester. Hospitals were set up in churches, public buildings, and suitable private homes, organized separately by each army corps. Because there was barely enough food for the wounded, the commanding officer in Winchester seized 8,000 rations from the army train heading to the front. Soon, over 4,000 wounded were transferred to their respective corps with competent surgeons in charge. The most experienced and expert surgeons were put in charge of operations.[54]

Twenty-two hospitals were established throughout Winchester, caring for both Union and Confederate wounded. By midday, regiments collected their wounded, transferring them from field hospitals to safer facilities within Union picket lines.

BATTLE AT FISHER'S HILL

Shortly after the Winchester engagement, General Early attempted a stand against the Union army three miles from Strasburg at Fisher's Hill. The enemy position was strongly fortified, their lines extended from the Shenandoah River on Early's right to the North Mountains on his left. The glaring heights of Fisher's Hill was a constant annoyance for Union commanders of the valley.

The Confederate's works were purely defensive. Their position was valuable only for defense, since it was as hard to get out of as it was to get in. Behind his old entrenchments, Early gathered the remnants of his army and

---

54 O.R., *No. 8. Report of Surg. James T. Ghiselin, U. S. Army, Medical Director*, Middle, Military Division, of operations August 27-December 31. Serial 090, p. 141.

began to strengthen himself with fresh earth works.[55]

Strong rifle pits were constructed by the rebels on either side while a strong earthwork was erected along the crest of the hill. This gave them full command of the road and approaches beyond. Between the Massanutten and the North Mountain peaks, the jaws of the valley were reduced to a width of about four miles.[56] This gave their position strength.

Sheridan wanted nothing less than the capture of Early's army. His plan called for General Alfred Torbert to push the Confederate cavalry through Luray, then take control of the valley pike road to the south in New Market and plant himself firmly behind Early's army—his only line of retreat. General Crook, hidden by the hills and woods, was to secure the back road and quietly approach Early's left flank and rear. The discovery of Crook's unexpected approach was planned to trigger the first sounds of battle, a signal for Wright and Emory to rapidly attack the Confederate front with everything they had.

On the morning of September 22, 1864, General Grover held the left position for the 19th Corps with his division forming two lines: Colonel Daniel Macauley with Birge, and Shunk with Molineux. Dwight would hold the right connecting to Major General Frank Wheaton. Grover moved the 128th NY and 176th NY regiments forward, driving back rebel skirmishers and occupying enemy ground, under the cover of brisk shelling from the guns of Captain Elijah Taft and Corporal Robert Bradbury. In the afternoon, Ricketts moved to the far right, seizing a wooded knoll on Fisher's Hill commanded by Confederate General Ramseur.

In preparation for the attack, Sheridan gave Emory the ground on the left of the railway, and Wright that beyond it; Molineux moved forward leading the advance of Grover. The sun was low, the noise of battle was heard to the right, seemingly far away. The noise came from Crook charging out of the forest onto the left flank and rear of the Confederate line, sending them running in retreat. The surprise was decisive with Wright and Emory now taking up the movement inspired by the presence and the commands of Sheridan. Descending rapidly on the steep and broken sides of the ravine to the bottom, then crossing Tumbling Run and charging up the rocky, almost

---

55 Irwin, Richard B. *History of the 19th Army Corps*. G. P. Putnam's Sons New York, (The Knickerbocker Press 1892), ch. 33, par. 1.

56 Tiemann, William F. *The 159th Regiment Infantry, New York State Volunteers*, In the War of the Rebellion 1862-1865, (Brooklyn New York 1891), pp. 103,104.

inaccessible sides of Fisher's Hill, the Union force swarmed over the strong entrenchments of the enemy taking line after line as they planted their colors upon the parapets and watched the whole army of Jubal Early scatter in retreat.

Leading the climb of the parapet was the 176th NY who had the good fortune of acquiring four pieces of artillery abandoned by the rebels in the panic caused by Crook's charge. At the same time, the 28th Iowa planted the colors of their regiment on the works completing the victory. The Union army now stood where the Confederates, formerly believing they had a secure position, abandoned sixteen additional pieces of artillery.[57]

Molineux described the engagement in his report:

"Finding the Third Brigade, Colonel Macauley commanding, throwing up works to protect the hill and the ravine to the right, I formed my brigade in rear and threw up a second line. I then placed two companies of the Twenty-second Iowa in the stone mill on the Shenandoah to guard against any advance in that direction and sent the 159th NY to throw up and held a rifle-pit in the orchard and on the road, to more completely guard against any attempt to outflank us on our left. The works held by the two brigades were strengthened, so as to enable the First Maine Battery to open on the enemy a very efficient fire. Our working parties and the battery were much annoyed by the enemy's sharpshooters in a line of rifle-pits directly in our front, and at 1 o'clock I was ordered to make a simultaneous attack in connection with a regiment of the First Division and one from my own line upon their rifle-pits."[58]

The 159th NY was detailed to take charge of the prisoners: over 200 in number. The regiment (with the prisoners) followed the brigade leading the advance in pursuit of the enemy force in full retreat.

The pursuit of Early's army continued until 3:30 a.m. on September 23, 1864. The advance halted just south of Woodstock, thirteen miles south of the mountainside engagement. A quantity of small arms, several pieces of artillery, horses, wagons, and an additional 800 prisoners were captured. The 159th NY was assigned to take charge of the large convoy heading back to Winchester. Delivering the prisoners on September 25, 1864, the men reversed direction in

---

57 Irwin, Richard B. *History of the 19th Army Corps*. G. P. Putnam's Sons New York, (The Knickerbocker Press 1892), ch. 33, par. 4.

58 O.R., Colonel Edward L. Molineux reports on the Fisher Hill engagement. Serial 090, p. 331.

charge of a supply chain heading back to the front. Completing their detail, they rejoined their brigade in Harrisonburg, Virginia.[59]

Leading the pursuit of the enemy, Colonel Molineux was extremely upset at the lack of support received by Sheridan, putting his men in harm's way. He shared his thoughts in a letter home:

"The fight at Fisher Hill I was under fire from 10 a.m. to 6 p.m. and commanded two brigades. We took two lines of works and four pieces of artillery—loss light and mostly from shell and shot. At 6 p.m. I was ordered to follow up the enemy all night. This was a most horrid thing and unmilitary as no cavalry was sent with us. Sheridan was at that time furiously drunk or he would not have done so. He grossly insulted me and other officers. The consequence was we fell into an ambush twice during the night and each time those in rear opened a furious fire on us who were in the front. It was a most trying horrid, horrid affair. God save me from another. Nothing but stern bravery saved us from being shamefully cut up in the night. Sheridan was ashamed of himself in the morning. He is a brave, excellent soldier but rash to madness if liquor is in him."

Further along in the letter, he wrote:

"Since writing the above, General Sheridan has publicly apologized for his insult to me, so I no longer have any hard feelings."[60]

General Sheridan's loss at Fisher's Hill was 52 killed, 457 wounded and 19 missing, a total of 528. General Early's loss in the infantry and artillery was 32 killed, 210 wounded and 995 missing (now prisoners of war).[61]

ALCOHOL AS THE COMFORTER AND DESTROYER OF MEN

On October 5, 1864, the Camden Confederate newspaper, printed an editorial with their opinion on why the Confederate Army was suddenly having major losses in the Shenandoah Valley:

---

59 Tiemann, William F. *The 159th Regiment Infantry, New York State Volunteers*, In the War of the Rebellion 1862 -1865, (Brooklyn New York 1891), p. 105.

60 James I Robertson Jr Civil War Sesquicentennial Legacy Collection at The Library of Virginia – Edward L. Molineux collection, *Letter Fisher Hill.*

61 Irwin, Richard B. *History of the 19th Army Corps*. G. P. Putnam's Sons New York, (The Knickerbocker Press 1892), ch. 33, par. 5.

CHAPTER SEVEN

"Cause of our Defeat in the Virginia Valley." "Correspondent of the Savannah Republican, writing from Richmond states the following as the reason of Early's defeat in the Virginia Valley: The Confederate arms have met with a fresh disaster in the Valley of Virginia. After his defeat at Winchester on the 19th, Early retreated up the Valley to Fisher's Hill, a strong position a short distance above Strasburg, and which, he would be able to hold. Not so, however. On the 22nd, Sheridan assaulted him in this strong position, turned his left, which soon gave way, followed by the entire line. We lost twelve pieces of artillery, though but few men. Such is Early's official report to General Lee. The Confederates were retiring up the Valley towards Staunton. Do you ask for an explanation of these rapidly occurring disasters in a portion of the State where the Confederates, until September 19th, never suffered defeat? It is simple and easily given: We have two enemies to contend with in the Valley, one of whom has never been beaten since Noah drank too much wine and lay in his tent. These enemies are the Federal army and John Barley Corp. Sheridan has been largely reinforced, and the Valley is running with apple brandy. Here is the key to our reserves. Officers of high positions-yes, of very high positions- have, to use an honest English word, been drunk- too drunk to command themselves, much less an army, a division, a brigade, or a regiment. And where officers in high command are in the habit of drinking to excess, we may be sure their pernicious example will be followed by those of lower grades. Shall I call names? Not now. The names are known to the authorities, and shall be to the country, unless there be a speedy reformation. Let us wait a little to see whether the guilty parties will not reform their habits, and especially whether the President, Secretary of War, Gen. Lee and Gen. Bragg will take hold of these men and punish them as they deserve. Just think of a drunken man in command of a body of men in battle?"[62]

This was not the first time alcohol caused issues for the military, and it wouldn't be the last. Both the Confederacy and Union suffered greatly from the flow of intoxicating beverages throughout the ranks. But more importantly, unnecessary harm to foot soldiers was caused by commanding officers who were too intoxicated to make sound decisions.

---

62 *The Camden Confederate, apple brandy*, Oct. 5, 1864. The Library of Congress, Natl. Endowment for the Humanities, Chronicling America.

HORRID SCENES FROM THE HOSPITAL

John was still in Winchester assisting the medical staff, caring for the wounded and tending to the dead. In his next letter, John describes for Junior, as vividly as he can, the horrid scenes in the hospital, how the wounded are suffering, and the many dead are literally piled up on top of each other. Still concerned that Junior might still be thinking of enlisting, John uses every tactic possible to deter him.

Letter #33 - Sunday, September 25, 1864

Winchester

To Mrs. John W. & Family,

I thought I would inform you that I am well except for the diarrhea which I have had since I left Charleston. It seems to wear on me. I have been at the hospital at this place since the first day's fight, I was some 300-400 yards from the enemy's breastworks when our Regt. made the first charge and was obliged to fall back. The roar of canon and musketry was awful and the way the shot and shell and many bullets flew was awful. If you had been where I was and seen what I have passed through millions of dollars would not be a temptation for you to enlist. But we have whipped the Reb's awfully, and we have got them on a keen run. Our men fetched in 21 canons which they captured from the Reb's and thousands of prisoners. In fact, we completely destroyed Early's Army after the first day's fight. I went back with one of our men wounded and the Doctor would not let me go back, he kept me to take care of the wounded and since then our wounded have been moved here and I have been here ever since. I think there are 150 wounded in this building. They are generally doing well, some arms and legs off. Our killed, wounded, and missing is large but some of our men who were detailed to bury the dead say they buried 3 dead Reb's to one of ours. I don't know whether they are done burying yet, it was awful. I received your letter No. 36 just as we started from Charleston on the evening of the 19th Inst. I was glad to hear that you were all well. I hope this will find you the same. Our Regt. was here last night and fetched with them between 8 and 900 prisoners and they went back again this morning to the front and left me here; the front is about 50 miles from here. I don't think I will be able to get any letters or papers here from home as long as I am in the hospital, but I can send letters to you, and you can hear from me. I can't tell how long I shall stay. My work is tiring and hard and disagreeable, dressing wounds and lifting on the chambers and off and only sleep from 4 to 6 hours out of 24. Last night was a very cold night, the wind in the north and our wounded suffered very much from cold, no windows in the building and 3 story high of brick, 8 nurses in my ward No. 2 and 75 wounded men and up night and day all lay on the floor on a little hay. Our Regt. has lost a large number, I won't give no names now the first day or 2nd day I helped load up 11 that died after they were brought to the hospital. If you could see how we threw them in, you would of fainted. Head over heels, some faces down to, some head first and feet first; all on top of one another officers and all. Well, you would be horror stricken. I forgot how many loads they drew away 4 or 5, I helped load one. I don't want John to enlist, or Eb for heaven's sake. If you don't feel satisfied just come down here and take one march and see one battle and you will have your belly full, and then go through the hospital to see the surgeons amputate arms and legs and see the wounded and you won't talk of enlisting. March long and heavy march, carry 40 to 60 lbs sleep on the ground and you won't say enlist. But I think it is right to enlist, but knowing what I now know, I would not enlist for all the money you could pile between here and New York 10 feet high if I did not know the future.

John W. Mambert [63]

---

63 *Letter #33, September 25, 1864, Winchester, To Mrs. John W. & Family.* John W. Mambert letters from the Civil War.

Chapter Eight

"I THINK I AM 10 YEARS OLDER THAN WHEN I ENLISTED, GETTING UP AND DOWN AND LIFTING THE WOUNDED MAKES ME FEEL OLD."

SHERIDAN'S FIELD HOSPITAL

The old building continued as a hospital for the 19th Corps' wounded. John was to remain at the hospital to care for the wounded. His regiment departed Winchester heading south to join Sheridan's army in pursuit of the enemy.

Surgeon John H. Brinton oversaw the construction, staffing and medical supplies needed to care for the great number of wounded men, establishing the largest temporary hospital of the Civil War. Constructed on the outskirts of Winchester, it was designated Sheridan's Field Hospital.

The following correspondence from Brinton to the Surgeon General on September 26, 1864, describes what he observed and administered:

"On my arrival here on the night of September 21st, I found matters in much confusion. Every exertion possible has been made by Assistant Surgeon General DuBois, U.S.A. in charge, of the hospitals, to systematize relief for the wounded, but the scanty supplies, the filthy condition of the town, and the number of the wounded, rendered the matter one of extreme difficulty. By the morning of the 22nd, all of the wounded were collected from the houses near the field and were brought to hospitals in or near the town, and within our picket lines. On the afternoon of the 22nd, 188 hospital tents arrived by wagons. I directed these to be pitched in an eligible situation, close to a fine spring near the town, and placed Surgeon Hayden, U.S.V., in charge. 108 tents have arrived since and are being pitched, -in all 296. 104 tents, the balance of the 400 sent, were by General Sheridan's orders pitched at Sandy Hook to increase the capacity of that hospital. The supply of medicine in hospital stores has been exceedingly scanty, only eight wagon loads having yet arrived, and these were rapidly exhausted. A train is now on the way hither. Of blankets and hospital clothing we have none. I directed Surgeon Shields, who has been appointed Purveyor of the Army, to obtain 5000 from the Quartermaster at Harpers Ferry. I also instructed him to make a requisition on the Purveyor at Baltimore for such additional as he might stand in need of. Cooking apparatus, etc., I obtained on application to the provost Marshal. Yesterday morning, the 25th instant, I sent 1200 wounded to Harpers Ferry, as the railroad to Washington was not yet opened. I last night received instructions from the Army Headquarters to send off light cases, Union and Rebel, to Martinsburg. Asst. Surgeon Ohhschloger has been assigned to duty in charge of transportation at that point. The severe cases I am instructed to place in the tents, and as fast as the latter become empty to transfer them to Martinsburg, the new army base. I telegraphed you this morning, requesting that the Quartermaster's Department may be furnished with 150 wood stoves, suitable for hospital tents. The weather is becoming cold, the town is overcrowded, and I must get the wounded into the tents as soon as possible. There probably will be 1500 men who cannot be moved for some weeks. P.S. Since writing the foregoing, a large train of medical supplies, and also 3000 blankets have arrived.[1]

The massive facility stood at Shawnee Springs, on the south end of town. Surgeon James V.Z. Blaney assumed command of the hospital on September 28, 1864, after Brinton completed its layout. The hospital quickly treated and evacuated more than 4,000 casualties from the Winchester battle, becoming a clearing and evacuation facility. Receiving wounded from engagements farther south in the valley, the United States Sanitary Commission personnel assisted patients, making sure they were clothed, fed, and processed; then transported to medical facilities in the North.[2]

The Union medical staff not only cared and treated their men, but they also cared for Confederate soldiers who were either left suffering on the battlefields, or may have been treated by their own surgeons in makeshift facilities throughout the valley.

A report from Surgeon James T. Ghiselin stated:

"At Strasburg and Woodstock a few Confederate wounded were found, destitute of all supplies and unable to bear transportation. These men were attended by their own surgeons, who were furnished with all the necessary medical and subsistence stores. We arrived at Harrisonburg on the 25th, where

---

[1] Briton, John H., *Personal Memoirs of John H. Brinton, Major and Surgeon U.S.V. 1861-1865*, (New York 1914), pp. 298-300.
[2] Ibid., p. 303.

there were several Confederate hospitals, containing 335 sick and wounded, attended by five Confederate medical officers. The surgeon in charge reported that he was in need of subsistence and a few essential medicines, all of which he was at once furnished with. One hundred and thirty-five sick and wounded were selected, who could bear transportation without injury, and sent to Winchester by a returning subsistence train. The medical officers here seemed to have some regard for hygienic principles in and about the hospitals, and their patients were probably as comfortable as they could make them with their restricted means; but at every other place, from Woodstock on, where Confederate wounded were collected by their own surgeons, the most extreme filth and positive indications of neglect were seen." [3]

On more than one occasion, John wrote of his intentions to move the entire family to Iowa. There were a few Iowa regiments with whom John and his regiment fought alongside throughout the war. John befriended men from those regiments who stirred an interest within him on what Iowa had to offer. In the future, some of John's family would move to Iowa, although John never did.

Letter #34- Monday,

September 26, 1864

> I feel better this morning. I expect to go to the front in a few days perhaps tonight or tomorrow. The news is good; such skedaddling of the Reb's you never seen before.
>
> I want you to keep this letter and put it away until I return home for there is an Iowa Regt. here I am familiar with and as soon as I get home, if ever, I do mean to go out there and try and buy you all a good farms. Make all the money you can and tell Eb to try and save all the money he can, and we will all go and see what we can do. They tell me of some good chances, here are their names etc. Elias Wolf[4], Benton Co. to go from Chicago to Clinton

---

3 O.R., *No. 8. Report of Surg. James T. Ghiselin, U. S. Army, Medical Director*, Middle, Military Division, of operations August 27-December 31. Serial 090, p. 142.

4 Elias Wolf, Private 28th Iowa Regiment Infantry: Age 30 a farmer in Cedar, Benton County Iowa. Lived with his wife and five children ranging in age from 5 to 13 years old.

on the Miss. River from there to Cedar Rapids 25 miles from there to Vinton which is the County Benton. Calvin Duke[5] Iowa City, Johnson Co., John Ladrop same as Duke. Charles Marton[6], Millersville, Marion Co. Indiana. Christian Macks. Fred Rowes[7] Claverack, Columbia Co., Stephen Brown[8], Orange County, Middletown, NY, B.A. Jack Benton Co. Iowa, Vinton. I shall get more names and you must be sure I keep these letters. Since I have seen this Country and traveled, I can see the roughness of our country. I think if the Blessed Lord will preserve my life, I think you can all have a good farm. If you had seen what I have you would be perfectly satisfied. I think before next June this rebellion will be over with. Politically I will say nothing about it, my impression is that Abe will be elected and if so, it is certain he will put down the rebellion, it is right that he should. If Mac is elected it will make a more united North, but the people ought to be united. If not, we'll have uphill work. I don't want you to think I have changed my coat. I am a Union man and fighting for it and the Rebs say that they would as soon have old Abe elected as little Mac, for they are not fighting for negroes but trying to establish a new government and to do that they say they must fight. I have talked with some of the Rebel officers, and they tell me so. But I think we will hoist them this fall, and I think this fall campaign will wind it up and by next June it will look different. If I knew how long I was going to stay here I would let you direct your letter here, at least a few, but I think you must direct as before. I feel old and stiff, and I begin to feel my age very fast. I think I am 10 years older than when I enlisted, getting up and down and lifting the wounded makes me feel old. Winchester, the place we are is quite a place and most every woman is secess and dressed in mourning for their killed friend, husbands, and brothers. However, there is some good union women here. But we have left

---

5 Calvin Duke, Private 22nd Iowa Regiment Co. I: Age 33 a farmer in Brandon, Jackson County Iowa. Lived with his wife and three children ranging in age from 5 to 13 years old.

6 Charles Marton, 4th Indiana Battery, Light Artillery: Age 20 from Crown Point, Indiana. Lived with his parents and five siblings.

7 Fred Rowe: Age 27, Hudson NY.

8 Stephen Brown, 18th NY Infantry Co. D: Age 34 from Middletown in Orange County, NY.

our mark here for the houses, barns, fences, and such are pierced with bullet holes. Oh, the horrors of war, the wounded Reb's are here in hospitals, and it is full. I was passing through the street with some hay for beds for the wounded and a little boy eight years old says to me "Hello Yank" and I replied, "Hello Johnny Reb." and some women say, "I don't like the Yankees." and I don't wonder but they have brought us to their doors and they must stand it or lay down their arms and submit to the United States Government. Today is pleasant, I must close. Continue to write and tell all our folks I am well and read this letter to the family. Tell them all to be good, Jobe and Walt, Martha and Permelia and Eb and such. I dream very often that I am at home now and sometimes I feel quite homesick, 13 months more and out, if I live.

Yours from
John W Mambert [9]

CARING FOR THE ENEMY

As efficient systems were put in place at the new tent hospital, beds became available more quickly. The wounded of the 19th Corps were transferred to more comfortable accommodations. This also reduced John's time at the makeshift hospital since he was transferred along with them to continue assisting patients there. John's experience assisting the wounded at the tent hospital was unusual; he not only cared for his comrades, but also his enemy. Confederate officers were cared for at Sheridan's hospital where John heard his adversary's perspective on the war, which likely created spirited debates. Straight from the enemy's mouth, and not the biased view of a newspaper, commander, or politician, John heard the enemy's honest thoughts: why they believed the country was at war. This opportunity gave John a more comprehensive understanding of the conflict to develop his own informed opinion. Under the care of the Union, the wounded Confederate officers were treated much better than the foot soldiers. John Brinton wrote about what he witnessed in his memoirs: "we had a large number of wounded rebel prisoners, some eleven or twelve hundred, I think. These were penned up, or confined in an open square, fronting the main street, and surrounded by a high iron

---

9 *Letter #34, Monday, September 26, 1864.* John W. Mambert letters from the Civil War.

railing, such as Washington Square, and the squares generally in Philadelphia used to have. These poor fellows were without covering or shelter of any kind, and what was far worse, we had scarcely any rations to give them. In fact, so closely did Mosby and guerillas watch all the roads and the outskirts of the town, that neither food nor provender could be brought, save under heavy escort. We had really almost nothing, and I know one or two mornings when the Commissary of Subsistence had only two or three hundred rations in store, and they were nothing among so many. Our poor prisoners were almost starved, and I have seen them struggle almost to the death for a biscuit or crust, thrown over the iron paling by their sympathetic friends." [10]

On September 30, 1864, the 159th NY regiment arrived in Harrisonburg, Virginia with the supply chain. Meeting up with the rest of the brigade, they established camp and remained there for the next six days until receiving orders to strike the tents. The brigade began a march continuing for twenty-three miles to Mount Jackson, Virginia. For the next seven days, they laid idle in camp. On the eighth day, the men were ordered into a line of battle that lasted the next two days. Shortly after the uneventful movement, the march resumed for another twenty miles until halted into camp two miles outside Woodstock.[11]

John's conversations with the enemy in his care stirred his emotions regarding politics in the country. He explains to family why it is important to support "Abe" and "give the president full authority to execute the laws of the United States over the whole United States." His adversary's attempt to justify why the country is divided does not sit well with him. John is not convinced by their explanation, but believes they rebelled, "to establish a new Government with slavery."

---

10 Briton, John H., *Personal Memoirs of John H. Brinton, Major and Surgeon U.S.V. 1861-1865*, (New York 1914), p. 301.

11 Tiemann, William F. *The 159th Regiment Infantry, New York State Volunteers*, In the War of the Rebellion 1862 -1865, (Brooklyn New York 1891), pp. 105-106.

Letter #35 - October 7, 1864

Winchester, Va.

To the Family.

I am well thanks be to God for his kind care over me. I hope when this reaches you it will find you the same. I received 2 letters from you yesterday. I was glad to hear that you were all well. I was quite homesick and saddened for I had not heard from home in quite a while and there is nothing that cheers me up so much as when I hear from home that you are all doing well. You say you can't find much news to write about seeing you write every week, but if it's nothing more than that you are all doing well, it is sufficient. But your news is of great importance to me, if you think it small it is great for me. You can gather up something of much importance to me. I have made arrangements so that I can receive my letters here now and you must write and direct your letters as usual. But we can't get our pay this time for our pay roll came back wrong so we won't be paid until I think the middle of Nov. If I am not with the Regt. I'll be mustered the last day of this month. I may have to wait until the middle of January. This is rather hard for me so I thought if you could spare a couple of dollars or even more you might send it in a letter, and it would be very welcome (green backs). I think it will come safe. I didn't receive my paper. I expect to get a furlough after we get in winter quarters. I don't want to come home until I am paid so I have some money to get me some decent clothes with it, and something to eat along the road. I am glad to know you received all the money I sent to you. Our Regt. is out to the front. I am attending to the wounded one man died last night, his leg was taken off. How long I shall be here I don't know. 5 or 6 of our drummers started to go to the Regt. with the wagon train and they were all taken prisoners by the gorillas. I refused to go, and I stayed here. They felt dissatisfied here to take care of the wounded and now they are, I suppose in Richmond and, or dead. Dad Wheeler and Solon Macy are here. I hope I may stay here until their Campaign ends which I think will be about the middle of Nov. or later. I presented some time ago to Col. Waltermire what you wrote me about, attending to our business, and he said he would see that I had a furlough, so I feel pretty sure of getting one. Although I don't put all my dependence on it, so that I won't be disappointed if I should miss. I shall press pretty hard for one. I am glad that Martha and Jobe has written to me. I must say that Martha has a very good mind to compose a letter, all the lack is in spelling, it is plain and easy to understand, Jobe doesn't seem to learn so fast to write, however, he does well, and I hope he and Walt will continue to write to me. The letters I read yesterday were 37 and 38.

Tell Permelia to write and Eb. I saw Eb's brother Bill[12] in the 128 Regt. About politics I have but few words to say and that I should note if I have a chance to and if I do, I will vote for Abraham Lincoln for several reasons. One is that the Southern Rebelled because he was elected President, and another is that they say they will fight as long as he is in. I approve the election of Abe because I feared the result, but if I am a Democrat he is the President of the United States and if the South won't submit, I say as General Jackson by the Eternal we will make them, if they had not rebelled I would do all in my power to change a vote for any good man for McClellan or any man, but for them to say we want our people of the North to say we will humbly bow and do as you want us, now, I say I never will. Abraham Lincoln is sure to be elected if the war is sure. I will say what I think about this war, if what we are fighting for is the unity, some of the people of the North say we are fighting to free the other people of the South, they say we are fighting for our negroes, but this is only ignorance. We are fighting to give the president full authority to execute the laws of the United States over the whole United States and while a part of the States has set up a president or bogus president to enforce laws upon a part of the United States of their own, to form a Confederate Government of the United

---

12 William Duntz, Private 128th Regiment Co. I: Age 22, promoted to Corporal September 1, 1863. From Gallatin, NY. He was a farm laborer who lived and worked for Robert Younghan and family. Ebenezer Duntz is his brother.

States. The Southern people are fighting to resist the laws of the Government of the U.S. It is not the negroes they are fighting for, or we either, for we never took their negro from them until the 1st Jan. 63 the 1st of Sept. 1862 President Lincoln issued a proclamation saying that if they would represent themselves in Congress they should have all their Negro's and property just as always and if not, he would issue an Emancipation Proclamation on the first of January. Their reply was all the representation they wanted, Stonewall Jackson, Beauregard, Lee and others, this gives the people of the South a choice to keep their negroes or rebel and lose them. The President, according to his word issued the Emancipation Proclamation but only for those States and part of States who were in rebellion, so you see it was only so far as the people were in rebellion against the Gov't. This was done because it was a war necessity because the negroes were used as a force in the hands of the Reb's to build fortifications, raise provisions for the Reb Army and such. We have now 500,000 Negroes aiders taken away from the Reb's that has weakened them so much and strengthened us so much. So, I think that the Southern people have emancipated their own slaves and seeing they were determined to fight the Gov't., it is the duty of the President to cripple and use every means to defeat their object as negroes is property used against us, if we can take them, do so just as soon you would take a cannon. The Southern people only wanted to establish a new Government with slavery and by it they have lost their Negroes, but they don't care for the negroes or else they would of kept them and returned to the union. They only hung the point on slavery because they thought that was the strongest point and cotton was king and that old Abe was elected President. Now I am hoping he shall stay there until the south will submit to the laws of the United States. I don't think I shall vote even if I have a chance, I might vote for J.C. Fremont, I think he has the best platform. The soldiers are for Abe Lincoln; some say they will fight three more years rather than to come down again, although some are for McClellan. I will say no more vote as you please. I write with a pencil for I can't get ink. Write every week and send my paper.

J. W. Mambert, Attend good to everything – trust in God. [13]

Letter #36 - (separate letter)

To John:

I want you to write on a separate piece of paper a letter stating that you want me home very much to attend to settling up our business which you can't get along without me, also stated the wishes from Lieut. Col. Waltermire soliciting him to grant me a furlough and perhaps it will have the desired effect.

Yours John W. Mambert

To John Mambert N.B. I also sent a letter with $1 to pay the express on the $25.[14]

WARTIME ELECTIONS

On June 8, 1864, the Republican Convention in Baltimore unanimously renominated Lincoln for president, and to "balance the ticket" Andrew Johnson of Tennessee, (a former slave supporting Democrat) for vice president.[15] When the announcement of Johnson's nomination was handed to Lincoln, he looked at the telegram for a moment, then said, "Well, I thought possibly he might be the man, perhaps he is the best man, but—" and rising from his chair, he walked out of the room. It was contended that Lincoln preferred Hannibal Hamlin and believed he should have been placed on the ticket for vice president a second time.

That election season had some interesting political movements. In Cleveland, a convention was held on May 31, 1864, where a large number of Republicans were critical of President Lincoln's management of the war. Forming a new political party, they nominated a rival candidate to run against him. Delegates representing ten states established the new radical

---

13 *Letter #35, October 7, 1864, Winchester, Va., To the Family.* John W. Mambert letters from the Civil War.

14 *Letter #36, (separate letter), To John.* John W. Mambert letters from the Civil War.

15 Plum, William R., *Military Telegraph during the Civil War in the United States.* Chicago, IL: Jansen, McClurg, 1882. p. 267-268.

party. John C. Fremont was nominated by acclamation as the party's presidential candidate with General John Cochrane of New York as vice president. The party platform was adopted, advocating the abolition of slavery by constitutional amendment, a one-term presidency, and direct election of the president by the people.[16] The new party was negligent regarding the constitutional provision against nominating both president and vice president from the same state.[17]

The thought of a radical candidate participating in the election worried Lincoln. Negotiations to persuade Fremont to withdraw began. On September 21, Fremont withdrew his name to ensure united Republican support for Lincoln's reelection. It was rumored Fremont's withdrawal from the election had a price: the removal of his old enemy, Montgomery Blair, Lincoln's Postmaster General. Ironically, the following day Blair announced his retirement.

Democrats in New York State claimed to uphold the constitution. Lecturing on the perversion of war by their opponent, they criticized violations of the government, corruption and extravagance in Washington, and the absurdity of forced emancipation. Bringing attention to the national debt, a special appeal was made to the taxpayer. The heading of an Argus[18] editorial was: "Half a Million Dollars a Week Something for Taxpayers to Consider!" This was their estimated cost to support "pauperized Negroes by the administration."

Another headline in Argus read: "Albany County Debt Taxes! Taxes! Taxes! More than half the value of the county and city absorbed. Can we afford to prosecute an 'Abolition War!'" The Democratic Party also made use of the impending draft with, "future call-ups predicted," bringing further debt and taxation necessitated by such demands. The Argus claimed, "a vote for Lincoln is a vote for more drafts."[19]

Meanwhile, the Democratic Party nominated George Brinton McClellan, a graduate of West Point and popular general who served in the

---

16 McKinney, Effie. *The Cleveland Convention* (1928).

17 Ibid., p. 269.

18 *The Argus* – semi-weekly Democratic publication, Albany NY.

19 Brummer, Sidney David, A.M. *Political History of New York State During the Period of the Civil War.* (New York 1911), pp. 425-426.

Union Army. The New York Herald favored the nomination of McClellan, commenting in its editorial:

"As for Lincoln, we do not think it possible that he can be reelected after his remarkable blunders the past three or four years."

The opinion was McClellan would win, (as he told them he would) he was sure to be elected and resign from the army on November 1st. However, the chairman of the Illinois Central Railroad Board, having dealt with both Lincoln and McClellan professionally, shared a different opinion. He spoke bluntly:

"No, Lincoln will beat McClellan, for he has the courage of his convictions and does things, but McClellan, while able and great in preparation, lacks confidence in himself at critical times. Even if elected, he would be a failure in the responsible position of President."

Advocates for peace at any price continued to spread lies and rumors about Lincoln, declaring the president was in favor of mixed-race marriages and many other political deceits, also labeling him a "widow-maker." In the North, the peace movement made some feel uneasy and dissatisfied with the handling of the war, and when events did not turn out as they expected, they blamed the administration.[20] However, most Unionists were concerned with the danger of changing the political party while at war and were quite vocal on not allowing the Confederacy an opportunity to remain in power. They declared McClellan's nomination a victory for the rebel armies. They felt that, if elected, McClellan was the only hope for Confederate victory, a tool in the hands of traitorous leaders.[21]

There was mutual distrust between Lincoln and McClellan. Lincoln was privately cynical of his past commander-in-chief, removing him from command in November 1862. This came after McClellan failed to pursue Lee's Army following the Union victory at the Battle of Antietam. Lincoln saw to it McClellan never received another field command.

---

20 Plum, William R., *Military Telegraph during the Civil War in the United States*. Chicago, IL: Jansen, McClurg, 1882. p. 270-272.

21 Brummer, Sidney David, A.M. *Political History of New York State During the Period of the Civil War*. (New York 1911), p. 427.

In a speech, Henry Ward Beecher[22] proclaimed:

"...and we have conquered half the rebel territory, hold the keys of the whole, and have nearly destroyed the military strength of the Rebellion in the field. All this in two years of war."

"Four years you mean," said a bystander.

"No," responded Mr. Beecher, "I said two years of war. In the first two, General McClellan was in command."[23]

As the campaign continued, it was common for Democratic Party speakers to threaten violence against any interference with an unrestricted ballot. One such situation occurred with (former) Judge Dean, a Democrat and past nominee for speaker of the assembly. Dean reportedly proposed, "in all the wards they should form white-boy clubs; and if any man came to the ballot boxes to prevent them casting their votes freely, let the white boys take care of him, put him where he belongs, hang him."[24]

By the end of October, revelations of alleged frauds in connection with soldiers' ballots were reported. Democrat voting agents were arrested and charged with impersonating US Army officers and soldiers, forging ballots and affidavits. One man pleaded guilty confessing to signing the names of a number of soldiers, and then accused a second person of affixing the required officer's name. The second confessed to having signed blanks with the name of "C. S. Arthur, captain and aide-de-camp," but claimed that no offense was committed since there was no officer by that name in the service of New York State. At the same time, Colonel Samuel North, New York State Agent at Washington, Major Levi Cohen and Edward Jones (two subordinate officials at the state agency) were arrested on similar charges. The office was closed, and the soldiers' ballots—ready to be deposited, were seized.

The press reported there were several dry-good boxes of forged votes for the (Democratic Party) national and state ticket nominees forwarded to New York. This was discovered after a letter was recovered addressed to General

---

22 Henry Ward Beecher (June 24, 1813 – March 8, 1887) was an American Congregationalist clergyman, social reformer, and speaker, known for his support of the abolition of slavery. During the Civil War, President Lincoln sent Beecher on a speaking tour of Europe to build support for the Union cause. Beechers speeches helped turn European popular sentiment against the rebel Confederate States of America and prevent recognition by foreign powers.

23 Ibid., p. 428. *Tribune, Oct. 25.*

24 Ibid., p. 432.

Farrell (one of Governor Seymour's staff). The letter read:

"I send you ... a number of ballots for your county. I have made out a number from the list you sent me ... I guess you have enough. Fearing that you might not, I enclose envelopes and powers of attorney sworn to; you can fill them out for Columbia or any other county." The Democratic press and stump speakers were indignant at the arrests stating it was all a "Lincoln Plot."[25]

On October 28, General Dix, commander of the Department of the East, confirmed receiving information regarding rebel agents in Canada planning to send large numbers of refugees, deserters and enemies of the government into Union states to vote at the approaching election. Dix was determined to guard the purity of the elective franchise against the threat.

The following day, Senator Morgan wrote to Stanton requesting three thousand troops be sent to New York immediately. The troops were sent, and General Butler was selected to oversee the troops and the mission. New York Governor Seymour felt Dix's order was unnecessary, so on November 2, he issued a proclamation:

"...the painful and exciting doubts in the minds of many with regard to the free and unrestricted exercise of the elective franchise, there are no well-grounded fears that the rights of the citizens of New York will be trampled upon at the polls. The power of this State is ample to protect all classes in the free exercise of their political duties."

In the end, the people of New York supported Lincoln. With help from the newly voting military, Lincoln won New York by less than 7,000 votes.[26]

BATTLE PREPARATIONS

On October 8, 1864, the men of the 159th NY began their march at 6 a.m., stopping twice to form a line of battle. They continued for nine miles that day, halting at 4 p.m. and going into camp in a ravine near Fisher's Hill.

The next day, Confederate General Thomas Rosser with a large cavalry force hovered about the rear of the Union encampment. He was being watched closely by the Union's cavalry commanded by General Alfred

---

25 Plum, William R., *Military Telegraph during the Civil War in the United States*. Chicago, IL: Jansen, McClurg, 1882. pp. 432-435.
26 Ibid., pp. 436-439.

Thomas Archimedes Torbert. Eventually the enemy cavalry got a little too close. Torbert's unit charged at them, effectively whipping the enemy. A number of rebels were killed, wounded, or taken prisoners and nearly all their artillery was captured. Two days later, the 159th NY moved their position crossing to the north side of Cedar Creek establishing a new camp and constructing earthworks.

General Early received new reinforcements and felt he could pick a fight with the Union army once again. On October 13, 1864, Early attacked in force the 8th Corps garrisoned at Hupp's Hill. The engagement was visible from the 159th NY's position. Forming a line of battle with two regiments from the brigade, they began marching to the front, strengthening the picket line. The rebels, viewing the movement, started falling back toward Fisher's Hill. (Their attempt was a reconnaissance to discover the Union army's position and force.)

Three days later at midnight, the regiment was awakened and put to work building additional breastworks. Working through the night under bitter cold conditions, the men constructed the breastworks, finishing the next day.[27]

Throwing Up Earthworks to prevent a Night Attack[28]

---

27 Tiemann, William F. *The 159th Regiment Infantry, New York State Volunteers*, In the War of the Rebellion 1862 -1865, (Brooklyn New York 1891), p106.

28 Illustration, *Throwing Up Earthworks to prevent a Night Attack*, artist Alfred R. Waud [Between 1860 and 1865] Photograph. Library of Congress, # 2004660724

BATTLE AT CEDAR CREEK

Colonel Molineux received orders on October 18, 1864, to hold the brigade in readiness; it was reported the enemy was on the move: the reporting proved accurate.[29]

At 3:30 a.m., hidden in the long shadows of the full moon, the enemy commander stood in readiness 1,200 yards above the mouth of Cedar Creek with General Joseph Kershaw behind him. They watched with restraint the sleeping ranks of Colonel Joseph Thoburn, commander of the Union's 1st Division. The rebel force was positioned directly in his front waiting for the appointed hour. At 4:30 a.m., General Early set Kershaw in motion crossing Cedar Creek unobserved and unopposed by the sleeping soldiers. Once the troops were on the north bank, Kershaw deployed his men to the right and left; the enemy now in position, stood-to-arms waiting for the signal.

General Gabriel Wharton formed a line of battle under the cover of trees on the edge of Hupp's Hill. His men crept down the slope to the front of the woods and remained in the shadows a mere 1,000 feet, also waiting patiently for the signal. To crown it all, as the dawn drew near, a light fog descended upon the river bottom and covered all the opposing troops with a veil.

Since the beginning of the 19th Corps' existence, it was their custom, when in the presence of the enemy, orders were to "stand-to-arms." At daybreak, they were. That morning, Molineux was to go on reconnaissance and by 5:30 a.m., his men had already eaten breakfast and were lying on their arms waiting for orders to march. Birge and Macauley were to be ready to follow in support after a proper interval, and Shunk was to cover the front of all three during their absence. Brigadier General James W. McMillan was also notified to support the movement of Grover's brigades.

General Emory himself was up and dressed. His staff's horses were all saddled, and while his horse was being saddled, from the left a startling sound broke the stillness of the morning air.

The sound was the roar of a tremendous volley of muskets by which Kershaw made known his presence before the sleeping camp of Thoburn. In an instant, before a single shot could be fired in return, Kershaw's men, with their loud and continuous rebel yell, swarmed over the parapet in Thoburn's

---

29 Tiemann (n1), p. 107.

front. The enemy seized his guns, sending his half-clad soldiers flying to the rear of camp. Kershaw, without artillery before the morning sunrise, suddenly found himself in possession of Thoburn's seven guns.[30]

At the first sound of battle, Colonel Molineux moved his men back into the rifle-pits they left only an hour ago. Emory ordered his corps to stand-to-arms as he rode swiftly to the left of his line toward the valley road to investigate this strange outbreak. Knowing that Molineux was near and ready, Emory drew from him two regiments, the 22nd Iowa and the 3rd Massachusetts, giving them orders to support the artillery planted left protecting the bridge. Barely had the regiments positioned when the enemy's shells began to fall among the battery, followed by a volley of gunfire directed toward the infantry lines.[31]

While all eyes were directed toward Kershaw, Confederate General John Gordon and his men remained hidden by the fog. The noise of the cannonade was not identified by the Union troops until the instant the enemy's line of battle attacked the ranks of General George Crook as they charged upon them. Still bleary-eyed after being startled from their sleep by the disordered rush of Thoburn's men running through their camp, Crook's old veteran soldiers poured a substantial well directed volley in response. Gordon and Kershaw began moving together against the exposed left, and rear of Emory's troops. At the same time, Early, who watched Kershaw launch his attack, rode back for Wharton and the recently seized guns of Thoburn, bringing them forward into position for a front attack against the Union line.

Additional Confederate troops headed for the valley road toward the sound of battle. Emory, seeing the enemy movement, sent General George Thomas across the road into a ravine and woods beyond. His orders were to stand fast at all hazards against the additional rebel soldiers. Unfortunately, time was too short. After a desperate resistance, Thomas's men were forced back by the overwhelming masses of the enemy, but not before this tried brigade left a third of its brave men on the ground attesting to their valor. The 8th Vermont, also in a desperate fight with Kershaw's men, again and again saw their colors brought down, as three flag bearers were slain in keeping the precious emblem protected.

---

30 Irwin, Richard B. *History of the 19th Army Corps*. G. P. Putnam's Sons New York, (The Knickerbocker Press 1892), ch.34, par. 7-11.

31 Ibid., ch. 34, par. 12.

It seemed there was nothing else the Union forces could do to delay the advance of Kershaw on the left and Gordon to the rear. Wharton, and the forty guns began their work on the Union front, beginning left and moving right continuously. Yet, as the Union soldiers moved toward the right and rear under that pressure, the men held well to their colors. Although there were many who fell, not a brigade nor a regiment lost its organization throughout the turmoil.

The enemy's pressure began to reach Molineux's 159th NY, along with Captain George Davis and his 10th Vermont regiment positioned on the opposite side of the entrenchments. Both brigades began moving under Emory's orders, taking position near the Belle Grove plantation, to the right of Brigadier General James Ricketts's division of the 6th Corps. Ricketts's division was extending its line diagonally in an effort to reach the valley road and cover the position. All of them were working to hold off the onward rush of Gordon's men.

Emory already posted the 114th and the 153rd NY on a commanding knoll 500 yards to the south overlooking the road. The detached regiments of Molineux and Lieutenant Colonel Neafie, came off the line joining the 159th NY with the 3rd brigade, consisting of the 156th NY and the 176th NY who just finished a hard hand-to-hand fight with the enemy as they pursued their colors. Birge withdrew along the works, opening ground beyond Meadow Brook where Shunk joined him.[32]

First Lieutenant Eben Haley's 1st Maine battery lost forty-nine horses killed in their harness; he abandoned three guns as they vacated their posts at the breastworks. Taft lost three pieces of his 5th NY battery during the difficult crossing of Meadow Brook west of Belle Grove. At the same location, and from the same cause, three guns of the 17th Indiana and two of the Rhode Island battery were abandoned.

The fierce combat continued and the losses to the infantry were now in the thousands. General Grover was slightly wounded, McCauley severely, Emory lost both his horses and for a time commanded the corps afoot, while Birge rode a mule. The enemy under Gordon's command, was on their flank and rear at every moment, getting closer to taking control of the valley road.

General Wright had to think quickly of the safety and success of the

---

32 Ibid., ch. 34, par. 13-15.

army he was commanding; he also needed to secure a plan for General George Custer and General William Powell to be able to hold off the enemy throughout the day. Time was needed, whether to bring up the troops, change front, or march to the rear past the faces of the advancing enemy. Whatever was to be done, it had to be done quickly. Wright, in predicting the enemy's move, made up his mind to retreat to a fresh position where his line of communications would be preserved, and his flanks protected.

The retreat handed Major General George Getty's division the difficult task of covering the exposed right flank of the army, while the left gradually swung in the direction of the new line to be formed. Getty, having perfectly performed the undertaking, crossed Meadow Brook and took position on the high, partly wooded ground rising beyond the brook toward the southern edge of the village cemetery. He planted his artillery while Grant held the immediate front. Under trees following the contour of the hill, Grant placed Wheaton with Colonel Joseph Keifer.[33]

General Emory needed to reach a position to the left of Getty while retreating. To accomplish this movement, he had to descend the hill from Belle Grove, cross Meadow Brook, climb the opposite slope, reverse his troops, and reform the line in good order on the crest of Red Hill. Successfully reaching his position, Emory stood at bay while Kershaw, who was following him closely, formed a line of battle facing his army. But now, seeing the fighting force of Emory's infantry, and General Wesley Merritt's cavalry in close support, Kershaw refrained from renewing the attack until Early could send Gordon to his aid.

As a result of the bold stand taking place at Red Hill, General Wright was provided the time he needed to reorganize while Kershaw waited for reinforcements.

Emory, following orders from Wright, crossed over to the cemetery and placed himself on the west of Getty while Thomas rejoined McMillan. Meanwhile, Torbert moved into position with Merritt on the left flank surrounding the cemetery. At 7:30 a.m., the unwavering strength of the Army of the Shenandoah was gathered, with every Northern eye looking once more toward the South.

The standoff continued. As the Union forces waited for the Confederates

---

33 Ibid., ch. 34, par. 16-17.

to attack, Wright strategically positioned his men on the battleground: From the south wall of the cemetery overlooking Meadow Brook on the left, he positioned a rough echelon[34] of divisions stretching to Marsh Brook, forming on the right—Keifer first, then Wheaton, Grover, and McMillan. Between Marsh Brook and the Old Forge Road on open elevated ground, Emory formed his corps in an echelon of brigades. Knowing this was where the decisive combat of the day would be fought, Emory began fortifying his front with the help of loose rails and stones.[35]

The increasing heat of the sun slowly dissolved the fog, giving Early a clear view of the Union army lying directly before him; looking so strong he decided not to have his infantry attack until after the concentrated fire of his artillery impacted their lines. As for Wright, he could see the fatal threat of Early's artillery and the weakness of his left flank broken by Meadow Brook.

The Old Forge Road, a country lane crossing the field from the north end of Middletown, was important only for its value to unite the wings of the Union army. Wright's objective in taking up this line was simply to gain time needed to enhance the fighting line developing in the far rear.

When Getty's troops reached the cemetery, Wright ordered him to slowly move in line of battle toward the north, guiding along the valley road. Merritt's cavalry, just beyond, would follow and cover Getty's operation. Emory would take up the movement in his turn and look toward Wheaton for his guide. Getty began his movement passing the left of Keifer, who followed on the left of Wheaton, and on his left Emory; all forces marching closely, shoulder to shoulder in a solid line. Keifer, now in position, formed the center line of battle, with Ball on his right and Emerson on his left. Passing over rough ground and among trees, the head of each line was guided by its right-hand neighbor. After five hours, the Union retreat reached its destination.[36]

The enemy, observing the retreat, moved forward to the crossroad beyond the cemetery and began positioning their troops. Early posted his troops behind stone walls, while Wharton extended his line on the east side of the turnpike and positioned three batteries between him and the road. Brigadier

---

34 Echelon – A formation of troops in parallel rows with the end of each row projecting further than the one in front.
35 Ibid., ch. 34, par. 18.
36 Ibid., ch. 34, par. 21-24.

General John Pegram covered the turnpike, with his left resting on Meadow Brook. Beyond Pegram were Generals Stephen Ramseur, Kershaw, and Gordon, who carried the line to the east bank of Middle Marsh Brook.

With the ranks of the Union infantry depleted by heavy battle losses of the early morning, and the loss of all the men forced from the encampment, there was no question the men standing by their colors on the Old Forge Road, planned to endure to the end. It was during this pause in the engagement many of the men (who hadn't eaten all day) munched on bread and meat from their haversacks while others made coffee, and many threw themselves in exhaustion upon the ground, falling fast asleep.

While both armies were preparing themselves for battle, farther down the road among the crowd of wandering men came a whisper of noise. The nearer it got, the louder it became until it swelled to an uproar: "Cheers!" It was the cheers of the stragglers. As the soldiers on the line turned toward the sound, they were amazed to see the lost men returning to the front.

It was soon realized they were returning because, among them riding into the field, was a small man on a black horse, and from the ranks arose a cry: "Sheridan!" Through all the ranks the message flashed as if it had been charged by an electric spark.

General Sheridan was in Winchester when he received reports of heavy firing twenty miles south, near the army corps' encampment. He thought little of it until, having finished his breakfast, he became uneasy at the continued sound of cannons. Accompanied by his staff, they mounted their horses and rode south to join his army where he left it—on the banks of Cedar Creek. Galloping at an easy trot, one mile from the battleground, Sheridan was shocked to see the tattered and disheveled column of stragglers. Every man was making his way toward the Potomac, without his arms, his equipment, or his knapsack, carrying in short nothing but what he wore after being shaken out of the ranks during Kershaw's raid on the camp of Thoburn.[37]

---

37 Ibid., ch.34, par. 25-27.

Sheridan's ride into the Battle at Cedar Creek, artist Alfred Rudolph [38]

Once Sheridan arrived on the battlefield, he lost no time in assuming command of his army. Establishing his headquarters on the hill behind Getty, he proceeded to complete the positioning of his armies already in progress. Sheridan, now content with the arrangement and appearance of his army and perceiving that his adversary Early was getting ready to attack his force, rode the length of the line of battle to show himself to his troops. A roar of cheers greeted and followed him as he passed in front of regiment after regiment; his presence emboldening confidence and assuring victory.[39]

Sheridan knew he needed to delay the counterattack long enough to allow Crook time to reform the stragglers under his command. After Crook accomplished his duties, Sheridan gave the signal for the whole Union line to go forward against the enemy. Getty began on the left as a pivot, while the whole right was to sweep and drive the enemy in front, sending them toward the valley road near the camps from the morning engagement. The movement was taken up in succession, and in a few minutes the whole line advanced steadily. From that moment until the battle's conclusion, the men did not stop for any reason.

Initially, resistance from the enemy was constant and here and there even

---

38 Illustration, *Sheridan at Cedar Creek*, artist Alfred Rudolph. Gift of J.P. Morgan, Library of Congress # 2004660854.

39 Ibid., ch. 34, par. 31.

spirited. For a few moments, the attack appeared to hang on the extreme right of the line. McMillan, pushing at a quick pace, suddenly found himself surrounded by the right wing of an opposing brigade commanded by General Clement Evans. Evans formed the extreme left of Gordon's division, and of the Confederate force. Although McMillan's troops were resisting the enemy, suddenly they were being swept away. Davis's force was directly upon the rebels' front and flank; he observed McMillian's plight and immediately changed his front to join McMillan's force. Together, the troops not only defeated Evans's attack, but also burst through Gordon's line and quickly swept the enemy brigade off the field.

Knowing this to be the critical point in Gordons's line, Sheridan ordered: "Stay where you are till you see my boy Custer over there,"

Almost like an introduction, General Custer appeared on the high ground leading his bold troops as they rushed down upon the broken wing of Gordon. At the same time, the whole right line of the Union charged, and while Custer rode down Gordon's left flank, Dwight, McMillan, and Davis began rolling up the Confederate line.[40]

Meanwhile, the left center of the Union attack trailed momentarily, as the 159th NY (positioned with their brigade, on a southerly slope of a wooded hollow) were about to be confronted by Kershaw's troops on the opposite crest. The only way Molineux's men could reach the enemy was to climb the steep side of the dirt hill. Molineux, seeing through the difficulties of the situation, organized his troops accordingly. When he felt the troops were ready, his men, along with Birge's, rose up, charging boldly together out of the hollow. Climbing the hill, facing a constant barrage of enemy fire, they crossed open ground and climbed the stone wall. Molineux overthrew Kershaw, sending a panic down Ramseur's line, spreading chaos through every part of Early's force; the enemy was now scattering in retreat.

"Back to your camps!" had been the motto since Sheridan showed himself on the field, and Dwight's men were the first of the troops to stand once more upon their own ground, while the infantry took up their original position.

The cavalry pursued the retreating enemy with intensity, but it was a single missing plank on a little bridge near Strasburg that made all the difference: purely by happenstance, that one open spot made it impossible

---

40 Ibid., ch. 34, par. 35-36.

for Early's artillery train to get the cannons across the bridge before falling into the hands of the Union army.⁴¹

Chasing the Rebs through Strasburg, artist Alfred Rudolph ⁴²

Confiscated from the battle were forty-eight cannons, fifty-two caissons, all the ambulances that had been lost in the morning, many more wagons and seven enemy battle flags. From every part of the abandoned battlefield, large stacks of rifles were gathered. According to reports from Sheridan's officers, the prisoner count was 1,200 men. Without stopping and with scarcely a show of resistance, the remnants of Early's army retreated back to Fisher's Hill. Merritt and Custer's cavalry kept in strong pursuit of the enemy well into the night. Their captures became so numerous in men and material that help was needed to supervise the custody.

Early reported his loss in killed and wounded without distinction, to be 1,860 men. He also reported capturing 1,429 Union prisoners from the early morning hours of the day.

---

41 Ibid., ch.34, par.38.

42 Illustration, *Chasing the Rebs through Strasburg*, artist Alfred Rudolph. Gift of J.P. Morgan, Library of Congress # 200466192.

General Ramseur was mortally wounded in the last stand by his division. He died a few days later at Belle Grove while under the care of his former West Point alumni who opposed him on the battlefield that day.

Sheridan's reported loss was 644 killed, 3,430 wounded and 1,591 captured or missing: in all 5,665.

6th Corps had 298 killed, 1,628 wounded—combined total: 1,926

19th Corps 257 killed, 1,336 wounded—combined total: 1,593

Crook lost 60 killed, 342 wounded—combined total: 402

Cavalry 29 killed, 224 wounded—combined total: 253

Missing: Wright 194, Emory 776, Crook 548, Torbert 43

The greatest proportionate loss of the day was suffered by the 114th New York with twenty-one killed and eighty-six wounded including seventeen mortally and eight missing, in all 115 men out of the 250 engaged that day.[43]

The total number lost in the battle for the 159th NY were two officers and two men killed, one officer and twelve men wounded (one mortally) and five men missing, a total of twenty-two.

Killed
Major Robert McD. Hart, A.O.O & I.,2nd Division

COMPANY A, LOSSES, WOUNDED AND MISSING

Killed
Private Lewis Morrin

Wounded
Private John S. Lown

Missing
Private Joseph Clearwater

COMPANY B, LOSSES, WOUNDED AND MISSING

Wounded
Private Michael Murray

Missing
Private Thomas Cavanagh
Private John Dailey (died in Salisbury, NC)
Private Gottlieb Schnepf (died in Salisbury, NC)

---

43 Ibid., ch. 34, par. 39.

COMPANY C, LOSSES, WOUNDED AND MISSING
Wounded
1st Lieutenant Ranson Barzillai
Private William Tator

COMPANY D, LOSSES, WOUNDED AND MISSING
Wounded
Private John J. Brown, mortally
Private Michael Furness
Private George H. Miller

COMPANY E, LOSSES, WOUNDED AND MISSING
Killed
Private James T. Perkins

Wounded
Private George Proper

COMPANY G, LOSSES, WOUNDED AND MISSING:
Wounded
Private Elliott Rowlinson

COMPANY H, LOSSES, WOUNDED AND MISSING
Wounded
Private William Herbert

COMPANY I, LOSSES, WOUNDED AND MISSING
Wounded
Private John Devlin
Private Almon Edson

Missing
Private James Morton (died in Salisbury, NC)

COMPANY K, LOSSES, WOUNDED AND MISSING
Killed
Captain Duncan Richmond

Wounded
James S. Leonard [44]

---

44 Tiemann, William F. *The 159th Regiment Infantry, New York State Volunteers*, In the War of the Rebellion 1862 -1865, (Brooklyn New York 1891), pp. 110-112.

Among the killed and mortally wounded were the following officers: Bidwell, Thoburn, Kitching. General Lowell, despite being severely wounded in the morning leading his brigade, held fast propped against a stone wall until his deathblow came in the last decisive charge. General Grover received a second severe wound early in the final charge when Birge took command of his division.[45]

That evening, the 159th NY moved down the valley to Strasburg along with the rest of the brigade. The men wished they had their blankets as the night was piercing cold.[46] The next day, the regiment marched back to their old camp at Cedar Creek.

The evening following the battle, Molineux found himself in a precarious position of his own doing, almost buried alive:

"After fighting hard for fourteen hours, during which retreating and advancing ensued, we found ourselves at nightfall sweeping over the old campground from the morning, following Jubal Early in his retreat down the valley of the Shenandoah. Gladly we left the task of gathering up the fruits of victory, to the cavalry division and returned to our original camp. Of course all such things as blankets. overcoats, shoes and shelter had disappeared during the fight; but some kind hands had deposited the body of the Major in one of the packing boxes, having first removed one end to allow for the length of his body. The night air was keen and sharp and as we bivouacked on the frozen ground; even the fatigue and exertions of the day could not overcome the effects of the cold. I threw myself down on the lid of the box containing the body of the Major, to get some sleep. As the night progressed, I could not help thinking that the inside the box would be more useful to me for an hour or two, than the brave unfortunate man within. Finally the impulse to change places became so irresistible that in a minute more I had accomplished the feat by carefully placing the body on the ground, on the cover. Then placing the box sideways I crawled in, where I was sheltered on three sides and fell asleep.

Some time later I was awakened suddenly by the box being tipped over so that I found myself on my face in a jammed condition. Then there was a hoist and I, in my wooden overcoat, was being carried on the shoulders of men, as carefully

---

45 Irwin, Richard B. *History of the 19th Army Corps*. G. P. Putnam's Sons New York, (The Knickerbocker Press 1892), ch. 34, par. 39.

46 Tiemann, William F. *The 159th Regiment Infantry, New York State Volunteers*, In the War of the Rebellion 1862 -1865, (Brooklyn New York 1891), p. 112.

as circumstances and the frozen ground would permit; to the place of burial. The wind shrieking and moaned and he howled aloud. There was an instant cry of dismay from the men, an awful thump on mother earth, splintered the wood and a most dilapidated Brigadier sprawling on the ground. Explanations were in order, made and received; then back I marched with my former pallbearers to point out the rightful owner of the packing case."[47]

The harsh weather in the valley was making life for the men of the 159th NY miserable. The rain and intense cold continued for days. Everything was wet, including the wood, making it impossible for the men to dry out their wet clothes and get warm. Uncomfortable as they were enduring the severe northern weather, how many do you suppose wished to be back in the southern lowlands with its poisonous swamps, filthy water, overpowering heat, and intolerable mosquitoes?

As for John, he was in Winchester, warm and dry, assisting the wounded in the ward; the cold and rain outside likely had no effect on his comfort as it did his comrades.

On October 24, 1864, congratulations from President Lincoln were read to the troops.

Executive Mansion
Washington, October 22, 1864

Major-General Sheridan

With great pleasure I tender to you and your brave army the thanks of the Nation, and my own personal admiration and gratitude, for the month's operations in the Shenandoah Valley; and especially for the splendid work of October 19, 1864.

Your obedient servant,
Abraham Lincoln [48]

---

47 From the collection of E. L. Molineux, *extracts from his diary*, NYS Military Museum and Veterans Research Center.

48 Tiemann, William F. *The 159th Regiment Infantry, New York State Volunteers*, In the War of the Rebellion 1862 -1865, (Brooklyn New York 1891), p. 112.

Sheridan's army during the 1864 Shenandoah Campaign[49]

The presidential election is just two weeks away. This letter (if received on time) will be John's last opportunity before the vote to share his thoughts on the current situation. He writes to his family that Lincoln needs to finish the job, or the South will persist independently, and the conflict will worsen.

Letter #37 - October 24, 1864

Today we are to be mustered for pay and if we don't then we won't be paid until November. If I can get a pretty good number of the boys to leave here with me, I think I will come home very soon, but I don't want to come home until I have been paid, so I have some money, and if we don't get mustered today then we won't be until the last day of October. Write and let me know where Jobe and Walt goes to school and write how the cow gets along and if you have got enough hay for her this winter, and how the sap run on and how the bees get along, if you have any, and how Mig gets along. You must get enough cabbage for winter, and potatoes and fish and onions, get a good supply, and get plenty of apples and pancake flour. As

---

49 Illustration, *Sheridan's army following Early up the Valley of the Shenandoah*, artist Alfred R. Waud. [between August and 1865 March] Photograph. Library of Congress #2004660195.

soon as this war goes down everything will go down too. I don't think it will make any difference who is President as it regards the war, for if the southern government for their independence, then no president will do different from what Lincoln has done and the war will be still more desperate. Give my love to all.

John W. Mambert
Remember me.[50]

NEW YORK AND THE PRESIDENTIAL ELECTION

On November 8, 1864, the 159th NY received orders to march. The regiment struck their tents at 4 a.m. and proceeded to lie in a cold rain all day, waiting. Eventually the men were instructed to put their tents back up as the rain continued throughout the night.[51]

This was also the day of the presidential election. Secretary of War Edwin Stanton agreed with Lincoln's principle that a democratic society fighting for its existence should include all legal voters who wished to participate in the election. Stanton implemented an absentee voting system regulated by each state; it allowed soldiers deployed to the front lines the opportunity to vote in their local, state, and federal elections. Twenty-five states changed their laws allowing soldiers to vote while serving, either at a field station at their military encampment, or by mail. The other nine states required the men to be present at their local polling place.

As many states passed measures to give the ballot to those in military service, the Unionists of New York wanted their soldiers to have that opportunity. But this issue was made more complex due to varying interpretations of state mandates.

UNIONISTS 'FOR'

The state constitution provided that "no person should be deemed to have lost residence by the reason of his absence in the service of the United States." This clause, the Unionists felt, was the means to allow soldiers the opportunity

---

50 *Letter #37, October 24, 1864*. John W. Mambert letters from the Civil War.
51 Tiemann, William F. *The 159th Regiment Infantry, New York State Volunteers,* In the War of the Rebellion 1862 -1865, (Brooklyn New York 1891), p. 113.

to vote. The members sought to avoid constitutional objections to a soldier's voting bill by authorizing volunteers in the army and navy to vote by proxy. In brief, the bill provided that a legal voter in the service might empower any voter who is a landowner of his election district, the right to cast a ballot for him by transmitting to such citizen a written certificate, duly attested by a witness, and acknowledged before a commanding officer (including suitable provisions to guard against dishonesty).

NEW YORK DEMOCRATS 'AGAINST'

New York's Democrats opposed such a measure as unconstitutional and liable to "fraudulent abuses." Democrats perceived that while they might safely fight the bill, it would be politically unwise to resist giving a ballot to the soldiers. When the senate bill came up in the assembly, there was a resolution offered to amend the constitution to allow those in the federal military service to vote. This amendment would require favorable action by two successive legislatures and ratification by the people. This would not give New York soldiers the ability to vote until 1864, and then only if a special election should be held to ascertain the will of the people on the proposition. Without submission, the soldiers could not vote until 1865, missing out on the important Presidential election in 1864.[52]

Unionists wanted soldiers to have the opportunity to vote. They likened the Democrats' declared constitutional scruples as nothing more than a political cloak behind which to hide their desire to deprive soldiers their absentee vote. While the bill was pending in the Assembly, Democratic Governor Seymour sent a message to both houses protesting it. In his communication, the Governor stated that absentee voting by soldiers "would be worse than a mockery" if soldiers were given the freedom to choose a candidate without being properly informed. There were uncertainties about news on the home front reaching soldiers away at war in a timely manner or at all.

"The conduct and policy of high officials have caused great distrust in relation to the freedom from restraint and coercion which should be accorded to the absentees in the exercise of this right [i.e. of voting]...It would be worse than a mockery to allow those secluded in camps or upon ships to vote,

---

52 Brummer, Sidney David, A.M. *Political History of New York State During the Period of the Civil War.* (New York 1911), pp. 279-280.

if they are not permitted to receive letters and papers from their friends, or if they have not the same freedom in reading public journals, accorded to their brethren at home, to aid them in formation of their opinions in respect to the conduct of those in power, the issue to be decided at the election, and the character of the opposing candidates…"

Despite strenuous Democratic opposition, the measure passed in the assembly with every Union member and the speaker voting aye, and every Democrat voting nay. Seymour promptly sent in a veto message, declaring the bill not only unconstitutional but also defective in that it afforded opportunities for wholesale frauds. The Unionists accepted the Democrats' proposal for the passage of concurrent resolutions amending the constitution. The Senate promptly passed the bill over Seymour's veto with the Assembly following suit. [53]

Republicans regained control of the Assembly in the 1863 election, and the legislature issuing soldier voting was taken up promptly. There was no opposition to the second passage of the amendment, making it through both houses without a dissenting vote. A bill providing for a special election to be held March 8 was passed, and the amendment was submitted as a referendum to be voted on by the people of New York. It passed unanimously.[54]

President Lincoln subsequently wrote to General Ulysses S. Grant:

"New York votes to give votes to soldiers. Tell the soldiers."[55]

McClellan was expected to win the popular vote among the Army of the Potomac. His platform to negotiate an end to the war and Lincoln's dedication to the Confederacy's defeat, the vote within the army could mean a choice between life or death.

JOHN YEARNS FOR HOME

The war continued on. For John, caring for the constant flow of wounded, dying and dead soldiers affected his attitude about the country. Conversations with enemy officers, people about town and comrades were beginning to wear on him. John's yearning to be home with family was stronger than ever—to be home, away from the unrest of the nation.

---

53 Ibid., p. 280-282.

54 Ibid., p. 360.

55 Putnam, George Haven, *Abraham Lincoln: the Peoples Leader in the Struggle for National Existence,* (The Knickerbocker Press 1909). p. 153.

Letter #38 - November 8, 1864

Winchester, Va

To the Family

I'm well and enjoying very good health, as good as I have since I have been in the Army, for which I am thankful to God, trusting in him that I may be blessed with the same during my stay in the Service. I have not received a letter since No 40, I suppose they go to the Regt. Our Quartermaster has gone to the front and so I don't get them. You better write and direct 1 or 2 or 'til I inform you different in this way (John W. Mambert, W.J. Hospital Winchester, Ward C. Va.-there is where I am) but that is not right. Direct in this way, John W. Mambert, in care of E.H. Smith, Agent U.S. Sanitary Commission, Winchester Va., that is all. And as long as I am here, I will get them, but I may not stay here long so you must be careful what you write for I may not get the letters. Don't write many in that way for I expect soon to go in winter quarters and where I don't know. It has been reported in this place that the 19th AC was to report to New Orleans again, but I think it was nothing but a camp rumor. As soon as I can, I shall get a furlough and come home. I have been detailed as Brigade carpenter for the hospital by the doctors. I don't like it much for I have steady work, yesterday I made 2 coffins. I have made a good many for this place. 7 died last Saturday, all wounded soldiers, and we lose more or less every day. I have a good shop to work in and in the warm and dry. If it wasn't for coming home, I would try to stay all winter here. I was paid 2 month's pay for July and August, so there is 2 month's pay due me and by the first of Jan. there will be 4 months due me that will be $64. I don't expect to be paid again until January for I was not mustered on the last day of Oct. I sent you $25 and $1 in a letter to pay the expressing, which will be 50 cents. I think I shall lose my fife coat and overcoat and blanket; they were boxed up with the company boxes and left a week or 10 days ago to the Regt. and I not being there may lose them all for which I feel sorry. I have a blouse and the elbows are all out. I bought a vest off a man yesterday second hand for $1 which will answer well for the winter. Today is the election in which I take no interest for which I feel better than many do who are quarreling every day about it. I shall not vote at all. Last knight all Winchester was in an uproar. They expected the Johnnie's, and soldiers were marching in all directions but this morning it is all quiet. I feel lonesome and wish thousands of times I was at home, but only to see all my family. Some friends, I feel as if I never could make that country my home again. I have lost all affection for it. It looks like a rough country full of hard labor and little gains. About our business affairs, see if Sylvester Bartle can pay that mortgage, and if it is impossible for him to do so; see if it makes no difference to J.W. Elseffer[56] and if not, let it be another year until I am home for I don't want any land there at the present prices and see that A. Decker keeps up his interest on the Elseffer mortgage; if not, you must try to pay it or see that it is paid as soon as this war stops everything will be down to nothing except taxes. If Ves Bartle can pay them, let Elseffer assign that mortgage to me and have the assignment recorded. You must not call the mortgage paid and satisfied; it is only transferred to me to hold against the place. You must be careful how you do business. Get Esq. Hauver to assist you and I will pay him for it if necessary. If that won't work let A. Decker pay up the $13.50 mortgage and take grandpop Mambert's mortgage of $1500, the interest and principal will amount to $1500 by the first of April next do the best you can. Maybe I will be there to attend it. Write immediately.

Yours, John W. Mambert [57]

ELECTION RESULTS

It is assumed that John did not participate in the election. He would have needed to send his affidavit home prior to election day, and it does not appear

---

56 Jacob W. Elsaffer: Age 38, a sawyer (timber man). Lived in upper Red Hook on the Dutchess and Columbia County line.

57 *Letter #38, November 8, 1864, Winchester, Va., To the Family.* John W. Mambert letters from the Civil War.

that happened. Truth be told, only about 25 percent of Union soldiers were given the opportunity to vote.

In states where absentee votes were not accepted, soldiers needed furloughs to return home and cast their vote. (Good luck getting that furlough.)

When election day came, votes for McClellan did not materialize, and Lincoln won seventy-eight percent of the military's vote and fifty-five percent of the popular vote. The former general won three states: Kentucky, Delaware, and New Jersey. The electoral college tally was 212-21 in Lincoln's favor. Lincoln also won areas of the Confederacy occupied by the Union Army, where citizen voting was allowed.

Lincoln's second-term win was propelled by recent Union victories on the battlefield. The Union achieved important military triumphs, capturing Atlanta and Mobile, dealing a major defeat to General Early's army in the Shenandoah Valley. These military victories boosted morale among civilian and military voters. Soldiers in particular seemed to agree with Lincoln's campaign slogan:

"Don't change horses in the middle of a stream."

Two days after the election, President Lincoln gave the following speech on the necessity of a national election:

> It has long been a grave question whether any government, not too strong for the liberties of its people, can be strong enough to maintain its own existence, in great emergencies.
>
> On this point the present rebellion brought our republic to a severe test; and a presidential election occurring in regular course during the rebellion added not a little to the strain. If the loyal people, united, were put to the utmost of their strength by the rebellion, must they not fail when divided, and partially paralyzed, by a political war among themselves? But the election was a necessity.
>
> We can not have free government without elections; and if the rebellion could force us to forego, or postpone a national election, it might fairly claim to have already conquered and ruined us. The strife of the election is but human-nature practically applied to the facts of the case. What has occurred in this case, must ever recur in similar cases. Human-nature will not change. In any future great national trial, compared with the men of this, we shall have as weak,

and as strong; as silly and as wise; as bad and good. Let us, therefore, study the incidents of this, as philosophy to learn wisdom from, and none of them as wrongs to be revenged.

But the election, along with its incidental, and undesirable strife, has done good too. It has demonstrated that a people's government can sustain a national election, in the midst of a great civil war. Until now it has not been known to the world that this was a possibility. It shows also how sound, and how strong we still are. It shows that, even among candidates of the same party, he who is most devoted to the Union, and most opposed to treason, can receive most of the people's votes. It shows also, to the extent yet known, that we have more men now, than we had when the war began. Gold is good in its place; but living, brave, patriotic men, are better than gold.

But the rebellion continues; and now that the election is over, may not all, having a common interest, re-unite in a common effort, to save our common country? For my own part I have striven, and shall strive to avoid placing any obstacle in the way. So long as I have been here I have not willingly planted a thorn in any man's bosom.

While I am deeply sensible to the high complement of a re-election; and duly grateful, as I trust, to Almighty God for having directed my countrymen to a right conclusion, as I think, for their own good, it adds nothing to my satisfaction that any other man may be disappointed or pained by the result.

May I ask those who have not differed with me, to join with me, in this same spirit towards those who have? And now, let me close by asking three hearty cheers for our brave soldiers and seamen and their gallant and skilful commanders.[58]

---

58 Collected Works of Abraham Lincoln. Volume 8. Lincoln, Abraham, 1809-1865. Concerning this speech and Lincoln's response to a serenade on November 8 (*supra*), John Hay's *Diary* records under date of November 11: "The speeches of the President at the two last serenades are very highly spoken of. The first I wrote after the fact to prevent the 'loyal Pennsylvanians' getting a swing at it themselves. The second one, last night, the President himself wrote late in the evening and read it from the window. 'Not very graceful,' he said, 'but I am growing old enough not to care much for the manner of doing things.'"

SOLDIERS BURIED FAR FROM HOME

When a soldier lost his life from disease or battle, family members back home feared that their dead loved ones would be buried so far away, they would never be able to visit the gravesite. For many families, the inability of holding a traditional final viewing of the body and burial in a family plot or local graveyard kept them from experiencing closure.

However, if the family of the deceased had the financial means, they could hire someone to locate, retrieve, and bring the body home. If the loved one died in one of the larger hospitals, for example New Orleans, a metallic casket could be purchased for $100 and the body could be shipped home. Or, a family member could retrieve the body, just as Corporal Kline's father recovered his son's remains. (He traveled nearly 1,500 miles deep into the South, mostly through enemy territory, miraculously escaping unnoticed.)

Families financially or physically unable to bring home their loved ones agonized, not knowing where or how their family members were interred. Many men who died on the battlefields were buried where they fell.

After the 159th NY made their gallant charge at Port Hudson, a burial party waved a flag of truce to locate their dead comrades. They had fallen close together, not far from the enemy works. The fallen were interred there, along with the piece of color-staff left on the field.[59]

Shortly after the war broke out and mortality numbers continued to rise, the government realized it needed accurate records of the dead. The War Department issued G.O. #75 in September 1861, making commanders of Union forces responsible for burying their dead, keeping records of where the men were interred, recording the soldiers last moments, and the location of interment.[60] In reality, under certain circumstances this order was impossible.

After the surrender at Appomattox Court House and the war's slow movement toward a conclusion, government agents began a thorough search on all battlefields and hospital burial grounds. They carefully located soldier graves, copying the writings etched into rough wooden grave markers.

---

59 Tiemann, William F. *The 159th Regiment Infantry, New York State Volunteers*, In the War of the Rebellion 1862 -1865, (Brooklyn New York 1891), p. 37.

60 Moss, L. (1868) *Annals of the United States Christian commission*. Philadelphia, J. B. Lippincott & co. [Pdf] Retrieved from the Library of Congress, https://www.loc.gov/item/02018786/. p. 375.

Initially the government repaired old grave markers (wooden headboards) as they rotted, making sure the grave was protected.[61] Eventually wooden headboards were replaced with granite stones for all dead soldiers (due to the long-term cost of replacing wooden markers). In 1867, the U.S. Quartermaster General established national cemeteries, and all known graves were relocated to further protect the soldiers' graves.

THANKSGIVING AND WINTER QUARTERS

On November 9, 1864, the 159th NY struck tents at 5 a.m. and began marching five miles to a new encampment situated on a hill two miles from Newtown, Virginia. Two days after arriving, the enemy attacked their cavalry. Hearing the heavy firing in the distance toward the battlefront, the regiment immediately began construction on breastworks. During the severe skirmish, the cavalry drove the enemy into a hasty retreat, ending the engagement. The regiment completed construction on the breastworks, and gave their new fortress a few names, "Fort Waltermire", "Bears' Retreat," or "A Rough Hug for the Rebs."

During the next eleven days the weather remained as it had been, rainy and cold. The men were miserable. On Thanksgiving Day, they were relieved of their duties, except those absolutely necessary. Thanks to family and friends back home, the 159th NY was provided with a bountiful feast. The regiment received turkeys, chicken pie and many other luxuries. General Emory ordered an issue of whiskey for all the troops in the 19th Corps that Thanksgiving Day.[62]

The men were elated. It had been a long time since they thoroughly enjoyed a full meal. If only the families back home could have seen the joy on the faces of their loved ones so far away that day.[63]

On November 29, 1864, the 159th NY received orders to go into winter quarters. They moved to Camp Russell, named for General David Allen Russell who was killed by a shell fragment during the battle in Winchester. The regiment was positioned

---

61 Ibid., p. 439.

62 G.O. *Emory ordered an issue of whiskey,* Regimental Company Books for the 159th NYSV, National Archives, Washington, DC.

63 Duffy, Edward B., *History of the 159th Regiment, NYSV* New York 1890, p. 38.

to the left of Front Royal Road near Strasburg and Carisbrooke Redoubt.[64] Their camp was on high ground anchoring the east end of the two-mile-long earthworks. The 159th NY guarded the southside of Camp Russell where it straddled the Valley Pike. They began constructing log huts as their housing for the winter. The men were supplied with plenty of wood to keep them warm, dry and as comfortable as possible. The winter was severe; harsh temperatures froze streams solid, and several inches of snow covered the ground often after frequent snowstorms.[65]

The following letter home was from an officer at Camp Russell on November 23, 1864:

"Colder than any huckleberry pudding I know of! Whew, how it blew and friz last night! I took my clothes off in Christian style last night. No enemy near for a week and more makes this the correct thing. It got windy, flue disgusted smoked, let the fire go out, then grew cold; put on pants, coat, and vest, in bed. Cold again, put on overcoat and in bed again. Colder than ever, built up the fire, [it] smoked. So I wanted to be cold, and soon was. Tent-pins worked loose from the wind flapping the fly; fixed them after much trouble; to bed again, and wished I was with my wife in a house of some sort! Today the men were to have had overcoats, stockings, shirts, etc., which they greatly need, but behold, we learn that the clothing couldn't come because all the transportation was required to haul up the turkeys and Thanksgiving dinner! We must wait until next train, eight days! And we all laugh and are very jolly in spite of it. 8 P. M.--The clothing has come after all. The turkeys are issued at the rate of a pound to a man. Very funny times we are having! When the weather is bad as it was yesterday, every-body, almost everybody, feels cross and gloomy. Our thin linen tents--about like a fish seine, the deep mud, the irregular mails, the never-to-be-seen paymasters, and "the rest of man-kind," are growled about in "old-soldier" style. But a fine day like today has turned out brightens and cheers us all. We people in camp are merely big children, wayward and changeable. Believe me, dearest, your ever-loving husband, R. (Rutherford B. Hayes)[66]

---

64 Redoubt: a temporary or supplementary fortification, typically square or polygonal and without flanking defenses.

65 Tiemann, William F. *The 159th Regiment Infantry, New York State Volunteers*, In the War of the Rebellion 1862 -1865, (Brooklyn New York 1891), p. 113.

66 Rutherford B. Hayes Presidential Library & Museums. *Rutherford B. Hayes Vol. 2 Chap. XXV In Garrison End of war*, p. 540.

CHAPTER EIGHT

PRESIDENTS FROM THE VALLEY

When the 159th NY (attached to the 19th Corps) was ordered to Shenandoah Valley, there were two additional corps accompanying them: the 6th Corps (Army of the Potomac) and the 8th Corps (Army of West Virginia); neither were newcomers to the valley. Part of the 8th Corps was the 23rd Ohio Infantry Regiment, commanded by Lieutenant Colonel Rutherford B. Hayes. Hayes was an attorney; he joined the volunteer regiment in June of 1861. The Governor of Ohio appointed several of the officer positions for the regiment with Hayes promoted to Major. The 23rd Ohio set out for western Virginia in July 1861 as a part of the Kanawha Division; he was then promoted to Lieutenant Colonel.

At the Battle of South Mountain on September 14, 1862, Hayes led a charge against the enemy position and was shot. One of his men tied a handkerchief on his arm above the wound, trying to stop the bleeding. Wounded, Hayes continued to lead his men in battle. The next day, in a letter to his mother, Hayes wrote:

"I was wounded in the battle yesterday. A musket-ball passed through the center of the left arm just above the elbow. The arm is of course rendered useless and will be so for some weeks."[67]

In 1864, the army command structure in West Virginia was reorganized and Hayes's division was assigned to George Crook's Army of West Virginia. Hayes and his brigade entered the Shenandoah Valley with the Army of West Virginia attached to Major General David Hunter's Army of the Shenandoah.

Retreating to Maryland, the Army of West Virginia was reorganized with Major General Philip Sheridan replacing Hunter. In August of 1864, as Early was retreating up the valley with Sheridan in pursuit, Hayes's troops fended off a Confederate assault at Berryville, Virginia. They were also engaged at Winchester, Fisher's Hill and Cedar Creek, fighting along with the 159th NY.

During the battle at Cedar Creek, Hayes was injured again (for the third time). Writing to his wife two days later, he explained:

"As usual with me I had some narrow escapes. While galloping rapidly, my fine large black horse was killed instantly, tumbling heels over head and

---

67 Rutherford B. Hayes diary entry: *Middletown, Fredrick County, Maryland, September 15, 1862.* Volume II [1861 – 1865], p. 353.

dashing me on the ground violently. Strange to say I was only a little bruised and was able to keep the saddle all day. I lost all my horse trappings, saddle, etc., including my small pistol.) I was also hit fairly in the head by a ball which had lost its force in getting (I suppose) through somebody else! It gave me only a slight shock."[68]

Hayes's leadership and bravery drew the attention of his superiors. General Grant later wrote of Hayes:

"his conduct on the field was marked by conspicuous gallantry as well as the display of qualities of a higher order than that of mere personal daring."[69]

Hayes was promoted to Brigadier-General in October 1864. Before the war ended, Hayes was promoted to Brevet[70] Major General.

Three of our nation's presidents have in common the fact that they were all superb leaders during the great battles of the Shenandoah Valley in 1864: Ulysses S. Grant, William McKinley, and Rutherford B. Hayes. Ulysses S. Grant became the 18th president in 1868, Rutherford B. Hayes succeeded Grant in 1877 becoming the 19th president, and William McKinley the 25th president in 1897. Hayes and McKinley both served with the 23rd Ohio regiment and both had their horses shot from under them.

The brave soldiers who fought under the command of those notable men likely never expected they were fighting alongside future leaders of this great nation.

REUNITED

With the battles in the valley winding down and the number of wounded in Sheridan's hospital decreasing, the number of caregivers and carpenters was reduced. The largest surge of patients at Sheridan's field hospital came after the battle at Cedar Creek; the hospital evacuated 3,400 men between October 23 and 31. Patient counts reduced at a rapid rate thereafter and they transferred the last patient on December 28, 1864.

John's appointment came to an end, and he was relieved of his hospital duties, returning to his regiment in November. (The next letter John wrote

---

68 Rutherford B. Hayes diary entry: *Camp at Cedar Creek near Strasburg, Virginia, Oct. 21, 1864.* Volume II [1861 – 1865], p. 527.
69 McFeely, Williams S., *Ulysses S. Grant, an Album.* (New York 2003), p. 564.
70 Brevet: (pronounced *brehv-it*) An honorary promotion in rank, usually for merit. Officers did not usually function at or receive pay for their brevet rank.

to his family back home is dated December 1, 1864. It's worth noting that a letter from John is missing, dated between November 8 and December 1. John writes this next letter after returning to his regiment.)

Letter #39 - December 1, 1864

Camp Russell

To my Family

I am well and enjoying good health. Things are at camp as they were when I last wrote with the exception that we have had orders to build winter quarters which we'll have to commence today. I received two letters from you today dated Nov 16 posted 23 or 28, I can't tell if the other one dated Nov 27th requesting me to come home to settle up our business. I must tell you that I am not in a good situation to come home before Jan. I have no coat, nor a blouse nor pants fit to come with nor no money. I applied for clothes, and they could not get them. We draw clothes on the first of every month and I missed, so I have to wait until the 1st of January before I can get them. I want an overcoat and at the end of Dec. we muster again for pay and perhaps we won't be paid until the middle of the month or later. So, I don't think I can come home before that. I have not shown Col. Waltermire your letter yet, but he told me a few days ago that he would see that I got a furlough so I will be home sometime this winter. About you or John coming down here, I don't think he can get to the Regt. but I will see and write in a letter and let you know as soon as I can. It is quite dangerous to get here for Mosby's Gorillas; however, I will let you know. I am glad to hear that mother is provided for the winter. She must have pancakes when I come home, but my mustache is so big I don't think she will want to eat with me I must say to Permelia that I received her money and was happy and all of the rest for which I was thankful, but at the time I had been paid and wasn't very much in need of it. She must not worry so much about me nor dream so much for I am doing well and soon will be home. Tell Jobe and Walt that I was very happy to receive their letters and to hear that they are well and catching rabbits and that Mig is so smart, but don't go on the creek on the ice and fall in and get under the ice and drown, be careful about that never go to the

trains, to break up the means of conveying intelligence and thus isolating an army from its base, as well as its different corps from each other, to confuse plans by capturing dispatches, are the objects of partisan warfare."[75]

While Sheridan's army was in northern Virginia, Mosby's Rangers were under the command of General Early. They repeatedly struck at Sheridan's invading force. Mosby's raids on the Union Army supply lines were so effective that Sheridan stated:

"Mosby has annoyed me considerably."[76]

Mosby's raids forced Sheridan to divert a large number of his troops to guard supply lines.

In September 1864, it was learned Mosby was in the vicinity of the Thirteenth New York Cavalry. Five men were selected for the duty of capturing him and his companions. Crossing paths, Mosby was recognized by the Union cavalrymen. They were within a few yards of each other, and all fired at the same time. One ball shattered the handle of Mosby 's pistol and another entered his groin. Severely injured, he managed to escape on his horse.[77]

On Friday, September 23, during Mosby's absence, a group of his men engaged with a unit from Sheridan's cavalry. During the brief battle, a Union soldier was killed and riddled with bullets from the rangers, who rode over him in their flight. The cavalry captured seven of Mosby's men, executing them. They were killed because Sheridan's men believed one of their cavalrymen was murdered while trying to surrender. Three were shot, and two were hung to a tree in sight of the town of Front Royal. A paper pinned on the breast of one read:

" Such is the fate of all of Mosby's gang."

Mosby decided he had to take firm action and put a halt to such executions.[78]

On Sunday, November 6, 1864, Captain A. E. Richards returned from the valley, bringing in fourteen prisoners. These, along with prisoners brought in by others, made up a total of twenty-seven men. It was decided that seven of them should be taken and executed, in retaliation for the six men of Mosby's

---

75 Ibid., pp. 26, 23.
76 John Singleton Mosby, Confederate Guerrilla Leader, *"Mosby's Confederacy."* Encyclopedia. com
77 Ibid., p. 233.
78 Ibid., p. 240.

command hung or shot at Front Royal.

The twenty-seven prisoners were drawn up in a single line. Twenty-seven pieces of paper, seven of which were numbered (and the remainder blanks) were put into a hat. After the hat was shaken, each prisoner was required to draw their fate. Numbered pieces meant death by hanging, blanks were a trip to Richmond or Libby prison. Among the prisoners was a drummer boy who drew a number. (In October, drummer boys from the 159th NY were believed captured by Mosby's men.) Mosby said the drummer boy should not have been allowed to draw, and that there must be another drawing with a substitute prisoner for the boy, who was released. The seven unfortunate prisoners were then sent off under guard, with orders to execute them on the valley turnpike as close to General Sheridan's headquarters as possible.

Mosby ordered the following letter be delivered to Sheridan's headquarters on November 11, 1864:

"Major-General P. H. Sheridan, "the execution of my purpose of retaliation was deferred in order, as far as possible, to confine its operation to the men of Custer and Powell. Accordingly, on the 6th instant, seven of your men were by my order executed on the Valley pike, your highway of travel. Hereafter any prisoners falling into my hands will be treated with the kindness due to their condition, unless some new act of barbarity shall compel me reluctantly to adopt a line of policy repugnant to humanity."

Sheridan took Mosby's warning letter seriously. [79]

MOLINEUX'S PROMOTION

The first two weeks in December were pleasant but cold at their quarters in Strasburg. The men of the 159th NY enjoyed some down time after two-and-a-half months of non-stop movement in endless skirmishes, and battles. For John, it was also a much-needed rest after spending countless hours performing his duties in Winchester.

On December 16, 1864, all the batteries fired salutes in honor of the victory obtained by General Thomas's defeat of General Hood in Nashville. Also on that day, Colonel Molineux was promoted to Brevet Brigadier-General, with rank from October 19, 1864, for "gallant and meritorious

---

79 Ibid., pp. 289-295.

conduct at the battles of Opequon, Fisher's Hill and Cedar Creek." His regiment rejoiced with him on this well-deserved promotion.[80]

Settled-in and rested, John was physically feeling better, but lonesome for home. He received sad news that by order of the General, no furloughs would be given. On the bright side, he believed the war was close to ending after hearing good news from Savannah, Georgia.

During the early morning hours of Wednesday December 21, 1864, the Confederate army defending Savannah during the siege, snuck out of the city on a pontoon bridge leading to the Carolina shores. By 4:30 a.m., the mayor of Savannah surrendered the city to Major General William Tecumseh Sherman.

The letter below reveals that John does not always finish a letter to his family in one sitting, or even in the course of one day. In this letter, dated December 19, John writes about the surrender at Savannah, which happened on December 21.

Letter #40 - December 19, 1864

Camp Russell, Va. 159 Regt. NYS

To John Mambert, (jr)

I recd. a letter this morning No 45 stating that you were all well. I am happy to hear that you are all well, and you will be glad to hear that I am well and enjoying good health. I begin to get my natural color again that is turning from a creole to a white man. I have not received a letter in sometime and I feared some of you were sick and did not want to let me know, and I felt somewhat lonesome, but I will excuse you under your business. I feel happy to think you are capable of filling my place. I hope you will make a carpenter of Eb. We have built our winter quarters, my tent

---

80 Tiemann, William F. *The 159th Regiment Infantry, New York State Volunteers*, In the War of the Rebellion 1862 -1865, (Brooklyn New York 1891), pp. 113-114.

mates are Greecy Hamlin or rather Augustus Hamlin[81] and Wm. Nichols[82] and Lewis Hart[83], all from Hudson, drummers. (But they are the Devil's Angels.) I have got my things of last Spring and in addition, or in all 3 woolen blankets, 1 rubber blanket, overcoate, my dresscoate and such, 1 shelter tent, portfolio, knapsack, haversack, canteen, housewife and such, housewife is a thing to carry pins, thread and needles and such, I don't want mother to think I have another wife to keep house. So, you see if I move, I shall have a load large enough for a jackass to draw. You talk of coming to the army for making money, there is great chances for making money, a man can make money as fast as he is a mind to in sutler business you can sell things as fast as 3 men can hand them out. I have frequently had to wait ½ hour to get a chance and then he was out, could not be waited on and at high prices at that. (But) there is so much risk to him of losing all that it is not safe. If the Rebs pounce on you it is all is gone, another is they will rob and steal. I have seen them steal whole cheeses, whole firkin of butter when the men stood almost on the firkin in mid day watching them. I seen them steal whole boxes of beer bottles, whole boxes of plug tobacco. They will steal and you will stand right by and won't see it. Another thing you want 3 or 4 in partnership and there ought to be more, you want a wagon and horses, tent, and such, but if guerilla comes when you are fetching your goods all is gone. Then on pay day there is such a rush that it takes 20 men to deal out and that lasts about 2 weeks then the money begins to get scarce with them, but it is impossible to keep enough to supply them but there is a great chance for making money in various ways. We have no sutler with our Regt. because they stole him naked, he trusted some of ½ or 2/3 of them and they did not pay him, so if anyone wants to be

---

81 Fredrick Augustus Hamlin, Private 159th Regiment Co. E: Age 17 at enlistment (actual age 15), a drummer. His older brother James Hamlin, Private 159th Co. E: Age19, appears also to have lied on age (actually 16), a drummer. They lived with their parents in Hudson along with five siblings.

82 William Nichols, Private 159th Regiment Co. C: Age 17, a drummer. From Hudson, NY. Lived with his parents and five siblings.

83 Louis S. Hart, Private 159th Regiment Co. C: Age 16, a drummer. Hudson NY. Lived with his parents and five siblings.

a sutler you have go to get permission from the Col. of the Regt. and he would want perhaps $500 or $1,000 for the privilege or a share, so you see all these obstacles are in the way. It will cost you $30 or more to come here to see, I would like to have you come but you better wait for I think I shall be home about the middle of Jan. You may then begin to look for me although I might not until 1st of February, I can't be home by the first of January. Gen. Sheridan issued orders that he could not issue furloughs until after the 1st of January for fear Old Early might pounce on us, so after that I expect one. Col. Waltermire said I should have one. You have heard all the good news of Gen. Thomas Sherman; the Rebellion has seen its best days we now begin to see the end. All is quiet here except for the firing of 200 guns in honor of the good news which was some fun. I can't write more – you must write, I mean all of you as often as you can.

My love to you all.
J.W. Mambert, Co. B 1st R. N.Y.[84]

DRUMMER BOYS

The "Devil's Angels" John refers to, are a group of boys from the regiment's drum corps. It appears they were returned to the regiment, possibly after facing death at the hands of Mosby and his men. This young group was inclined to stay safely within the regiment's camp instead of venturing outside the breastworks.

The minimum legal enlistment age to join the Union army was eighteen for soldiers and sixteen for musicians, although younger men could enlist as musicians with parental permission. When the Columbia/Dutchess 167th was enlisting men for the regiment, a number of the drummers enlisted from Hudson: James Hamlin and his brother Augustus, William Nichols, and Louis S. Hart, all of legal age to serve. Once the boys were mustered in, their position as musicians came with a number of responsibilities, not just drumming. If needed, they would become mounted couriers or runners,

---

84 *Letter #40, December 19, 1864, Camp Russell, Va. 159 Regt. NYS, To John Mambert, (jr).* John W. Mambert letters from the Civil War.

hospital attendants, guards, orderlies, chaplain assistants, water carriers or even barbers.

One can only imagine what was going on in the minds of these four juveniles who were in the midst of a horrid adventure. Youth was likely the reason for many poor decisions made during this group's military musical career. John made a few references to this quartet in past writings, so it is understood why the title he gave them was one deserved.

THE SUTLER

Prior to the breakout of the Civil War, the number of U.S. army soldiers (regulars)[85] was under 20,000 men, by the end of 1861 there were over 700,000. The war department was unprepared to supply all these soldiers with their needs, sometimes even the most basic to sustain life. To fill the need, U.S. Army established governmental sutlers appointed by the secretary of war. Each post and regiment were allowed one sutler selected by the post or regimental commanders.[86]

"Sutlers" were men who stepped in and filled the shortage by providing soldiers with provisions either the army didn't have enough of or did not provide for every soldier. They were civilian merchants following the army throughout their campaigns, selling provisions to the men in the field, camps, or quarters. Selling wares from the back of a wagon or a temporary tent, they provided the soldiers with things they were not issued but needed, such as paper and envelopes for writing home, and beer. Pies, cakes, cheeses, and other novelties provided a valuable food supplement and relief from the monotonous army rations, and their molasses cakes or cookies, sold at the rate of six for a quarter, made a pleasant and not too expensive dessert when hardtack got to be a burden.[87]

Because the sutlers were civilians, they were not subject to army discipline, but they were subject to some regulations and implemented rules to govern them and the cost of their merchandise. But regulations didn't stop them from overcharging their customers. Soldiers loathed them because they sold goods at exorbitant prices; they were viewed as thieves taking advantage of a

---

85 Regulars: Full-time active professional component of the United States Army.
86 Billings, John D., *Hardtack and Coffee*. Published 1888. p. 224.
87 Ibid., p. 226.

soldier's suffering for easy money. A soldier's thirteen dollars a month would not go far patronizing a sutler. When the paymaster came round, hundreds of soldiers signed away their entire pay settling accounts with the sutler.

On the contrary, the sutler got the short end of the stick dealing with the 159th NY. That merchant was either financially forced out of business due to outstanding credit, or he got wise, cut his losses, and found a new regiment to follow.

John D. Billings, author of <u>Hard Tack and Coffee</u> wrote about prices and merchandise offered by the regiment sutler:

"butter, warranted to be rancid, one dollar a pound, cheese fifty cents a pound, condensed milk seventy-five cents a can, navy tobacco of the blackest sort one dollar and a quarter a plug."[88]

In November of 1864, Mosby's Men scouting in Fairfax, lay in the woods watching the road between Vienna and the Fairfax Court House. Stopping a sutler's wagon, the rangers emptied his goods in a pile—gloves, calico, buttons, cakes, crackers and canned goods. Trailing behind the sutler was a wagon carrying a supply of milk, also seized (going well with the sutlers cakes and pies). Cans of oysters and turkey were broken open with stones.

As the men feasted, the sutler looked on and said:

"Now you've taken everything, what are you going to do with me? You are not going to take me to Richmond?"

"Yes," said one of the rangers, "you'll have to go."

His face appeared as if in pain and he limped around, saying,

" Oh, I'm sick. I know I'll never live to get there. Gentlemen, I'm really sick."

Once the sutler realized the men were amusing themselves at his expense (they did not think him of sufficient importance to take to Richmond) his health miraculously recovered.

Before Leaving, each man put on a pair of new buckskin gloves from the sutler's stock and happily shook hands with him. As the sutler shook their hands, feeling his merchandise slip away from him, his eyes slowly scanned the scattered remains of his stock.; He stood speechless. As they trotted off, he could hear one singing:

---

88 Ibid., p. 225.

"When I can shoot my rifle clear, At Yankees on the roads, I'll bid farewell to rags and tags and live on Sutlers' loads."[89]

A Sutlers Tent Near Headquarters, artist Arthur Lumley [90]

THE ROUGH RIDE OUT OF THE VALLEY

Christmas Day, December 25, 1864, was celebrated by the men of the 159th NY regiment with great enjoyment, for the men received boxes from home. The following day, a salute was fired in honor of the capture of Savannah by General Sherman.

Four days later, the brigade received marching orders to be ready to depart the next morning.[91] The following day, the regiment started marching from their comfortable quarters at Camp Russell to Stephen's Depot (at the southern terminus of the railroad from Harpers Ferry). Once the men arrived, they established camp. The next morning, the regiment was awakened by a

---

89 Williamson, James J. [1834-1915], Mosby's Rangers, A Record of the Operations of the Forty-Third Battalion Virginia Cavalry. The Polhemus Press, New York 1895. pp. 106-107.

90 Illustration, *A Sutlers Tent Near HQ*, artist Arthur Lumley. Gift of J.P. Morgan. Library of Congress #2004661305.

91 Tiemann, William F. *The 159th Regiment Infantry, New York State Volunteers*, In the War of the Rebellion 1862 -1865, (Brooklyn New York 1891), p. 114.

frigid cold, one they had not experienced since arriving in the Shenandoah Valley. There was no wood for fires. Freezing in the cold, the men waited all day for the teams pulling the supply wagons (loaded with boards from their winter quarters) to make fires, and floors for their tents.

The brigade remained camped at the depot until January 5, 1865, when they received orders to be ready to move. The next morning at 6 a.m., under pouring rain, they struck their tents and readied their belongings. Loaded onto rail cars, they departed from the depot at 9 a.m., heading north to Harpers Ferry. The men were packed tightly in dirty boxcars. Some in the brigade were not so fortunate, traveling on open platform cars until their arrival at Harpers Ferry.[92]

THE LOSSES OF 1864

The year 1864 proved to be another time of tremendous loss for families of the men serving in the 159th NY. The regiment lost twenty-two officers and soldiers killed in battles, an additional seventeen succumbed to sickness, one from alcohol. Seven soldiers suffered, then perished in a North Carolina Confederate prison camp.

The regiment's losses in officers and soldiers who were discharged totaled fifty-one, with twenty-nine having disabilities. Additionally, twenty-four men hightailed it north for home, deserting their comrades.

Total losses were 121 men, bringing the regiment numbers to 518 active soldiers—entering the year 1865 at half their original strength.

---

92 Ibid., p. 115.

Chapter Nine

"CAPE HATTERAS IS ONE OF THE STORMIEST PLACES ON ALL THE OCEANS VERY SELDOM DO YOU PASS WITHOUT A STORM AND THE WATER ALWAYS PILED UP IN MOUNTAINS AND THE VESSEL STANDING FIRST ON ONE END AND THEN UPON THE OTHER"

POLITICAL CHANGE IN NEW YORK

The end of 1864 brought change to New York politics, as Democratic Governor Horatio Seymour lost his bid for a second term after being in office for two years. The Democrats, who claimed to be the genuine war party, won the election of 1862, with Governor Seymour appearing the patriot. Although he tried to make the impression of being a loyal Union man, Seymour appeared to be the opposite, since his sympathies leaned toward the Copperheads. (To say you support the war, but at the same time condemn it by excessively criticizing the administration and its measures for subduing the enemy was pulling in contrary directions.)

The climax came with the draft riot in New York City. Democratic leaders, Seymour included, along with the press, evoked a fierce resistance to the act for enrolling and drafting. But the conclusion took a different direction from what Seymour and the Copperheads intended. In the controversy over the conscription, Seymour was on the wrong side, while Lincoln's tact and firmness gave renewed proof of his statesmanship. The Unionists (Republicans) won back the state, fortifying their hold by giving the ballot to the soldier. Reuben Eaton Fenton, one of the original organizers of the republican party, became the state's twenty-second governor, serving the next two terms.[1]

---

[1] Brummer, Sidney David, A.M. *Political History of New York State During the Period of the Civil War.* (New York 1911), pp. 445-446.

THE SLOW BOAT TO SAVANNAH

The 159th NY arrived in Baltimore, Maryland, on January 7, 1865, at 2 a.m., where they quartered in barracks at Camp Carroll Hill, one mile from the city. At the barracks, men suffered great discomfort because the accommodations were overcrowded and excessively dirty. The regiment remained there until January 13, 1865, when at 2 p.m., they eagerly departed and marched through Baltimore to the base of Fell Street. Their mascot, Bruin, attracted great attention at the head of the regiment as they marched to their destination; their bear was constantly surrounded by an admiring crowd. When the men arrived at the wharf, they embarked on the transport ship *Servo Nada*, and began their journey south toward the Chesapeake Bay arriving in Hampton Roads. The transport dropped anchor near Fortress Monroe on January 14, 1865. The regiment waited for their orders, and once received, they were to proceed to Savannah, Georgia, and report to General Sherman. General Grover and his staff, along with the 128th NY and 22nd Iowa were also on the steamer. The next morning, the transport hauled in along the pier taking on provisions before departing for their destination. At sundown, the transport left the pier. After a 700 mile trip from Baltimore, they found themselves anchored off Tybee lighthouse at the mouth of the Savannah River.[2]

They lay anchored for quite some time while trying to find a pilot capable of taking the vessel (with a nineteen-foot draft) up the river. Unable to find a capable pilot, the decision was made to steam north to Hilton Head to continue the search; again with no success. Heading south again, they arrived in coastal waters south of Tybee and dropped anchor. Molineux sent Lieutenant Handy ashore to locate a pilot qualified for the mission; finally, success.

Two steamers—the *Fountain* and *Sylph*, arrived alongside their transport, taking all but 100 men left to guard the commissary stores. On January 20, 1865, under a pouring rain, the *Servo Nada* proceeded up the Wilmington River to where it joins the Savannah River. The remaining men and stores were transferred to the steamer *George Leary*. The three vessels proceeded together an additional fifteen miles while avoiding obstructions littering the river. Making their way toward the city of Savannah, they landed at 3 p.m. Upon their arrival, General Grover took command from General Sherman

---

2 Tiemann, William F. *The 159th Regiment Infantry, New York State Volunteers*, In the War of the Rebellion 1862 -1865, (Brooklyn New York 1891), p. 116.

(the post and defenses) informing the 159th NY they would be doing garrison duty. The men marched to the Georgia Central Railroad Depot where they were assigned comfortable quarters.[3] The regiment's bear weighed-in at 206 pounds on a large scale in the depot. He grew very large and strong, his temper not quite as good natured as it had been in his youth.[4]

John's return to Dixie reminded him of the unjust treatment that commoners, as well as slaves, received at the hands of a powerful few. In John's eyes, it was clear what the aristocracy had created throughout the South.

In his next letter to his family, John expresses his belief that there needs to be a dramatic restructuring of the Confederacy to raise up the poor, ignorant people of the South from the miserable life created by a wealthy minority: "I think I can see and judge for myself if I am honest and willing to own the truth. It is sufficient to say that the Southern States need revolutionizing very much."

Letter #41 - January 24, 1865

Savannah

To: Mrs. John W. & Family:

I must say that I feel quite well with the exception of a cold and cough. You will see by this letter that I am in the State of Georgia at the City of Savannah and that I have had another sea voyage again of 3 days and 3 nights, tossed to and fro upon a stormy sea and I was so seasick I thought I could hardly overcome it. Cape Hatteras is one of the stormiest places on all the oceans very seldom do you pass without a storm and the water always piled up in mountains and the vessel standing first on one end and then upon the other and that makes me puke and shit very soon, but I am all right now. I received your letter No 47 as I was going aboard the Steamer Survo Nada at Baltimore, and I was glad to hear that you was all well. I hope and trust that this may find you the same. I read a letter from Permelia and Jobe and Walt

---

3 From the collection of E. L. Molineux, extracts from his diary, *assigned garrison duty*. NYS Military Museum & Veteran Research Center.

4 Tiemann, William F. *The 159th Regiment Infantry, New York State Volunteers,* In the War of the Rebellion 1862 -1865, (Brooklyn New York 1891), p. 117.

containing a part for Wm. Duntz which I gave him. Savannah is a very curious place; it appears to lie right in a swamp and finally the whole country is swamp and marsh. Savannah is one of the dirtiest places I have ever seen, everything is dreary and desolate. The people have made themselves miserable. You have no idea of the difference between a slave country and a free country. If you could but see the evil of the institution you would think it could not be possible that people would make themselves so miserable, but they, the common class, are very ignorant, and the aristocracy keeps them so. I have traveled a good deal since I have been in the army by sea and by land and I think I can see and judge for myself if I am honest and willing to own the truth. It is sufficient to say that the Southern States need revolutionizing very much. The weather here is rainy and damp. We arrived here yesterday; we went aboard the ship on the evening of the 13th and started the next morning. It is quite warm here and the trees are all green and the rose bushes begin to look as if they were going to bloom in a short time, the birds are all here, etc. As for a furlough I can't tell whether I can get it now or not seeing there has been such a great change moving from the Shenandoah Valley to Baltimore, MD and from there here so far away again and those who got furloughs at the Valley are up today. They started away about 10 o'clock Jan. 5th and before 4 o'clock p.m., same day, we had marching orders. Sometimes I think I might almost as well stay 9 months more and then have my own furlough, for it will cost me $20.00 for me to visit home and 9 months is but a short time in the army, however, I will see if I can get one and I will be home. We have not been paid yet. I hope we may be paid in a short time. We expect to stay here perhaps the rest of our time. Gen. Grover is in command of this place, and we may stay here and do provo duty. I don't think this war can last long. I think it is about done I suppose you are looking for the draft, I hope you can raise your quota without the draft but try and shun the draft if you have to go west or somewhere else, don't never allow yourself to come in the army for you stand 100 chances to die or be killed where you stand, but one chance to go clear. I mean John and Eb, money is no object and there is no honor about it. You are deprived of every right or privilege, you have no right what so ever, it is all given over to others who have shoulder straps, and you must now do just as they say, those who have been picked up out of the gutters, rum shops, thieves and cut-throats. So, get around it, if there is any such thing. Let me know how much mother is in debt for her groceries, etc. and all and try to settle up everything and if A. Decker pays his interest to J.W. Elsaffer let it be and remain if he can't do anything more and if Ves Bortle can't get enough to pay that mortgage, try to let it be until I come home, but tell them all will have to be settled up when I come home, if ever I do, and that I am going off to Iowa or some other place and it will all have to be paid, either next fall or in the spring without fail and so they will have good notice of my intentions. Try and save the interest cleared this year, and put it at interest with good security, and if they will, Ves or A.D. I will give good notes that is all that is necessary.

Now I want to say that A. Decker must pay that $200 note in the spring, and you must take $100 of interest and pay Mag, take up those 2 notes and let granpap send it to her. I want this attended to and done. I want you to let Ves Bortle pay all he can on the mortgage, that is if he can. If he can pay it all then take up the $700 mortgage of J. W. Elsaffer and there will be $150 left which you must pay out at interest, times are good for the farmers and now is the time to get the things straightened up. You must try and save $100 besides $100 of interest. I will send you some more money soon. On the last day of every other month we are mustered for pay on the last day of Dec., Feb., April, June, Aug. Oct., so you know just when we are mustered and about 2 weeks after the paymaster comes along and pays us, and then you can judge how long it takes to reach you, so you know just what we do on the last days of every month. I see a bill is presented to the Congress to increase the pay of soldiers 50 per cent, this will make quite a difference. I want you to take care of my two little boys and see to them. Tell Martha that I have to laugh at her letter about Ebeneazer. I have not gotten Martha's miniature and don't expect to get it. It must be in the 128th Regt. Tell Martha I get enough to eat, such as it is, but I live principally on bread and butter and coffee. Tell her to write and tell Permelia to write and

Jobe and Walt. Let John write at least once a week and I would like to see your handwriting and a few lines from you and send me your Miniature and all the rest and ask my father if he would send me his, tell him not to put his arm so far through his coat sleeve and brush up his hair, and Uncle Al Mambert, to send his too with Johns. You wrote that James Sheldon[5] was drafted the second time and Daniels[6] also drafted. I am sorry but I hope John escapes the draft for then there would be nobody to see and care for the family except you. He must be clear on account of his eye. Harm is clear on account of his physical disability. How is it with Rams? I hope he may go clear. I see the increase in your neighborhood has been great and mostly all soldiers good luck to them. I think you had better not buy a farm yet. Things are high and there will be a fall in things, that is as certain as there is a rise better wait until I come back and then I will see. Content yourself only settle up Mag's affairs as I said. You said that people are afraid of greenbacks, take all you can get. Don't be scared of greenbacks. There is no money down here, they only have postage money. It is quite a site to see a silver piece here, there is none, no pennies, no 3 cts except postage stamps, and postage stamps are as scarce as can be. Dr. Dedrick told me he had never seen Obadiah Miller[7] in this country, he said he had not seen him in a number of years and told me how he and Obe used to go to school. Obe lived with Charley Miller[8], old Aunt Lang's husband. So, you may believe Obe Miller or Dedrick just as you please. You say the war will not last long 8 or 10 months, that is long enough. I think it will be worn out before that time, if any of you get drafted, don't come down here for if you do you will stand 10 chances to die, where you stand one to live. I always said I would

---

5 James Sheldon: Age 31. Moved to Manchester, Wisconsin. Married with one son. No record of draft or serving during the war.
6 Ashbel Daniels: Age 46, had a draft record, but not listed in any N.Y.S regiments. A farmer in Claverack, NY. Lives with his wife and one daughter.
7 Obadiah Miller, Private 159th Regiment Co. C: Age 38, deserted at Staten Island N.Y. in November 1862. A farm laborer in Claverack, NY. Married with nine children ranging in age from 4 to 16 years old.
8 Charley Miller: Age 63, farmer in Claverack, NY. Married. (Parents to Obadiah).

come back I think to get, I have good faith and I know whom to trust. I will write as often as I receive letters to answer if I don't receive letters, I get lonesome. I have no company to suit me, but I make the best of it. The time will soon pass, and I will be home once more, if I have a chance I shall come by the way of Chicago.

To my family J.W. Mambert [9]

CAPE HATTERAS

Throughout the war, navigating the treacherous waters of Cape Hatteras was dreaded by soldiers on both sides of the battle. Unpredictable currents and rough waves caused havoc for many captains, terrifying soldiers facing the high seas, possibly for the first time. John, along with most of the men, never experienced sailing open seas before becoming a soldier. Sailing the waters of Hatteras was a harrowing experience they would never forget.

After his voyage through Cape Hatteras, Private Andrew Sherman wrote:

Barring the usual seasickness, the first few days of the voyage to the southward were pleasant, and to most of the boys the novelty of being on the great, blue ocean was fascinating, but when off Cape Hatteras, a terrific storm burst upon us, "The vessel," I now quote the words of another "with its freight of a thousand men, refused to obey the helm, and wallowed helplessly in the tough of the sea, shivering under the mountamous waves, while flash after flash of lurid lightning revealed the terrors of the situation." Men trembled who never trembled before, men knelt in fervent prayer on the sea washed decks, who had not, perhaps, prayed since the innocent days of "now I lay me down to sleep. "and many who had been far from exemplary vowed future obedience if only the storm would abate and the imperiled vessel reach her destination in safety.[10]

---

9 *Letter #41, January 24, 1865, Savannah, To: Mrs. John W. & Family.* John W. Mambert letters from the Civil War.

10 Sherman, Andrew M. (Andrew Magoun), 1844 – 1921; *In the Lowlands of Louisiana in 1863*, pp. 9-10.

Steamship in Storm, man overboard, Naval operations 1861-1865 artist Alfred Rudolph[11]

PROVO DUTY IN SAVANNAH

On January 26, 1865 at 8 a.m., the 159th NY marched through Savannah to fortifications one mile from the city, on the west side near Ogeechee Road. General Molineux was in charge of the works and the two brigades. The provost and fatigue duty were demanding on the men, and gave little time to complete their work on the fortifications. Two days after the men settled into their new camp outside the city, there was a great fire. It was caused by a burning magazine with powder and shell exploding continuously. The 159th NY was ordered to fall-in and be ready to assist if needed, but their services were not required. The fire was contained without their aid, fortunately doing little damage to the surrounding buildings. On February 1, 1865, fresh bread was issued with the men's rations; a treat for the regiment after eating hardtack throughout their military career. Fresh bread was considered a luxury, a simple staple which prior to service, the men would have taken for granted. For the next six days, the regiment dealt with steady rainfall, feverishly adjusting tents and reworking trenches in their attempts to stay dry.[12]

---

11 Illustration, *Steam Ship in Storm, Man Overboard*, artist Alfred R. Waud [Between 1860 and 1865] Photograph. Library of Congress #2004660690.

12 Tiemann, William F. *The 159th Regiment Infantry, New York State Volunteers*, In the War of the Rebellion 1862 -1865, (Brooklyn New York 1891), p. 117.

John often states in his letters: "...tell Webb..." when giving instructions where to send his newspaper. Reading the next letter, it appears John has a personal connection when calling out Alex Webb, the publisher of The Daily Star:

"Let John tell Webb and be sure to let him know to send my papers to Savannah, G.A. instead of Washington and see that he does it, he may say yes and forget it."

Letter #42 - February 6, 1865

G. A. Savannah

I received 3 letters last night, the 1st mail since we have been here, and 4 papers. Your letter, one from mother, had no date, only Jan. 1865. One was from Albert Sheldon, one from John, they were dated, Albert Jan. 18th and the other Jan. 23rd I said from John, but it was from Martha. I must say that I have to cough a great deal at night and my throat is very sore. Otherwise, I feel good. If you send a box, put in some cough medicine or some red salts, I can't get any here. When we will be paid, I can't tell and I suppose you are in need but don't be discouraged, try to get along. When it comes it will be good. I will send home all I can. I want you to write and let me know how much you owe so I can make some calculations. I would like to have all my interest this Spring and add $50.00 if I could; if I send you $75.00 this pay, will that do you. I shall send that if they pay me 4 months pay for, I have been speculating some. You can tell A. Decker that you are in no want of that money until a year from this Spring and if he persists in paying it, you and John and someone else go to Hudson to the Farmers Bank and there let him pay you and put the money in the bank for what interest you can get for 6 months, or 3 months or so and if that would do try to add the interest which will make $1500. and see if you can't get grandpap Mambert's mortgage and then you are alright. If I can possibly come home on a furlow I will. But if A. Decker will wait one year more to accommodate you until I return, I would like it, but on the 1st of April 1866 that is a year from next April all must be paid up. If I live and all is well.

About Martha going West I don't know what to say. As for my part she can go but if she gets homesick what then if mother won't be lonesome and it is but a short time and I shall be home, and if I come home and she is gone, I know if she knows I am home she will be homesick and it won't be but a short time after I get home before I shall go out there and then if we all go together it will be better. I don't want to pay her passage twice and if she goes, she will have to stay out there until we come out there or to Iowa. About Walt's rabbit, tell that man that he must give you your rabbit or pay you for it, that you catched it in your snare and that he had no right to rob you. If he didn't give you your rabbit, I would hoist him when I come home and tell him you don't want him to bother your snares or traps either, and you must not touch his. I must say to mother, since I have been in the army I have learned to live saven and if a person doesn't want half so much room as he thinks he does when at home, if you could live with as little furniture as I do, you would do well. When I move, I carry all my bedding, dishes, clothing, and my grub for 4 days and water and in fact I could live in a log shanty with the greatest of pleasure. I have got good schooling and just fit to go west. It is warm here and nice, so warm that we don't need fire in tents and the flies are flying about like in Oct. up there. I wrote you a letter a few weeks ago. The reason why you did not get a letter from me was we were on the move, and I had not received one from you to answer. Let John tell Webb and be sure to let him know to send my papers to Savannah, G.A. instead of Washington and see that he does it, he may say yes and forget it. I have got so homely I don't like to look in a glass and what is worse than all I am turning dark again; I don't know if I shall ever be white again. If I get paid, I will send you my photograph in my full uniform if I can get it taken. You may send me a few postage stamps in a letter. Excuse me for this time. I will direct your letters to Hudson in care of E.A. Rorraback Mrs. John W. Mambert from John W. Mambert [13]

---

13 *Letter #42, February 6, 1865, G. A. Savannah.* John W. Mambert letters from the Civil War.

HYGIENE

John wrote about his "homely" appearance, although in reality, he was no different from the rest of the men in the regiment, or army for that matter. Undoubtedly, all the soldiers felt ugly after living in squalor for the past two years. They were required to bathe at least once per month, but would be lucky to find that opportunity. If the men were encamped for a longer time period, they might find a river or some other body of water nearby to take that much needed bath. The men were also required to wash their hands and face daily, often a challenge due to their constant troop movements and lack of clean water.

It was relatively easy for the men to keep their teeth clean with brush and paste on a regular basis. Toothbrushes were simple in design, sometimes even fashioned from sticks, and the paste was mixed from a powder.

When the men enlisted into the Union Army and the recruits underwent their physical, it included an examination to determine "whether he had sufficient number of teeth in good condition to masticate food properly or tear his cartridge quickly and with ease." As regulations were revised, guidelines became more specific, "total loss of all the front teeth, the eye-teeth, and first molar even if only of one jaw" was cause for rejection.[14]

Though John never mentioned lice in his letters, it was another battle the men skirmished with daily. Due to their close quarters and movements, the louse easily transferred from one soldier to another, triggering infestation throughout the army. The men worked diligently to free themselves from the pesky little wingless parasites, using a comb, picking them off themselves or a comrade, or even boiling their clothes if chance were granted. Amid the unkind circumstances of war, nothing seemed to completely rid the soldiers of them, so they dealt with it, turning an overwhelming situation into a favorite sport: louse racing.

John's next letter is upbeat for the most part, maybe because he often dreams of being home.

---

14 Bumgardner, E., *Disqualification for military service in the Civil War on account of loss of teeth.* [Dental Cosmos 1894]. p. 429.

Letter #43 - February 14, 1865

Tell Jobe to direct his letters in this way.

Mr. John W. Mambert
CO. B. 159 REGT. NY. VOLS.
Savannah, GA
2 BRIG. 2 DIV 19 A.C.

Keep this and lay it away and direct all your letters and papers and all boxes until I write to you to write differently. I think Jobe improves in writing very much. I dream very often of being home and always there come so many to see me and everyone is so glad. We lost one of our men since we have been here. I think his name was City[15]. You must all try and get along as well as you can next year. And after that I will try and get you all settled together, if you all live so you can all enjoy yourselves. I have got so used to living outdoors, and in a tent, that I think I could live with half the room I used to. I could live and travel all over the world and steal my way and cook anywhere by the side of the road. We live here pretty much now; we may stay here our time out and we may also move very soon but I think we will stay here. The war news is good, and I think we will soon hoist the Johnnie's, I am looking soon from Richmond and the siege which I don't think will be long. If you were here now you would be for making gardens. I saw the roses in bloom yesterday. Oh, how nice they were, and it is nice here, yet I don't like it a good hard winter is worth all this fine dixie. There everything tastes natural, here it doesn't. I must close by saying goodbye, write as often as you can and trust in God and he will take care of you.

Yours sincerely,
John M. Mambert [16]

---

15 Edward Citty, Private 159th Regiment Co. E: Age 28, died February 16, 1865 of disease in Savannah GA hospital. From Ghent, NY.

16 *Letter #43, February 14, 1865, 'Tell Jobe to direct his letters in this way.* John W. Mambert letters from the Civil War.

WALTERMIRE TAKES COMMAND

Molineux was detached from the 159th NY regiment and their brigade on February 21, 1865. He received orders to report with two of his officers, Handy and Wilson, to Fort Pulaski on Cockspur Island in Georgia. Molineux was placed in command of the fort—a temporary prison for 300 Confederate officers and 100 privates. There were also 250 Union soldiers garrisoned there. Molineux could not understand why he was taken from his soldiers and put in command at Fort Pulaski. It was later discovered that General Grover requested a "general" officer for the command, but the telegraph operator understood the request to be a "good" officer, turning Molineux into a jailor.[17]

Lieutenant-Colonel Waltermire took command of the regiment while Molineux was assigned to his new position. Shortly after taking command, Waltermire issued a series of orders to assure that the men and camp remained clean and orderly while under his authority.

The first order related to passes soldiers needed to leave the encampment. This was tightened up, allowing two passes per day only to men with proper authority. He then established a policing force within the camp that began at 8 a.m. each day (unless there was inclement weather).

Because of his concern for the health and well-being of the soldiers, Waltermire established rules addressing potential health problems within the camp. He ordered the policing force to dispose of all refuse matter accumulated at the mess, bring it to a pre-dug hole and cover it daily. Also, all sinks were to be covered. The officers in charge of companies were held strictly responsible to make sure that all men had clean, properly fitted clothing for their comfort.[18]

JOHN YEARNS FOR A TASTE OF HOME

Hoping that the regiment would remain in Savannah for a while, John's next letter includes a request for some of his favorite foods in a food box from home. Food boxes from home were not only a great treat for the men

---

17 From the collection of E. L. Molineux, extracts from his diary, *Molineux becomes a Jailor*. NYS Military Museum and Veteran Research Center.

18 G.O., *Waltermire establishes a policing force*. National Archives Washington DC, Regimental Company Books for the 159th NYSV.

because they might contain cookies, cakes, pies, and bread, but also staples for cooking such as butter (which was very expensive to acquire locally). Other items they received were cheese, smoked meats, fresh vegetables, and fruit. During the holidays, families sent turkeys or hams to their loved ones, also popular were canned items such as lobster, pickles and fruit.

John was yearning for a taste of home.

Letter #44

Savannah, Ga.

I think if I get a chance, I will send home by express my dress coat, scales and 2 or 3 blankets. If you can in your next letter send me a postage stamps for, I put the last one on this letter. Direct your letters the same as before (except Washington DC) put Savannah, Ga. or elsewhere, like this: John W. Mambert Co.B, 159th Reg't. N.Y. Vols., 2nd Brig. 2 Div. 19 A.C., Savannah Ga. or elsewhere and then they will come. Tell grandpap Mambert[19] that I wish he would write me a letter, he can write one very easy by writing how his new house suites him, and how much grain and hay he has to sell, and how many potatoes and apples and cider he has got, and how his stock comes on and what he has got and what his prospects are for getting along, and he can easy fill a whole sheet which will be interesting to me. I think he writes as good a hand as the great poet John Bunyon[20] did who was one of the great men. If we get settled here, then I want you all to clump together and send me a box. Cheese and butter are the two great principles. Uncle Steve Silvernail[21] promised to send me a cheese I was to let him know but we have not been settled yet and so I have not let him know. Tea and some fruitcake and a few good sound apples, nuts etc. would be a great treat. I have got, I think, an $18.00

---

19 John (Grandpap) Mambert: Age 64, married to Hannah Silvernail and father to John W. Mambert.

20 John Bunyan: November 30, 1628 – August 31, 1688, an English writer and Puritan preacher best remembered as the author of the Christian allegory, The Pilgrim's Progress.

21 Steve (Uncle) Silvernail: Age 68, farmer in Pleasant Valley, Dutchess County. Married with five children ages 10 - 20.

clothing bill due me so that will help some. I have been saving in clothes? for the last year we are allowed $42 a year but I have lost a great many clothes? Tell Grandpap Sheldon[22] that I send my love to him and Granny and hope to see them perhaps before next fall. I hope you will all give me your prayers and I know that God will hear and answer prayer. Write as often as possible.

Yours, John W. Mambert [23]

THE SOLDIERS UNIFORM

The clothing allowance for a Union soldier was forty-two dollars per year. Examples of what it cost the men to replace required clothing, becoming tattered over time was: pants $4.70, overcoat $12.00, flannel shirt $2.32, cotton drawers $1.00, cap $1.00, socks .40¢, woolen blankets $7.00, rubber blankets $4.40, shelter tents two pieces $ 9.80, shoes $2.70, blouse line $4.80, much needed haversack .95¢ or canteen .65¢.

Corporal Lawrence Van Alstyne with the 90th U.S Cavalry infantry described the whole ensemble the men received when they were first enlisted to serve:

"We are all togged out with new blue clothes, haversacks and canteens. The haversack is a sack of black enameled cloth with a flap to close it and a strap to go over the shoulder and is to carry our food in, -rations, I should say. The canteen is tin covered with gray cloth in shape it is like a ball that has been stepped on and flattened down, a cork stopper and a strap to go over the shoulder. It is for carrying water, coffee or any other drinkable. Our new clothes consist of light blue pants and a darker shade of blue for the coats, which is of sack pattern. A light blue overcoat with a cape on it, a pair of mud-colored shirts and drawers, and a cap which is mostly fourpiece. This, with a knapsack to carry out surplus outfit, and a woolen blanket to sleep on or under is our stock in trade. I don't suppose many will read this who do not know from observation how all these things look, for it seems as if all creation was here to look at them, and us." [24]

---

22 Grandpap Sheldon: Father to Job Sheldon, and grandfather to Mariah, married J.W.M..
23 *Letter #44, Savannah, Ga.* John W. Mambert letters from the Civil War.
24 Van Alstyne, Lawrence, *Diary of An Enlisted Man*, (New Haven Connecticut 1910), p13-14.

OFF TO NORTH CAROLINA

General Molineux's career as a warden was short, he was relieved of command March 7, 1865, when all the prisoners were sent North for prisoner exchanges. He returned to Savannah the next day and received marching orders for the regiment; preparations for a move were made. However, Molineux would not be going with his men; instead, he remained in Savannah. He gave Lieutenant-Colonel Waltermire orders to continue command of the 159th NY. At 9 p.m., the regiment's baggage was loaded on wagons and sent forward. At 4 a.m. the next morning, the men started marching through the city to the wharf, embarking on the side-wheel steamship *USS General Grant*. Setting sail, their destination was Hilton Head Island in South Carolina. Upon their arrival, the men disembarked from the transport and marched to the barracks at Fort Welles, Headquarters for the Department of the South. It rained steadily the entire time in Welles, thankfully the men had dry, comfortable barracks.

On March 15, 1865, the regiment received orders to move. At 6 p.m. they marched to the wharf and climbed aboard a small tug ferrying them to the transport *New York*, anchored just off the coast. Early the next morning, the steamer weighed anchor, and began its voyage. The regiment arrived in Charleston, South Carolina, passing Forts Sumter, Moultrie, Wagner, Gregg, Johnson and Castle Pinckney as they entered the harbor. The transport continued sailing northwest until dropping anchor in the Cooper River. The following day, the 28th Iowa, 52nd Pennsylvania and a detachment from the 54th NY were taken on board, all waiting for orders. The next morning, with sealed orders, and 1,600 men on board, they pulled anchor and headed back toward the harbor entrance. While at sea, the orders were opened by the Lieutenant-Colonel. The instructions were: report to Major-General Schofield in Wilmington, North Carolina.

On March 19, 1865, the transport dropped anchor offshore from Fort Fisher. A boat sailed out to meet the crowded transport with new orders: proceed to Morehead City. The following day, the troops sailed into Beaufort channel dropping anchor off Fort Macon. The packed transport sat offshore for two days. The men longed to be ashore as they stared at the wide-open coastline in front of them. The steamer *H. M. Wells* arrived alongside the transport taking the 159th NY, and the 54th NY on board. The regiments were taken to the railroad wharf in Morehead City where they disembarked and marched inland one mile setting up their encampment.[25]

The 54th NY regiment received its designation on October 15, 1861. Predominantly soldiers of German descent, the men were mostly from Brooklyn, and New York City. They were a sharpshooting regiment commanded by Eugene Kozlay. Also known as the "Hiram Barney Rifles," "Barney Black Rifles" or the "Schwarze Yaeger." Their uniforms were black and silver, hence their name. They were engaged in many battles in the northeast including Gettysburg.[26]

MOREHEAD CITY

Shortly before the war, North Carolina Governor, John Motley Morehead, decided the state needed a major commercial port town. He realized the newly developed North Carolina railroad could assist in its creation—if a railroad terminus was located there. Consequently, the terminus of the Atlantic and North Carolina railroads were constructed in 1857 with Morehead City officially incorporated in 1861. It soon became one of the largest ports in the region, mainly because the Newport River was eighteen to twenty feet deep, and a mile wide, allowing easy access to shipping.

Regrettably, just as Morehead City was starting to experience growth, the Civil War brought a halt to the development as Union forces invaded it early in the conflict. With the North occupying the city, many of its citizens fled the region for the duration of the war. Shortly after occupying the port town, Union troops ousted the Confederates from Fort Macon, taking possession. The fort, across the sound from Morehead City, was constructed in the 1820s to protect the coastal region against both pirates, and foreign invaders.

In the heart of the city, between Eighth and Tenth Streets, was the fenced encampment for the newly delivered Union soldiers. Occupying a number of town structures for barracks, the army also took over buildings needed for essential use. The home of Elijah M. Dudley, at 810 Fisher Street within the encampment, was used as the hospital. The Methodist Church on the 400 block of Bridge Street served as the regiment's bakery, which accidentally

---

25 Tiemann, William F. *The 159th Regiment Infantry, New York State Volunteers*, In the War of the Rebellion 1862-1865, (Brooklyn New York 1891), pp. 117-118.

26 *54th New York Infantry Regiment Civil War Historical Sketch*. NYS Military Museum and Veterans Research Center.

burned down during the occupation.[27] A fire also broke out in the state-founded salt works near the railroad freight wharf adjoining the government storehouse. The men of the 159th NY aided in removing the stores and extinguishing the fire. While the troops were stationed in Morehead City, the regiment was assigned to guard the commissary stores and provide details to guard the trains carrying the stores and other materials to Goldsboro and Raleigh, North Carolina.

The regiment's camp was adequate, and the seaside location helped make the stay comfortable. But on occasion, their comfort was unhinged by strong gusts of wind sweeping sand into the air, blinding the men.[28]

John received his food box from home. And with a bay full of fresh shellfish within reach, he ate like a king. Still, John dreams of home more than ever, as he explains in his next letter.

Letter with no date # 45

Sent from Morehead City NC

The weather here is somewhat gloomy and last night it rained. It is warm and pleasant. This country is all sand, white sand covered with the richest yellow pine timber you ever saw. Morehead City is 35 miles from New Bern N. C. A railroad running from here to New Bern, Beaufort is right across from Morehead City like Athens is from Hudson. The Bay is full of clams and oysters, and I am in for the clams for the oysters are small, the clams are large and the sand clams are the best I ever saw. But I have eaten so many I don't care for them. My box which I got from Uncle Steve is very nearly used up although I have a large piece of cheese and some butter yet, and over half of my tea. It has been a great blessing to me. I have a bad cough; I have to cough all night. If we should stay our full term in the Service, our time will be up when the 9 month men who work on the farm will be. I have dreamed about home and very often more so than I have before. I dreamed also that the doctor

---

27 Little, Ruth M. Phd., *The Historic Architecture of Morehead City, N.C. First Coastal Railroad Resort.* p. 11.

28 Ticmann, William F. *The 159th Regiment Infantry, New York State Volunteers,* In the War of the Rebellion 1862-1865, (Brooklyn New York 1891), p. 118.

took off my left leg above the ankle and I did not know anything about it until I looked down and saw it was gone, but I hope this may never happen. I dreamed Grandpap Mambert's barn was afire, and I was in a terrible stew and so I have been some homesick of late. My face pains me some. I begin to think now I shall soon be home. Our Regt. is pretty healthy, but there is 2 went away from here with the smallpox, or 4 I would say, and we lost a man by the name Frank Kertz[29] and I think we shall lose another man by the name of Jerome McDonnell [30] from Claverack.

Give my love to all the folks. John W. Mambert [31]

THE BASEBALL GAME

Since departing from Stephen's depot in the Shenandoah Valley three months earlier, the regiment did nothing but be transported from one location to another crammed into train cars or ships. No longer on the move and situated in their quiet seaside encampment the men needed to find activities other than searching for clams, playing card games or drinking. Boredom could easily set in followed by trouble, Waltermire needed to engage the men in activities. He and Briggs encouraged physical activities in an effort to keep the men fit and healthy. John mentioned he participated in boxing on occasion, but "only if he had to."

There were additional physical activities the men could enjoy participating in, for instance wrestling, or a favorite for the men from New York, baseball. One day, Lieutenant-Colonel Waltermire sent a challenge to Lieutenant-Colonel Lewis of the 166th NY, (organized in Orange County, NY) to play a game of baseball between picked nines of the regiments. The 166th NY won 19-17 over the 159th NY. A few days later, a rematch was played resulting in a victory for the 159th NY, 17-16.

As popular as baseball was, it was only mentioned one other time. Likely

---

29 Frank Kertz, Sergeant 159th Regiment Co. G: Age 21, died from wounds received in action at Halltown, WV. on April 28, 1865. From Hudson, NY. Left behind his parents and two younger siblings.

30 Jerome McDonald, Private 159th Regiment Co. F: Age 35, died at Fortress Monroe, V.A. May 1865. From Claverack, NY.

31 *Letter with no date # 45, Sent from Morehead City NC.* John W. Mambert letters from the Civil War.

due to the energy draining southern heat and humidity or the regiments constant movement in the Shenandoah Valley. Morehead supplied the men with the right climate and conditions to enjoy some "pick nine."

THE END IS NEAR

On April 7, 1865, the regiment received notification of General Grant's success taking Petersburg, Virginia. The news created great rejoicing among the men because they knew Richmond, the Capital of the Confederacy, would soon fall. Within the week, news was received of Lee's surrender to Grant at Appomattox Court, Virginia on April 9, 1865. No words could aptly express the deep feelings and intense enthusiasm this news brought to the troops. The men cheered until they were hoarse. The drums beat, and cannons were fired all night. It was the "dawn of peace."[32]

John could see the war nearing its end with the latest news of Lee's surrender. In a letter home, he wrote:

"I was informed that President Lincoln ordered five hundred thousand blank discharge papers to be printed."

John was hopeful that soon, he would be home.

Letter #46 - April 14, 1865

Camp 159th Regt. Morehead City, N.C.

To Mrs. J.W. and Family.

I thought I would write to you to let you know that I am well as usual, but I have been looking for a letter every mail and did not get one and I can't think what the reason is. I do really hope you will write for I am anxious to know how you have got along with our business this Spring. I received a letter from John, No. 51 came from Savannah dated Feb. 26th all other letters are back (if any there be). I received a letter from Permelia and wrote back immediately in John's letter I was informed that Mary's house was

---

32 Tiemann, William F. *The 159th Regiment Infantry, New York State Volunteers*, In the War of the Rebellion 1862 -1865, (Brooklyn New York 1891), p. 119.

burnt. I am very sorry to hear such bad news but thank God it is not worse. I don't know whether you have moved or not, but I hear some bad news from the neighborhood about Ms Hawver and Jones W. Rockefeller[33] and others and I suppose you know about it. I have heard other news and such. I suppose you have heard all about the fall of Richmond and the Capture of General Lee's Army. I think Gen. Jackson will soon be captured if not already. I think now I shall be home in a couple of months or sooner if the Blessed Lord preserves my life. The war is at an end, I was informed that President Lincoln ordered five hundred thousand blank discharge papers to be printed and I think there is one for me. We have not been paid yet nor I don't think now that we will be until we are discharged, and I think that will be soon. I suppose John lives at Jacob Nivers[34] old place near Churchtown. Permelia lives in Owens house now where you live, I don't know. I have waited, and looked for a letter until I am homesick wanting to know how you all do. Uncle Steve's box is about all used up except some tea and I must say it has made me feel good and strong and I begin to feel natural again and I expect soon now to be home again if I can call it home. I suppose there will be a great change in produce and prices. It is Sunday night, and I can't write much, the weather is warm and the punkies bite like fury. We expected to go away from here, but it doesn't look like it now. I have been informed that we have been put in the 10th Army Corp., 2 Brig., 1st Division, but I don't know for sure. Direct your letters the same as you did to John W. Mambert Co. B., 159 Regt., N.Y.V – 2 Brig, 2 Div. 19 A.C., Washington D.C. or elsewhere until I tell you to alter it. I will write at least once a week; I hope you won't have to write many months and that I may be with you. This letter is intended for you all. You must not expect that I can write you all separate letters. Write what Eb is doing and how he gets along and how John and Jobe and Walt gets along, and if they go to school. I know Permelia and Martha get along and how you get

---

33 Jonas William Rockefeller: Age 45 from Livingston, NY. Married with two children.

34 Jacob Niver: Age 48, a farmer in Livingston, NY. Married, son and daughter in law all reside at residence.

along. Tell John to write, so I must bid you good knight, hoping to see you all soon.

My love to you all.
John W. Mambert

P.S. I suppose by the time this goes in the mail I will receive a letter from you.[35]

FROM ELATION TO SORROW

On April 14, 1865, Brigadier General Molineux attended ceremonies at Fort Sumter. He wrote the following in his diary:

I have been placed in charge of the transport, Blackstone, to take a large party to Charleston, to be present at the ceremonies of restoring the old flag on Fort Sumter. The fort itself stands a splendid ruin; stronger against attack than ever before, and a monument to be admired by all military engineers. I am very glad I was able to attend the celebration, but I did not enjoy the trip so very much. So many of the officers insisted on rejoicing over the recent victory in a rather rough way that it was quite troublesome for me trying to make matters agreeable to the ladies.[36]

That evening, a singular act turned elation to sorrow for the Union cause. The North's elation at feeling the war's end in sight due to recent Union victories, came to an abrupt halt: The 16th President of the United States was assassinated.

While the president and his wife attended the play, "Our American Cousin" at Ford's Theatre in Washington, he was assassinated by well-known stage actor, John Wilkes Booth with a fatal shot to the back of his head. President Abraham Lincoln's died on April 15, 1865 at 7:22 a.m. at the Petersen House, opposite the theater. Lincoln became the first U.S. President to be assassinated.

The assassination of the president was the beginning of a scheme to

---

35 *Letter #46, April 14, 1865, Camp 159th Regt. Morehead City, N.C., To Mrs. J.W. and Family.* John W. Mambert letters from the Civil War.

36 From the collection of E. L. Molineux, *extracts from his diary*. NYS Military Museum & Veteran Research Center.

murder a number of government officials, including Vice President Andrew Johnson, Secretary of State William H. Seward, and General Ulysses S. Grant. Secretary of War Edwin M. Stanton ordered War Department agents to apprehend the conspirators. It was rumored there was involvement by top Confederate officials; however, beside Booth the actual conspirators included, Dr. Mudd who treated Booth's wound, David C. Harrold and Lewis Payne, Edward Spangler (of Ford's theater), George Atzerodt a German immigrant, Michael O'Laughlen and Samuel Arnold, both were Booth's childhood friends and Mary Surratt a boarding house operator. Eight individuals were arrested for treasonous acts, with a military commission finding them all guilty. Four were hanged, of the remaining four conspirators one died in prison in 1867, the three remaining received presidential pardons in 1869.[37]

On April 23, 1865, General Grant (on his way to General Sherman's headquarters) arrived at the Morehead City encampment, greeted by enthusiastic cheers from the troops.[38] Just six days earlier, General Sherman and the Confederate general, Joe Johnston, met at Bennett's farmstead in Durham, North Carolina, signing the surrender papers for the Confederate armies in the South. After Grant's short stopover, he boarded a rail car and proceeded to meet with Sherman in Raleigh, North Carolina.

In John's next letter, he expresses his concern for family and business matters, things that are often on his mind as head of the household, whether in times of war or peace. He writes how his experience as a soldier has changed him.

---

37 Herold, David E, United States Army. Military Commission, and Alfred Whital Stern Collection Of Lincolniana. *Trial of the assassins and conspirators for the murder of Abraham Lincoln, and the attempted assassination of Vice-President Johnson and the whole cabinet: the most intensely interesting trial on record: containing the evidence in full, with arguments of counsel on both sides, and the verdict of the military commission: correct likenesses and graphic history of all the assassins, conspirators, and other persons connected with their arrest and trial.* [Philadelphia: Barclay & Co., i.e. 1865, 1865] Image. https://www.loc.gov/item/12002579/.

38 Tiemann, William F. *The 159th Regiment Infantry, New York State Volunteers*, In the War of the Rebellion 1862 -1865, (Brooklyn New York 1891), p. 119.

Letter #47 - April 24, 1865

Morehead City, N.C

To John,

I am glad that you have plenty of work, but you must try and save your dollars for the time will soon come when everything will be very low again. You pay a large rent, and you will require considerable work. You must write once a week so I don't get homesick. I am, you may say, all alone and no such company suits me and sometimes I feel very lonesome, but I think I shall soon be home now. I think I will go right out to Iowa; we have 2 Iowa Regt's in our Brigade 22nd and 28th and the 24th Iowa is in the 3rd Brigade and one of old Herm Duntz's Boys is in the 24th Iowa, his name is Peter George[39] he lives in Iowa, but his wife is now at his father's. These 3 Regt's represent almost every County in the State of Iowa. There are soldiers here from most every State; Mich., Wis., Ind., Penn., NC, NY, Ill, Mass, Conn., Me., and Ca., so I hear from all over the United States.

I must close.
John M. Mambert

To Walt and Jobe they must go to school and learn all they can. To Ebenezer Duntz I am glad you have written a few lines to let me know how you get along. To Permelia I suppose that you feel better since you are keeping your own house and I hope you will enjoy yourself, and you have not written one word about Martha, where she was. I suppose John's Martha is the same as ever, all well. To Mr. J. Olin[40] I am pleased with your little note very much. I really wish you could all be here for a few days to live a soldier's life with me. When I was at home, I could not sleep with a window open during the night, but now I can take the ground for my bed and the sky for my cover and not get cold at that. I will say that I have seen the soldiers lay down and it

---

39 George P. Duntz, Private 24th Iowa Infantry Co. B: Married in 1855 and moved from Gallatin, NY to Tipton Iowa. His father J. Herman Duntz was a farmer in Gallatin, NY.

40 James Olin: Age 42, a stock drover who bought livestock to market. Lives in Taghkanic, NY. Married, no children.

would rain, and the water would stand around them and they would not wake up until the water became too deep around them that it runs in their ears. Now this seems strange to you, but nevertheless true. I have one boy with me in my tent, a drummer boy. My tent is 6 by 7 feet, the sides 2 feet high, our floor is the ground. In this we have a table, we sleep on the ground and put our legs under the table, and we sleep badly. I have 3 woolen blankets, 2 under and one over us. We have quite a good lot of furniture, a frying pan, tin cup, knife, fork, spoon, plate and accidentally 1 chair (Cup T) or T Cup, so you see we are well off. Now I wish you could come live with me for one week. I think now I understand how to live in the west. I am fleshy and feel pretty good. As for boxen I don't box unless it is absolutely necessary for my own welfare and then I think I do my share. One thing I shall never forget about one of my boys from our Regiment in the Battle of Winchester, he turned around to go back a short distance, I looked around and just then there bursted a large shell a short distance from me and him and it seemed to drive his backsides or ass almost from under him, it was like hitting a dog with a scoop shovel against his tail, he looked around and his eyes was like 2 white onions. Well, it was no fun, but I could not help but laugh, the smoke and iron flew so I kneeled, and it would drive most any man's backsides in. I wish you could have been here, but you might have been killed for it was a bad place. The nicest battle ever was on the plains of Mansura. I must now close hoping that I may soon see you face to face together with all the rest. To Grandpap Sheldon and Granny I send my love hoping to see them soon, and to all the rest our bear is well. I hope you will all hear when we return to meet us at Hudson. Our Regt. is small but simon-pure. I have as much to carry as a mule can draw.

Yours,
J.M. Mambert [41]

---

[41] *Letter #47, April 24, 1865, Morehead City, N.C., To John.* John W. Mambert letters from the Civil War.

TELEGRAPH OPERATORS: SOLDIERS AND CIVILIANS—AN UNEASY MIX

The following letter by John is dated thirteen days after President Lincoln's assignation. It is John's first mention to his family of the sad news. His regiment was 350 miles from Washington, yet it appears some news traveled slowly to the soldiers' encampment in Morehead City. It's possible the regiment may not have received news as quickly as expected. Most military communication traveled by telegraph. Although strategically and tactically important during the war, such communication was not always reliable. The United States Military Telegraph (USMT) was essential for the army's success. The system was busy everywhere, every day. The ebb and flow of Union military expeditions communicated through the telegraph provided the most vital and up-to-date information by the hour. Newspapers depended on these telegrams to relay the latest news on the war to their reading public.

Government messages were "supposed to" have the right of wire, and the lines were "red-hot" with military dispatches, but ordinary business and reports for the press were delayed. However, once the army was silent in sleep, and the bribing of "civilian" operators began; the news hit the wire.[42] This behavior exposed the USMT's major drawback—it functioned, to an extent, independently while employing both military and civilian operators. The result was a troubling combination: a telegraph system that served the military but was not in total control of it. Stanton, the Secretary of War, designed the USMT to function in that manner. It allowed him to supervise the telegraph system and maintain "control" over both military operations and the flow of news throughout the army.

Perhaps the two regiments were not considered a priority at their quiet encampment on the Carolina seashore. If that was the case, important information flowing from Stanton's "controlled telegraph system," might have been inadvertently minimized.

Soldier operators were militarily disciplined, understanding the importance of speaking only positively about military life (to which they could at any time be returned if disobedient). Soldiers were more reliable

---

42 Plum, William R., *Military Telegraph during the Civil War in the United States.* Chicago, IL: Jansen, McClurg, 1882. p. 14.

than civilian operators who could resign at will and find employment elsewhere with any of the private companies.[43]

Civilian operators were sometimes accused of creating conflict. (They were not subject to military authority, which was a concern to field commanders.) Though civilian operators performed their duties, they often refused to conform to military standards of discipline. The civilian operators felt patriotic, performing a military duty equal to soldier operators, but were often treated to disparaging remarks by military officers. To remedy some of the petty ills, it was proposed that the operators wear a uniform without distinctive marks representing rank. General Order No. 51. was issued: Telegraph Operators are hereby authorized to wear a formal uniform: Blouse, dark blue trousers, dark blue cloth, with silver cord one-eighth of an inch in diameter along the outer seam; vest, buff, white, or blue; forage cap, like that worn by commissioned officers, but without any distinctive mark or ornament; buttons, like those worn by officers of the general staff.[44]

Though civilian operators were often accused of engaging in conduct a military officer would have not tolerated from subordinates (drinking while on duty or taking bribes to transmit non-military messages ahead of military traffic) once the war concluded, Lieutenant General P. H. Sheridan gave his opinion of the USMT operators during the war:

"as a general thing, well performed, and the men were almost universally trustworthy. In my own experience, I found them invariably active, brave and honorable."[45]

Abraham Lincoln found the military telegraph indispensable and for four years during the war he spent more of his waking hours in the War Department telegraph office than in any other place, except the White House. While in the telegraph office, he was free from official cares. Out of the eye of the public, he was more comfortable to be himself.[46]

---

43 Ibid., p.107.
44 Ibid., pp. 110-111.
45 Ibid., p. 353.
46 Bates, D. H. & Alfred Whital Stern Collection Of Lincolnian. (1907) *Lincoln in the telegraph office; recollections of the United States Military Telegraph Corps during the Civil War.* New York, Century Co. [Pdf] Retrieved from the Library of Congress, https://www.loc.gov/item/07032385/.

In the next letter, John mentions the 159th NY is no longer part of the 19th Corps, but now attached to the 10th Corps. The 10th Corps was disbanded in December 1864, and then restored under the command of General Alfred Terry in March 1865. It was attached to the Department of North Carolina, and part of General Sherman's Army.

Letter #48 - April 28, 1865

1st Part of this letter to all the family.

159 Regt. N.Y.V.
Morehead, N.C.

To Mrs. John W & family.

I must say to you all that I am well and enjoying good health for which I am thankful to God. I received two letters today, half an hour ago and I was very glad to hear from you that you were all well. John's letter 54. The last letter before 54 was 51. So, all the rest are back, and I think letter 46 or 7 up to 51 I have never received. I may receive them yet. I have not heard from home before today since the letter of Permelia dated March 25th. So, you see these letters are a great treat and welcome messengers. I am glad to hear that all matters are all as they were and settled.

I suppose you have heard of the sad affair at Washington, the assassination of President Lincoln. This created the greatest ferocity in the soldiers you ever see and if they would have gone in a battle just at that time, after hearing the sad news, there wouldn't have been a reb left to tell the tale. The news of Lee's surrender did not create so much excitement, but of the assassination. I must say that is a black mark for our nation of which I am ashamed. This shows the principle of the southern people. I don't want any man to praise the southern planter or the aristocracy of the south or of the institution of slavery unless he goes and lives in Dixie between 2 planters a couple of years and then if he talks it will be out the other side of his mouth. Our people are as ignorant of the welfare of the common people of the south as the common people are of us when they come

8 or 10 miles to see if the Yankees had horns, and yet our northern people know all about it because they read it in papers, and yet they won't believe one word they read in the papers because they print all kinds of rumors to fill up the papers. I must here say that my passion has increased greatly since I have been in this climate, and I have all I can do to control it. I become much easier irritated than I used to, and I am afraid if I come home that I don't want them to tell me what I know or tell me what they know if they never saw it for themselves. I must stop. I expect to be home soon as the war is over as I suppose you have heard. I expect to be discharged in May although I may be mistaken, but I think I will be home by the 1st of June, and then, if we have good luck with the help of the Lord, we will soon be with you. Our Regt. is in good health, the weather here is very warm, I just came in from the seashore. I raked up 28 clams as much as I could carry. They were very large, the largest I ever saw and the best. I boiled them open and then put them in a pan and fry them; 12 or 14 makes as much as I can eat. I feel good and strong, Uncle Steve's box has done me good. You told me where you all were except Martha, let me know where she is. I am sorry to hear about the death of Mrs. Row. here we are my tent is about 10 paces from a hospital, about 10 paces from 3 graves. yesterday one died and the day before another died in the hospital tent, so you see we live in the graveyard among the dead and against the hospital. These two died with the same disease I had in Savannah, which I supposed to be a cold, a rattling in the lungs and a tickling as if there were worms in your lungs and I have a little of it yet, but I begin to feel good. April 29th this morning is quite cool, so I thought I would finish my letter to you. I write a few words to you all on the other sheet you must excuse all mistakes I can't write much and amongst it all a short letter to Olin I am glad A. VanDusen[47] is gaining. I don't expect to be paid until I am discharged. Try to make yourselves all comfortable and feel contented and thankful to God for his goodness to you all. There is no express office here or I would send home all I could spare of clothing. I must write her a little anecdote about one of the Dukes who was riding out in this country and it seems that he never

---

47 Abram R. VanDusen: Age 43. Farmer from Claverack, NY. Married with one son.

heard of a skunk or smelt one and while riding out in a country road in his carriage a skunk came from the side of the road and was hit by the horses when he let off his essence. The Duke began to hawk and spit and asked the driver what that was that smelt so awfully, the driver said it was a little nasty animal they had in this country and that he had only peed, well said the Duke, thank God he didn't shit. Thinking if he smelled so awfully if he only had pissed, how must he smell if he should shit. If you should have a little leisure and could write a few lines to me, it would be very welcome. If you write direct: To John W. Mambert, Co. B 159 Reg't-N.Y. Vols., 2 Brigade, 1st Div. 10th A.C., Morehead City, N.C. You will see that we have been changed from the 19th A.C. to the 10th A.C. If you see John tell him or show him, let him have a copy of the Directions. We all feel very good. If we should all reach home well, we will say that we fetch peace with us. I am very sorry about Marge's loss and about that time I dreamed your barn burnt up, you lost all you had, and I told you to never mind I had a little left yet, and that you should not want as long as I had anything, and then I woke up. I did not like my dream and I felt as if all was not well at home.

John W. Mambert [48]

SOUTHERN SYMPATHIZERS

For John, the taking of Lincoln's life was an unthinkable action. He was not alone in his anger. His Union comrades surely felt the same. Their smoldering rage could have potentially caused the men to attempt a bloody massacre against the enemy.

John learned, and witnessed while in Louisiana, that the aristocracy controlling the Southern people created the country's division, but he also understood that sympathizers in the North were just as responsible. Those sympathizers, known as Copperheads, were Northern Democrats who did not support the war, despised Lincoln, and were sympathetic to Southern aristocrats. Some sympathizers were so extreme, there were accounts of them

---

48 *Letter #48, April 28, 1865, Morehead, N.C., 1st Part of this letter to all the family.* John W. Mambert letters from the Civil War.

celebrating the death of President Lincoln. Ironically, in the South, many people were saddened by the cowardly act that took the President's life.

An excerpt from the Confederate paper, The Wilmington Herald, dated April 20, 1865 reads:

> "There can be but one feeling in the minds of all, of every section and party who are not depraved as the miscreants guilty of the foul crime,--grief for the dead, and humiliation for the dishonor brought upon the whole country. There was a strong personal attachment to Mr. Lincoln among the mass of the people, an attachment which increased as the true character of the man became better known. Even in the south, where so little could be known of him, where for years so strong a feeling had been excited against him, these same kindly feelings were beginning to manifest themselves."[49]

Despite Booth's success in murdering President Lincoln, his desire for the southern states to remain as they were, allowing slavery to continue, was not going to materialize. Andrew Johnson became President, and though he was a southern democrat, he did attempt to carry out plans for reconstruction of the South, including the abolishment of slavery.

NEWSPAPER SENSATIONALISM

John wrote about the people's mistrust of newspaper articles. (Apparently some things never change.) Dishonest reporting by journalists who weaved their personal opinions into news stories on the war compounded the unrest in the North and South. Editors supported their correspondents' reporting, expressing their own biases on the editorial page. The view of most editors at the time was perhaps best expressed by the New York Times, June 9, 1863:

> "While we emphatically disclaim and deny any right as inhering in journalist or others to incite, advocate, abet, uphold or justify treason or rebellion; we respectfully but firmly affirm and maintain the right of the press to criticize freely and fearlessly the acts of those charged with the administration of the Government; also

---

49 *The Wilmington Herald, Assassination of President Lincoln*. April 20, 1865. North Carolina Newspapers, p. 2.

those of their civil and military subordinates...[50]

The demand by the public for an uninterrupted flow of news created separate morning and evening editions from the same paper, inspiring humorous comments, "they issue those evening editions to contradict the lies that they tell in the morning."[51]

One of the first journalists employed as a Washington correspondent for a New York paper explained to William Russel from the London times, that initially he merely wrote news no one cared for, but then he spiced it up, got sarcastic, and made-up stories. He admitted that government officials called out his lies. But the public was captivated by his stories, so the paper encouraged him to stay on as a feature writer.[52]

There were numerous accounts of misinformation reported by journalists under pressure from editors. Time was of the essence for field correspondents more concerned about being first to report the story than about the facts. Many times, leaving an engagement before it ended, reporters created their own finale as they rushed to the telegraph office to dispatch their stories before their rivals.

Telegraph companies instituted a policy of "first come, first served," which soon resulted in abuses. In an effort to hold the wire until they were ready to send their dispatches, reporters would occasionally instruct the operator to tap out various chapters of the Bible (not surrendering the wire until they were reasonably sure of having an exclusive story). As might be expected, competing correspondents retaliated, resulting in a reporter unable to "get" the line, because the wire was somehow cut. This forced telegraph companies to establish the "fifteen-minute system" allowing no reporter to hold the wire for more than fifteen minutes at a time.[53]

Stories printed in newspapers revealing sensitive military information became a major issue for government and military officials. Careless reporters frequently transmitted intelligence, betraying army movements to the

---

50 *New York Times, June 9, 1863; THE LIBERTY OF THE PRESS.*; A Conference of New-York Editors A Platform Adopted.

51 Andrews, J. Cutler, *The North Reports the Civil War (Pittsburgh: University of Pittsburgh Press, 1955)*, p. 34.

52 Ibid., p. 35.

53 Ibid., p. 7.

enemy.[54] To stop sensitive military information from leaking to the press before the Battle of Bull Run, General Scott issued an order forbidding telegraph transmission of any army dispatch movements. Reporters were forbidden to telegraph arrivals, departures, movements of troops or predictions of movements to come. Correspondents who failed to comply were excluded and the newspapers that employed them were censored.[55]

Regardless of which side the newspaper sympathized with, journalists knew that irresponsible field reporting created problems for the military. A reporter from the Savannah Republican acknowledged the problem:

> "The truth is there are correspondents who invariably magnify our successes and depreciate our losses, and who when there is a dearth of news will draw upon their imaginations for their facts. The war abounds in more romantic incidents and thrilling adventures than poet ever imagined or novelists described; and it would be well if the writers of fiction from the army... would remember this fact." Peter Wellington Alexander, June 16, 1862.[56]

Correspondents understood the damage caused by reckless, inaccurate or false reporting. But they were insistent on reporting from the battlefield and camps, frustrating military leaders with damaging stories. After the battle on September 3, 1864, when the 159th NY was on the front line at Berryville, Virginia, fighting Early's rear guard, Sheridan had an argument with a New York Times correspondent. The newspaper man, using erroneous information from an unreliable source, reported that Sheridan's army captured an entire ambulance train from the enemy. He went on to report the Confederates recaptured the ambulances. In reality, the Union lost one ambulance back to the enemy:

"So you have been making fun of me in your damned newspaper!" were Sheridan's opening words.

"Fun, General?"

"Yes, you told all about those confounded ambulances and paid no sort of respect to the commander of the army in which you are suffered to live."

---

54 Ibid., p. 359.
55 Ibid., p. 95.
56 *Savannah Republican,* June 16, 1862.

"There was no exaggeration in my story, sir; you must admit that!"

"Admit hell!" cried Sheridan. "This business has got to stop. You are ordered to leave my department within twenty-four hours!"

"Well, General," retorted Williams, "you have just been made Commander of the United States Military Department. Even if I go back to New York, I shall still be within the lines of your command."

"Oh, go to the devil if you like!" shot back Sheridan.[57]

TIEMANN AND MOLINEUX RETURN TO THE 159TH NY

Captain Tiemann returned to his regiment on April 29, 1865 after five months of being held as a prisoner. Tiemann, captured at the battle in Winchester, was likely held at "Castle Thunder" prison in Richmond, Virginia. The prison, a converted tobacco warehouse, was notorious for prison guards with a reputation for brutality.

General Molineux remained in Savannah while his troops were in North Carolina. He contemplated resigning from service, partially due to health, but he also missed his regiment. He argued the point strongly with General Grover, but still did not receive orders to return to his troops. But on the morning of May 1, 1865, General Sherman arrived in Savannah ordering Molineux and his regiment to proceed to Augusta. Sherman wished for Molineux to receive the surrender of the Confederate forces in the region and take charge of the country.[58]

This new command satisfied Molineux, giving him renewed spirit as a commander. He wrote home the following:

On the 3rd I received orders to proceed to Augusta, Georgia and take command of that place. Before getting away, I was besieged by a host of speculators, sharps, cotton thieves, and sutlers, with applications to be allowed to go with me. Augusta being the El Dorado of such People. My orders being not to permit trade, they did not go. Since my arrival, I have been busy from 6 A. M. to 12 M. paroling prisoners, feeding the populace,

---

57 Ibid., p. 601.
58 From the collection of E. L. Molineux, *letter to Tiemann May 1, 1865*. NYS Military Museum & Veterans Research Center.

giving interviews to planters and merchants, securing Confederate States property to our Goverment, censorship of the newspapers, and ordering the cleaning of our streets. It requires constant watching to prevent outbreaks between citizens, paroled soldiers, negroes, and our men. I estimate I have seized for the United States, property worth over a million dollars and some three thousand bales of cotton. There are two great cares on my mind. Providing for the starving poor and the slave question.[59]

On May 2, 1865, the 159th NY was ordered back to Savannah. The next morning, they struck tents at 3 a.m. but never moved until 8 p.m., finally marching to the pier embarking on the transport, *Star of the South*, a screw steamer. Brigadier-General Birge with his staff and the 131st NY were already on board. When the regiment loaded on, it became so packed the men could barely move: The steamer didn't leave port until the following day.[60]

US Transport Star of the South, artist Alfred R. Waud [61]

---

59 From the collection of E. L. Molineux, *extracts from his diary*. NYS Military Museum & Veteran Research Center.

60 Tiemann, William F. *The 159th Regiment Infantry, New York State Volunteers*, In the War of the Rebellion 1862 -1865, (Brooklyn New York 1891), pp. 119-120.

61 Illustration, *Star of the South*, artist Alfred R. Waud. Gift of J.P. Morgan, Library of Congress #2004661370

The steamer sailed toward the mouth of the Savannah River, arriving at 5 p.m. Taking on a pilot, May 6, 1865, they proceeded sailing upriver, becoming grounded within a mile of the city. The transport remained stuck until the next morning, when help from the tide broke the vessel free. Shortly after, they grounded a second time, just off the city wharf. A steamboat came alongside and the packed troops loaded on, disembarking at the base of Bull Street in Savannah. The men of the 159th NY marched to the railroad depot camping near their previous quarters.

ORDERED TO AUGUSTA, GEORGIA

On May 11, 1865 at 2 a.m., the men were ordered to strike their tents, receiving orders to guard the wagon train. Seven hours later, the regiment along with the rest of the brigade commanded by Colonel Graham, departed Savannah.[62] The 159th NY also became part of the "Military Division of West Mississippi." (President Lincoln created the division with the purpose of reorganizing the troops in the western theater making General Ulysses S. Grant commander.) The Division consisted of the Departments of Ohio, Tennessee and Cumberland and embraced all Union armies situated between the Mississippi River and Appalachian Mountains. Major General William T. Sherman took command on March 18, 1864, with the Departments of Kentucky and North Carolina added in April 1865.

The brigade and wagon train covered seventy-seven miles in four days. On the last day, the troops received orders from General Molineux's headquarters to press forward quickly to Waynesborough, twenty-eight miles north. Moving rapidly, the troops arrived at the town's depot at sunset and loaded onto railcars. The brigade reached Augusta at midnight, 135 miles from Savannah. The men set up camp in a field alongside the railroad depot. Brigadier General Molineux arrived earlier and was in command of the region.[63]

---

62 Tiemann, William F. *The 159th Regiment Infantry, New York State Volunteers*, In the War of the Rebellion 1862 -1865, (Brooklyn New York 1891), p. 120.
63 Ibid., pp. 120-121.

Troops being transported by rail

MOLINEUX DINES WITH THE CONFEDERATE PRESIDENT

After the surrender of General Lee, Confederate President Jefferson Davis, along with his cabinet, fled south in an effort to continue the Confederacy, hoping better terms could be negotiated. Davis was captured May 10, 1865, by the Union army in Irwin County, Georgia. He was brought by rail with his family and other high-ranking Confederate officials to Augusta where they were detained on a steamboat. While in Augusta, the Confederate prisoners were allowed to visit friends in the community. President Davis, with his guard Colonel Pritchard, dined with General Molineux. The next day, the Confederate prisoners were transported down the Savannah River, then up the east coast to Fort Monroe, Virginia, where Jefferson Davis would be held for the next two years.[64]

COMFORTABLE QUARTERS

On the morning of May 20, 1865, tents were struck and the 159th NY marched through the city to the Old City Hotel in Augusta. The men were provided comfortable quarters with rooms assigned by the regimental companies.[65] Lieutenant-Colonel Waltermire was appointed Supervisor of

---

64 From the collection of E. L. Molineux, *Thomas Woodrow Wilson*. NYS Military Museum and Veteran Research Center.

65 Tiemann, William F. *The 159th Regiment Infantry, New York State Volunteers*, In the War of the Rebellion 1862-1865, (Brooklyn New York 1891), p. 121.

Trade. He issued General Order No. 5 on May 23, 1865, stating the daily schedule started at 5 a.m. with Reveille and ended with Taps at 8:30 p.m. Waltermire also directed: "under no circumstances were the men allowed to eat their meals in their rooms, but must be taken in the dining room which was set apart for that purpose." His concern for cleanliness was a priority, ordering First Sergeants to make sure the rooms of their respective companies were cleaned out daily with the refuse matter carried away by the policing party. He also noted that no Negroes other than officer servants, or enlisted cooks, would be allowed in the yard or buildings.[66] On May 29, 1865, there was a regimental review by General Molineux.

THE 128TH NY GOES HOME

From the Headquarters of the Military Division of the Mississippi, the following general order was released by Major General W. T. Sherman on May 30, 1865:

No. 3: The General Commanding announces to the Armies of the Tennessee and Georgia, that the time has come for us to part. Our work is done, and armed enemies no longer defy us. Some of you will go to your homes, and others will be retained in service till further orders.[67]

Sherman's orders were followed up on June 6, 1865, by headquarters of the Department of the South with General Order No. 11:

In compliance with orders the following regiments whose term of service expire before the 30th September, will prepare to rendezvous at Savannah, with a view of being mustered out of service – 22nd, 24th and 28th Iowa, 128th and 131st New York – Edward L. Molineux, Brevet Brig. Gen. U.S.V.68

Sadly, after surviving thirty-two months at war, John and the 159th NY were not returning to their families in the North at this time. They remained in the South, doing provost duty until their full three-year commitment came to an end.

---

66 G.O., No5, *May 23,1865 the daily schedule*, National Archives Washington DC, Regimental Company Books for the 159th NYSV.

67 G.O., No. 3, *Major General W. T. Sherman on May 30, 1865, go to your homes.* From the collection of E. L. Molineux at NYS Military Museum and Veteran Research Center.

68 G.O., No. 11, *Brevet General Molineux on June 6, 1865, mustered out of service.* From the collection of E. L. Molineux at NYS Military Museum and Veteran Research Center.

The veteran soldiers watched as their old friends and neighbors from the 128th NY headed home, back to their families and civilian life in Columbia and Dutchess counties, while the 159th NY remained in the South at the mercy of the military.

The 128th NY regiment served alongside the 159th NY for the majority of their enlisted time. They became part of the 19th Corps in May 1863, and were involved in almost every engagement together. Beside major battles, they also performed guard duty in Alexandria, Savannah, Bailey's Dam, and Morehead City. They also served together throughout the Siege at Port Hudson where the 128th lost twenty-two men killed in action, 100 wounded and six missing. Throughout their time served, the 128th NY deaths totaled 269 men, sixty-three killed in action and 206 by disease.[69] Between the two Columbia and Dutchess County regiments, 455 men would not be returning to their families, leaving a tremendous void in their communities.

One month after General Sherman spoke to Molineux regarding Augusta, he received orders from commanding major general, Q. A. Gillmore on June 5, 1865 providing complete instructions on what was expected when he takes control of the country:

"you have been assigned, the District of Northern Georgia, with your headquarters at Augusta. I have selected you for this command with sole reference to your fitness for it, and in the firm belief that my confidence in your zeal and ability is not misplaced. I desire you to take immediate steps to occupy all the important towns within your district. One company, or at most two, will probably be enough at each place except Augusta, where a larger force will perhaps be required. The object of this disposition of your forces in very general terms is to preserve good order, administer justice, settle disputes, and ratify contracts between freedmen and the planters, and encourage the inhabitants, whites as well as blacks, to resume the avocations of peace. I desire you to inform the people far and wide within your command that slavery is abolished, and you will take special care to inform the free people that they are expected to labor for their own support. You should establish mounted patrols between Augusta and your several posts, and you are authorized to hire horses for that purpose. The men selected for this purpose should be of such

---

69 *128th NYSV Infantry Regiment, Civil War Historical Sketch,* NYS Military Museum & Veterans Research Center.

intelligence that in addition to carrying dispatches simply they should be able to give you correct information of matters transpiring in the country through which they pass. An order will soon be published defining more definitely the status of the freed people, establishing provost courts and schools, and for other purposes. I enclose copy of General Orders, No. 82 to which I desire to have executed at once."[70]

BAD BEHAVIOR

At this time, it appeared the men of the 159th NY were becoming relaxed with their duties as soldiers. Excessive drinking and disorderly behavior were frequent occurrences. There were continuing issues with men not properly performing duties within the encampment. One instance was so appalling, Lieutenant-Colonel Waltermire issued an order on June 11, 1865, giving only officers and non-commissioned officers use of the washroom; due to the filthy conditions the subordinates left it in.[71] The following day, he issued another order pertaining to bad behavior. General Order No. 9 stated:

"Frequent complaints have reached the commanding Officer in relation to men of this command stealing and pilfering from one another. Men going to wash scarcely dare take off their caps, blouses or coats for fear they will be stolen. This state of things must not be allowed. Here after any man found stealing or interfering with the property of another or having in his possession stolen property will be severely punished. This order will be strictly enforced".[72]

In John's next letter, he paints a troubling situation for aristocrats in the South refusing to live with the freed people they exploited. John shares with his family thoughts on a solution for the noble planters.

---

70 O. R. *sent from Q. A. Gillmore, Major-General commanding, to General Molineux, June 5, 1865.* Ser. I, Vol. XLVII, Part III. p. 629-630.

71 G.O. No. 8, *Lieutenant-Colonel Waltermire, washroom filthy conditions, June 11, 1865.* National Archives Washington DC, Regimental Company Books for the 159th NYSV.

72 G.O. No. 9, *Lieutenant-Colonel Waltermire, stealing and pilfering, June 12, 1865.* National Archives Washington DC, Regimental Company Books for the 159th NYSV.

CHAPTER NINE

Letter #49 - June 19, 1865

Camp 159 Regt.
City Hotel, Augusta, Ga.

To Mrs. John W. Mambert,

Having a little leisure, I thought I would write a few lines to let you know that I was well and enjoying good health, thank God. I hope you are the same. I dream frequently about home, and I dreamed I had cut Walt with an ax. I near about killed him in his bowels and stomach, and he said Oh don't Pa and it has been harassing my mind very much, it makes me feel homesick. I hope you are all well and I long to hear from you I can hardly wait to start for home. I have thought more of home of late and felt more worried than ever. I would apply for a furlough but then I would have to cross the ocean twice and that I don't like. However, if you say I must apply for one I will risk it. But we may soon start for home although all prospects indicate that we shall have to stay very near our time out and that is about 4 months or a little better. I sent a letter with Ed Hollenbeck[73] a few days ago with some Confederate money in, I forgot how much, I think over 200 dollars and 2 papers left in care of Ed Roraback. I also sent you a box addressed to Churchtown, this was wrong, I should have sent it to Hudson. Soon after I sent a letter with the receipt in. If you get the receipt then you can go to Hudson to the express agent and find out where it is if you have not got it, it was valued at $20. I received your letters 56 & 57 dated respectively May 9th and May 16th. I was very glad to hear from you that you were all well for I have worried about you very much. And when I don't hear from you once a week, I feel lonesome and homesick. My health is pretty good I think, but we don't know here when we are in good health or not. I want you to tell Walt as I forgot to mention in my other letter that I thank him for his $5.00 bill to come home with if it is not worth much. I think it is worth as much as the Confederate

---

73 Edgar Hallenbeck, Private 159th Regiment Co. C: Age 17, promoted to 1st Sergeant September 19, 1864. From Hudson NY. Lived with his parents and six siblings ranging in age from 10 to 26 years old.

money is and you may give him a $5 or 10 dollar bill of the Confederate money but you take care of it for him, and give Jobe one also and John, Eb, and Permelia and Martha and keep the rest. Keep Martha's, Job's, and Walt's so they don't lose it. I expect to be home soon, but you have seen the order from the war department and that just leaves us, however, I think there will soon be another, the war is over. There has just been a boat that arrived here from Savannah but no mail. You stated that you had some wine, yet I hope soon to help you drink it at home. I will first come to John's when I return or to R. Decker's Hotel Churchtown but think we will first go to Albany to be mustered out of the Service and paid. You will hear, I think, when we will arrive at Hudson. Ed Hallenbeck is orderly sergeant of Camp C 159 and if he comes back to the Regt. send me one dollar with him not more (i.e.) if you have it if you see him, he lives I think in Hudson I have just inquired, he lives in <u>Johnstown</u> instead of Hudson in the Livingston House, so I am informed. John H. Allen I suppose is home; Jerome McDonnel of Claverack is dead; I have been informed. We are very much scanted here for diet, a piece of dry bread and a cup of slush coffee for breakfast and supper is our meals for dinner about 4 to 6 ounces of pork and a cup of cold water is our dinner and every time I go to get it, I feel like some poor beggar and if I can keep my soul and body together, I think I will do well. We have had a few small pieces of stinking codfish instead of pork, and a couple of times half mackerel, they were good if we only had enough. However, I can live but it is rather hard. I have not tasted butter since I ate last of Uncle Steve's, I don't want you to worry about it.

I saw something this morning which made me laugh. I heard a terrible noise in front of our hotel this morning of cheering. I went through the hall to the balcony and behold there some 5 or 600 negroes old and young singing Kingdom is Coming and going down streets throwing their hats and caps in the air. I look up street and sure enough Kingdom was coming, a United States Negro Regiment regular, apparently 1,000 strong they marched in companies the music was 10 paces in front and ahead of them was the squad shouting and singing Kingdom is Coming. To see them all dressed in uniform it looked like a Blue Black and in such precision order, and every foot up and down at the same time, every Negro as straight as a stake and as proud as a peacock. It was the best discipline I ever saw. They had beautiful colors, and it was enough to do a patriot's heart good. But what the fun was to see the citizens look at them, who they thought could not be soldiers, but the Negro's say, "how are you old master?" This is the most abhorrent to the white Reb's, but I don't pity them one bit. Now they say if they will only take the negroes away then it is alright. They won't live where they are, but I say they have brought that kind of stake in the country and they always lived with them and they amalgamated with them and they made war for them, and the Negro's made the South what it is, and God knows it is not much compared with the North. And now they won't live where they are, but my answer is to them is to just let the Negro's have the lands of those who don't want to live amongst them and let them move off to some other government they think better than the United States, to Africa or Asia, or Mexico or the West Indies Island. I don't care where, for we can spare such characters of disloyalty. They have been a curse to us as a nation and we can spare them to get better loyal people to fill their black hearted places (Black or White) I care not who. I can buy a plantation 3 miles from here for $2500 (500 acres). It is cheap but I don't want to live here in this society. The land is pretty good, pretty sandy, poor house, but good out-buildings and about half cleared the rest in timber, principally yellow pine, you could sell wood enough from it to pay for it. Carpenter's work is good here 3 or 4 dollars per day. Country's produce is reasonable enough, but no money. If you send with Ed Hallenbeck, send 5 and 10cts postal currency or silver, there is no charge except a little. Greenbacks are worth all the gold here. The 128th Regt. will soon be home now if they have good luck. I don't envy their happiness, but I live in hopes soon to follow them, and if not, then I think by time we leave the Railroad we will be done, and we may come by Railroad to Charleston or Savannah. Marching is what hurts. Carrying from 20 to 40 lbs 15 or 20 miles a day through a hot sun requires a mule to stand it. I wish you could be here till you got tired of seeing how different things are here, from what they are there, you would not like it long here. The general mass of people 8 out of 10 can't read or write. They come to town from the country with a cow or steer hitched to a small cart with fills a single yoke.

On the end of that is the team of a great many, an old woman and a boy generally, a few apples, a few cucumbers, radishes, onions, eggs, apples, pears, peaches, etc. and the soldiers steal –sometimes half and sometimes all. This is what turns my feelings and perplexes me very much. The women, 7 out of 10 chew snuff rolled up in a white muslin patch and the tobacco juice running down over their chin. This is abominable yet, it's a general thing. Some have a mule before a cart, and a boy riding the mule. I must stop. Write often.

John W. Mambert [74]

THE REALITY OF RATIONS—REPULSIVELY BAD

The government established the portion sizes of rations for the military: twelve ounces of salted pork or twenty ounces of salt beef, twenty-four ounces of flour or one pound of hard bread, and some dried vegetables. The vegetable could be navy beans or peas or 1.6 ounces of rice, and three pounds of potatoes a week per man. The meat should be fresh rather than preserved "when practicable."

Unfortunately, the reality of the situation was quite different. Throughout the war, staples such as meat, and hardtack were often repulsively bad. Quartermasters were challenged, finding it impossible many times to supply food, leaving the men hungry.

Ideally, coffee, sugar, salt, vinegar, candles, and soap were also part of the ration, but more often than not, the men feasted on dry bread, slush coffee, four to six ounces of pork and cold water.

At this time, the 159th NY was feeling hunger pains from a "lack of substance" being supplied to the regiment. But John did write about opportunities for foraging to help supplement their diet.

NEGRO ABUSE

As impressed as John was with the Negro regiment for their discipline and obvious pride in being U.S. Army soldiers, there were still many individuals

---

74 *Letter #49, June 19, 1865, City Hotel, Augusta, Ga., To Mrs. John W. Mambert.* John W. Mambert letters from the Civil War.

who found issue with Black soldiers; and it wasn't just people from the South. Information was delivered to Lieutenant-Colonel Waltermire on a scheme that was brewing, one that may have included some U.S. Army troops, a plan to violently attack Negroes. Once he learned of the situation, he issued General Order No. 10 on June 21, 1865, stating:

"The Lieutenant- Colonel commanding has been informed that enlisted men of this regiment and the 13th Connecticut Volunteers intend this evening to commit violence upon Negroes in this City. It is ordered that any soldier of this command who threatens to use violence to negroes in Colored Troops in this City will be severely punished. Interference with troops whether white or black, so long as they are soldiers of the U.S. is mutiny and any man committing this crime of guilty of mutiny and are liable to be shot by sentence of G. L. Martial.

If the white troops persist in abusing the colored race whether citizens or soldiers, they will be sent from the City to relieve Colored Troops, in order that they may do duty in the City.

It is therefore ordered that no enlisted man of this command under any pretense whatever shall leave their quarters to go around the City without a pass approved at these headquarters".[75]

THE NATIVE GUARD

Members of the Native Guard were originally part of the Confederate Army. Established on May 2, 1861 in New Orleans, they were free men of color, who enlisted to defend their homes from the enemy. Prior to Admiral Farragut sailing up the Mississippi and taking control of the Crescent City from Confederate General Lovell, the Black troops were disbanded because they served as nothing more than public display and were never intended for military use.

After General Butler took control of New Orleans, he planned to organize a special brigade to capture and occupy western Louisiana and control the salt mines in New Iberia. After unsuccessful requests for additional reinforcements, he decided to remedy his shortcomings, writing back to Washington:

---

75 G.O. No. 10, *Lieutenant-Colonel Waltermire, abusing the colored race,* June 21,1865. National Archives Washington DC, Regimental Company Books for the 159th NYSV.

"I have determined to use the services of free colored men who were organized by the rebels into the Colored Brigade."

Butler located the original twenty Black officers from the Confederate Native Guard, offering them the opportunity to become part of the United States troops. They unanimously agreed with one condition, to be officers as they were prior. Banks supported their request, and the men committed to raising two regiments within ten days. The 1st Regiment of Native Guards consisted of 1,000 free Black men. They were mustered in for three years of service on August 22, 1862, becoming the first officially sanctioned regiment of Black soldiers in the Union Army.[76]

In a short time, three regiments of infantry and two batteries of artillery were ready for service. Butler was impressed with the men describing them as intelligent, obedient, highly appreciative and dignified. He also stated the men learned quickly and "they learned to handle arms and to march more readily than the most intelligent white men."

When Butler departed his command in New Orleans, his replacement Banks, continued to enlist more Black troops. But Banks removed the goal of their ambition: equality to the White soldier. Banks removed the commission from the Black officers, calling them the Corps d'Afrique.[77] He formed this cadre of Black soldiers through General Field Order No. 40:

HEADQUARTERS DEPARTMENT OF THE GULF, NINETEENTH ARMY CORPS,

OPELOUSAS, May 1, 1863

GENERAL ORDERS No. 40. - The Major-General commanding the Department proposes the organization of a Corps d' Armee of colored troops, to be designated as the "Corps d' Afrique." It will consist ultimately of eighteen regiments, representing all arms - infantry, artillery, cavalry - making nine brigades of two regiments each, and three divisions of three brigades each, with

---

76 Butler, B. F. (Benjamin Franklin). (1892). *Autobiography and personal reminiscences of Major-General Benj. F. Butler: Butler's book: a review of his legal, political, and military career.* Boston [Mass.]: A.M. Thayer & Co. pp. 491-492.

77 Ibid., pp. 493-495.

CHAPTER NINE

appropriate uniforms, and the graduation of pay to correspond with the value of services, will be hereafter awarded.[78]

In the beginning of their service, Black soldiers were used primarily as labor detail, chopping wood, digging earthworks and guarding rail lines. Although it was reported many of the soldiers were not treated in a manner equal to the White soldiers, the army did officially have Corps d'Afrique posts, which included hospitals, and officers responsible for training the Black regiments. In June 1863, the three Native Guard regiments were re-designated the 1st, 2nd, and 3rd Corps d'Afrique

André Cailloux, a free man of color and cigar maker by trade, was one of the many who accepted a commission from General Butler, joining the 1st Regiment and becoming Captain of Company E. Cailloux's regiment saw combat for the first time when they assisted during the first assault on Confederate forces at Port Hudson. As part of the attack the first day, Cailloux was ordered to lead his company of 100 men in a suicidal assault against a high redoubt manned by two regiments of Confederate troops with heavy artillery support.

Despite his company suffering heavy casualties, Cailloux continued to lead the charge, shouting encouragement to his men. Then, a mini ball tore through his arm. Severely wounded, Cailloux continued undaunted, until a Confederate artillery shell struck and killed him. His decomposing body lay on the ground for forty-seven days until Port Hudson surrendered.

Within a year, the Corps d'Afrique was dissolved and the soldiers were placed in the newly organized 73rd, 74th and 75th Regiments of the United States Colored Troops.

---

78 G.O. No. 40, *May 1863 The Major-General commanding the Department proposes the organization of a Corps d'Armee of colored troops*. National Archives Washington DC, Regimental Company Books for the 159th NYSV.

Assault of the 2nd Louisiana (Negro) Regiment on the Rebel works at Port Hudson May 27 [79]

KIND WORDS FROM A CONFEDERATE NEWSPAPER

On June 28, 1865, Lieutenant-Colonel Waltermire was paid a visit by a local newspaper, The Augusta Chronicle. The Lieutenant-Colonel was interviewed regarding the 159th NY and their service to the Union. The following day the article below appeared on page three:

The Flag of the 159th New York – Having occasion to visit Col. Waltermire's office yesterday we were shown a flag carried by that celebrated regiment of which he is the popular Colonel and which for workmanship has scarcely if any equal. It does not show too much advantage now, having been through the dust and smoke of many battles, and bears the marks of having been very near the opposing forces, as several bullet and cannon ball holes will attest. The 159th have seen much service in the past two years, being engaged in the following battles: Irish Bend, La., April 14, 1863, Port Hudson, 1st assault, June 14, 1863, 2nd assault June 14, 1863, Manassa Plains, May 16, 1864, Halltown, Va., August 24, 1864; Barryville, Va., September 3, 1864, Opequan, Va., September 19, 1864, Fishers Hill, September 22, 1864, and Cedar Creek, October 19, 1864.

In all of these battles the regiment distinguished itself, but more especially, at Fisher's Hill and Cedar Creek. Col. Waltermire has succeeded in bringing this regiment to a degree of perfection in drill

---

79 Illustration, *2nd Louisiana Colored Troops*, special artist. Library of Congress, # 2003668336

and discipline, rarely if ever excelled. It is worthwhile to see the flag alluded to above, as such workmanship as that which is upon it, is scarcely ever put on flags intended for military in active service.[80]

PROVOST DUTY IN AUGUSTA

Time passed uneventfully in downtown Augusta for the 159th NY, with only a few exceptions noted during their two-month stay. The men were kept busy with drills or provost duty, but there was downtime, when boredom led to trouble for some men. July Fourth was patriotically observed in 1865, a day when many of the men drank more peach brandy than was good for them, causing serious disturbances quelled by the provost guard.

Guard mount, better known as changing of the guard, became a great attraction with large crowds of citizens watching the ceremony. The people applauded with great enthusiasm at the appearance of the regiment displaying their perfect drill and discipline. The men took great pride in looking neat—their clothing, arms, and accouterments most carefully looked after. The thought of going home was in each soldier's mind, and that anticipation motivated the extra care of their belongings.

The 159th NY was relieved of their provost duty in Augusta by the 19th U.S. Infantry (regulars) on July 21, 1865.

The regiment marched a short distance north of the hotel to the foundry where the Confederates manufactured cannons and ammunition during the conflict. They were encamped only a brief time at the foundry before being marched a few miles outside the city to the Sand Hills where they camped in the old United States Arsenal grounds.[81]

---

80 The Augusta Chronicle, June 28, 1865, page 3: *The Flag of the 159th New York*.
81 Tiemann, William F. *The 159th Regiment Infantry, New York State Volunteers*, In the War of the Rebellion 1862-1865, (Brooklyn New York 1891), p. 122.

Chapter Ten

"We are very anxiously waiting and expect every day to start for home."

MOLINEUX GOES HOME

On July 29, 1865, Brevet Brigadier-General Molineux resigned, and Brigadier-General J. H. King was appointed to command the department. Before his departure, Molineux issued the following letter to the regiment:

> Officers and Soldiers of the 159th New York State Volunteers:
>
> Having tendered my resignation from the United States service I shall soon no longer bear the honored title of your commanding officer.
>
> The war is ended, and it will not be long before you return to your Northern homes; but in parting from you at the present time, my feelings impel me to say a few words in farewell greetings, both as an officer and a brother soldier.
>
> We enlisted together at a time when the fortunes of our country were dark; many brave comrade have fallen from the ranks on the field and by disease, but we, the survivors, may return thanks to God that their deaths have not been in vain, and that our labors have been crowned with success.
>
> As a regiment you have won an enviable reputation for steady, persistent bravery on the field, and for good order and conduct in the discharge of any and every duty. Since my connection with you I have kept firmly to the rule that my conduct should do you no discredit and the reputation of the 159th New York state Volunteers should be second to none.
>
> I am proud of you, and am grateful for the cheerful obedience which you yielded to strict discipline. It is this which has rendered you successful, and those of you who have at times

thought the discipline severe and rigid must remember that it was necessary, and that although at times the voice may have been harsh yet the heart has always been warm towards you.

Recollect, comrades, whenever and wherever we may meet, I shall always be glad to greet you, and be ready to assist you in any way in my power.

Edward L. Molineux,

Colonel, One Hundred Fifty-Ninth New York State Volunteers, Brevet Brigadier-General United States Volunteers, Commanding.

On August 1, 1865, General Molineux bid goodbye to the men of the 159th NY; the men were saddened at his departure. He was the regiment organizer and commander, and he endeared himself to all his men by his concern for them. By his careful discipline and personal bravery, he proved himself to be one of the best soldiers the war had produced. His resignation was followed by the regrets and well-wishes of every man in the command.[1]

General Molineux was going home, returning to his family and the life of a private citizen. Edward was married to Harriet "Hattie" Davis Clark of Middletown, NY, on July 18, 1861. Hattie gave birth to their first son Leslie in 1863, while Edward was commanding his men in Louisiana. After the war they had two more sons, Roland, born in 1867 and Cecil born 1876.

Before the war, Edward was a partner in Daniel F. Tiemann and Co., a paint and color factory manufacturing pigments and paints. After the war, he had a long career with C. T. Raynolds and Company of New York, one of the largest paint houses in the United States. In 1880, he was appointed Brigadier-General of the New York State National Guard, and in 1885 was elected Major-General. In 1886, Edward was elected commander of the military order of the Loyal Legion. He was a contributor to literature, writing various papers on military subjects and physical culture for public schools. He also wrote on the suppression of riots in cities and on railroads.[2]

---

1 Tiemann, William F. *The 159th Regiment Infantry, New York State Volunteers*, In the War of the Rebellion 1862 -1865, (Brooklyn New York 1891), pp. 122-124.

2 *Edward Leslie Molineux*, The Virtualology Project. http://famousamericans.net/edwardlesliemolineux/

Edward Leslie Molineux [3]

In John's next letter, he mentions his pay:

"I expect to be paid (they say 8 months) pay."

Eight months, turned into 10 months—no income. When the men enlisted, they expected to be paid at the latest every two months. Thankfully, John's family were homesteaders, living in the country with other means to survive, mainly by growing and harvesting their own provisions. But imagine how the families of the men residing in the cities suffered with no income.

Remembering the young wife of the soldier in the ambulance corps; she was distressed shortly after the regiment left New York. Many families must have been destitute.

Letter #50 - August 6, 1865

Camp 159 Regt. Sand Hill Arsenal Augusta, Ga.

To the family:

I am well and enjoying good health for which I am thankful to

---

3 E. L. Molineux Photo Collection. *Edward Leslie Molineux*. NYS Military Museum and Veterans Research Center.

God. Hoping and trusting in God that this may find you the same. I received your letter with the money this morning. I was very glad to hear from you, I thank you for it, although our paymaster has arrived here last night for the purpose of paying us they say tomorrow, so if they do so, I will send it to you as soon as I get it, and you must go to the post office (the Adams Express, I mean) and inquire without further notice, for I have no reason to doubt our being paid tomorrow. You stated that I had given a Confederate bill to all the rest except Martha, John's Wife I don't know as I gave a Confederate bill to Eb. I thought I had given one to each of my children. I expected, I suppose, that John composed one family by giving to him was to his family and by giving to Permelia was to her family and the same to the rest should they become the head of a family, so that each family should have one. For that purpose, I said that mother should take care of them for those who were not capable of taking care for themselves. If I have forgotten to mention here it was because I forgot it, as you many times do in letting me know about Jobe and Walt, you forget it which I sometimes am most anxious to know. So, you see that if I have neglected to send one to her (that is) Martha and even Eb it has been because I forgot it. I don't want any of you to think that I think more of one than another. I think well of all who do well and do what is right.

Tomorrow, I expect to be paid (they say 8 months) pay, but I would almost as well not be paid until discharged if it wasn't for you being in want of money and if I am paid, if you get it, I want you to pay all you owe and deposit the rest in the Farmers Bank of Hudson until I come home for safekeeping.

Our Col. Molineux has resigned and gone to Nashville Tenn, and we expect to get orders to be mustered out very soon, at least by the 15th of this month and maybe by the time this letter reaches you, we may be on our way home We are very anxiously waiting and expect every day to start for home. I don't suppose we shall stay our time out. Our Regt. is pretty healthy. We have moved twice since I last wrote, 1st, from the City Hotel to the foundry in the city where they made cannons and munitions of war. We stayed there a little better than a week. From there we moved to this place 3 miles out of the City called Sandhill arsenal the building in which we are encamped is 500 feet long, our Regt. and the 13 Conn. Vols, each occupy a room on each end. It is a beautiful building (our Regt. is pretty well) The lot in which the building stands contains some 20 acres or more, all level and green grass with here and there a shade tree with benches around the foot of the tree. It lies on a hill I should think 150 feet above the level of the city. It is surrounded by groves of trees with gentlemen's residences if I may so call them. It is a beautiful place. The water is poor, the place is healthy and good, and while I was in the city, who do you think I found? I came to my quarters in the city hotel, and someone said on my way there was a man there waiting to see me very much. I stepped in and there he was, they asked me if I knew him and said he was a jayhawker, anyway I could not tell who, and later I recognized him, and it was Steve Lawton[4]. He had been in the Rebel Army, 26th Alabama Regt. and fought us all through the Shenandoah Valley and he had been taken prisoner and was paroled and was on his way to Alabama again. He said he had been pressed in the army but he said if he had known we had been there he would have deserted if he knew he would have been shot, he did not know he was fighting his own boys! but oh how glad he was to see us and I got the Col. to give him transportation with us and he said he was going home with us, I divided rations with him and sometimes some of the rest does and he stays with me and expects to come home with me. You can tell his folks. He is dreadfully heavy built and wonderstout, you would not know him. We had an accident in our Regt. with one of my associates, his name was John W. Coons.[5] He was from Southwestern part of Gallatin. He and I went out one day

---

[4] Steve Lawton: Confederate soldier, 26th Alabama Regiment. Originally from Columbia County NY.

[5] John W. Coon, Private 159th, Co. I: Age 27, died shot by an assassin at Augusta Lane Hill on August 1, 1865. A farm laborer in Gallatin, NY. Left behind a wife and four children ranging in age from 3 to 11 years old.

after watermelons and we got 4. We went 2 miles out and we did everlastingly eat, and we had quite a good deal of fun or sport. The next night he thought he would go and get some corn to eate with a couple of other fellows and when he came out the cornfield a man raised up about 20 yds off and shot a whole charge of shot right in his arm and side. And then the gunman fled, but his companions were unable to catch the man, so they sent for a stretcher and fetched him in, he lived two days and died. He was shot right through, and his inwards were cut so you see this was mortifying to us and we are after the man and if caught hang him or shoot him. Coons was a good man not accustomed to stealing, good company, always lively and sporty and is quite a loss to me as he was good company for me.

I must tell you that this is a great country for peaches, grapes and watermelons, sometimes they come to the camp with a load and one day there come two loads and the man went to sell some to the Col. and the boys stole the whole 2 loads in less than a minute and a half, and after the man was gone such eating, you never see. I see them steal a whole load from a negro the next day. Well, I live on melons, peaches and grapes. Before I got melons my belly was small, I could not eat more than half of my rations but I have stretched it so I can eat all and sometimes more. I think they are healthy here, the healthiest thing we can eat, they make me feel good and I think peaches have brought me back to good health again. If I get home soon, I will just be home in fruit time so I will see 2 fruit seasons. I forgot to tell you I have plenty of figs to eat, they grow here, plenty of sweet potatoes are coming soon. I was out yesterday on a scout beyond the place where Coons was shot, it is about 2 miles and I got melons, corn, figs and peaches and such. Al Van Deusen's brother Elias and 4 others we all feel pretty good, and we love to eat good things, but we are always prepared for any assault. After this you must direct your letters to John M. Mambert 159th Regt., N.Y. Vols. Co. B, 2 Brig, 2 Div 19 A.C. Augusta Ga. instead of Washington D.C. They are delayed too long in Washington. This is letter No. 53 July 9th and here it is Aug. 6th, so you see the difference.

This is a private letter, and this is the 5th page. I made a mistake and wrote on the wrong page, I will number the pages. August 11, 1865,

I postponed sending this letter until today and today I am going to the city to send you some money. I was paid yesterday, and they gave me 10 months pay and more besides. I think they, the paymaster, made a mistake, his payroll stated that I was paid up to the 30th of April and there was only 8 months due me up to that time and instead of what he paid me 184. and 61 cts, but I had 18.61 due to me last year up to Jan. last. So, if the mistake is not discovered the next pay day, then I will have better than 2 months pay ahead. Now when you write, direct your letter as I said, you can go to Webb and let him send my paper to you to Churchtown for I don't get on an average every 4th one and then 3 or 4 weeks old. If you can send me now and then and direct as I said about the letters. I can't write much this morning. I am well and enjoying good health. I live with an expectation of getting home soon at least in 90 days time and that is soon, time begins to look short now.

To John's Martha, I expect to come to your house first if I come out on the Claverack Road, and if I come out on the other road, I expect to stop at Rams Decker's[6] first, and from there I could not tell, but I think to your house first and then home. I dreamed of being home last night.

To Mrs. Mambert I want you to pay all you owe and put the rest of this money I send to you in the bank and you had better put $100 in the bank and not take it home with you and take the remainder and pay up your debts, and if you have any left put it in the bank til I come home for I am barely saving. You must excuse all nonsense in this letter and all mistakes and write as soon as you get this letter and the money.

John W. Mambert [7]

---

6 Rams Decker: Age 42, proprietor of the R. Decker Hotel in the hamlet of Churchtown, Claverack NY.

7 *Letter #50, August 6, 1865, Sand Hill Arsenal Augusta, Ga. To the family.* John W. Mambert letters from the Civil War.

John's story about Steve Lawton, the Confederate soldier originally from Columbia County, NY, is remarkable. Why Lawton fought on the side of the Confederacy is not currently known. What is known, is that before the war (and after) many Northerners relocated to the South for financial reasons; those with an education or a profession could make more money in the South. Lawton might have been seeking to gain financially, leaving the North and settling in Alabama.

Steve told John he had been "pressed into the army," so we can assume he did not purposely go South to fight against the Union. Most likely, he was the victim of the Confederate Conscription Act, passed April 16, 1862, making any white male between eighteen and thirty-five years old liable to three years of military service. Interestingly, there are no records of Steve serving the 26th Alabama regiment. He is not listed on the roster—probably due to his being a jayhawker. This regiment likely caught him and forced into service.

AN ACCIDENTAL DEATH

As time drew closer for the men of the 159th NY to be mustered out of service, their life was becoming more normalized. Light duty, good sleeping quarters, better food, and freedom to explore their surroundings; this was the daily routine. Keeping the men in check and behaving like responsible soldiers was becoming more difficult for Colonel Waltermire. Quartered at the arsenal with no official duties, the men had little motivation to follow through on their routine assignments. They began purchasing, and wearing, civilian clothing when wandering throughout the city—it seemed that the time for wearing government uniforms had passed...if not for the death of soldier John W. Coons.

The Colonel may have attributed the unfortunate death of John W. Coons to the fact that he was not in uniform, possibly seen by his assailant as a thief stealing his property. The event leading to this tragedy occurred the day after John and Coons went foraging for fruits and vegetables. The next day, Coons went by himself to gather provisions when he stumbled upon the infuriated farmer. This could indeed be the reason his death was listed as an accident, because only a few days later Waltermire issued the following:

"It is hereby ordered, that here after no enlisted man in this command shall be permitted to wear any outside clothing, such as pants, dress coats, blouses and hats other than that required by regulations. The only exceptions to be made in the above are these – beside regulation cap the men may wear black soft felt hats, but of no color, linen dusters may be used when not on duty, other than that army dress must be used.

All enlisted men having light-colored hats, citizen pants and blouses must dispose of the same within 3 days. Anyone after expiration of that time found wearing outside citizen apparel will be punished and the property confiscated. Commanding officers of Companies and First Sergeants will be held strictly accountable that this order is enforced."[8]

Colonel Waltermire's concern for the safety and welfare of his men was ongoing, as evidenced by reading past orders he issued. This latest one was no different. He took his position as commander of the regiment seriously, and the health and wellbeing of his men was important to him; he wanted all his soldiers to return safely to their families.

Striving to keep the ranks in order continued to be difficult during this gradual lessening of military duties. During the first week of August, after the officers received their pay, many of them went into town for the evening to indulge in libations. These became days of celebration extending over a week before the men found their way back to camp—likely penniless. Upon their return on August 16, 1865, one Captain, one 1st Lieutenant, the 2nd Principal Musician, six Sergeants and three Corporals from the regiment were reduced in rank for breach of discipline after being absent without leave.[9] The Captain and 1st Lieutenant (for reasons unknown) were accused of a crime and tried by a military Judge, both receiving dishonorable discharges.

Since the incident left the regiment short on officers to manage all the companies, they were marched to the Augusta Arsenal where the 159th NY could be overseen by a reduced staff. The men were surrounded by high walls and limited exits. By the middle of September 1865, the officers "reduced" were returned to their former ranks.

In his next letter, John feels thankful to God and dreams of home. He also shares with family the regiment's big plans for their return:

---

8 G.O. *Colonel Waltermire, to regiment, proper attire*. National Archives Washington DC, Regimental Company Books for the 159th NYSV.

9 Tiemann, William F. *The 159th Regiment Infantry, New York State Volunteers*, In the War of the Rebellion 1862 -1865, (Brooklyn New York 1891), p. 124.

"to Hudson and march through the streets and such, at least we are making calculation upon it…"

Letter #51 - August 24, 1865

159th Regt.
Augusta Arsenal, Ga

Feeling thankful to God for the preservation of my life and health and the privilege to write to let you know that I am well. I hope and trust you are the same. I sent you $175 on the 11th day of this month by the Southern Express Co. which you will find at the Express Co. in Hudson (the Adams) probably you have got it before this time, write and let me know. I told you what to do with it in my letter of the 11th last.

I feel somewhat lonesome and sick of a soldier's life, and I dream often about home. I dreamed last night that Job went away from home. None of us knew where he had gone to, and we searched everything, and could not hear from him and I thought he had gone to look for me, or was somewhere dead and I have felt very strange ever since. I hope this may not be so. I have been looking for a letter in the last 2 mails with great anxiety. I hope you will all take good care of yourselves for I think I will soon be home if the Lord permits, and then our hearts will be filled with joy and gladness. I also dreamed a few nights ago about you (Maria) I mean which has embarrassed my mind. What it was I won't say at present. I hope you will all keep good courage; time seems long, but 2 months more if I live and I will be home. We expect every day to get marching orders for home and we are looking on the bright side of the picture. We are living in tents in a yard of one acre surrounded on all sides by a high wall and buildings something like State Prison with a guard at the gate and no one is allowed to pass out or in, this was done on account of our men on being paid off went down to town and got drunk and did not return in a week or more. We did not have enough men for guard, so we were moved out of our big house in this place. We are laying about 100 yards from the Magazine with some 5 or 6 tons of powder, if that should explode there won't be a grease spot left of us. We have the news today about camp that we will move about 1 ½ miles from the powder house nearer the City, this is Camp life but a short time in one place. The 18th Ohio Regt. does patrol duty in the city, and one company of the 19 Regulars has arrived in the city and the rest are on their way here, and we think to relieve us. The 13 Conn. Lays in the building where we moved from to here, about 400 yards distance. One company of our Regt. has gone away to guard a small place they said 150 miles from here and it is thought that we will all be distributed by Comp. over the Country until our time is out, but this is only supposition. The 176 NY Vols. are downtown, they came yesterday from Waynesborough, 29 miles from here on the Augusta Railroad. The 156 and 175 N.Y. vols. are also about the Country here, I don't know where.

The weather is somewhat cooler than it has been, we have frequent showers here. I and Thomas B. Miller were out peach hunting this forenoon. We can pass the guard most anytime. I have more privilege than others. We eat about a peck of peaches and watermelons, I have ate until I am quite pot-gutted. I was down to the city the day before yesterday and I passed a large looking glass and I saw myself and I was quite scared, I thought I was the humblest man I had seen in a good while. Daddy Wheeler, Thomas Miller and I live in one tent together and daddy is laying on the bed playing baby and trying to get his big toe in his mouth. Steve Lawton is downtown.

I suppose the 128th Regt. is at home and no doubt you have seen them before hearing from me. I gave you some information about Philip Friss[10]. Our Regt. is enjoying very good health; our Bear is getting along fine. If we come to Hudson as a Regt. I hope you will all come to meet us. I think we will come as a Regt. to Hudson and march through the streets and such, at least we are making calculations upon it, and then I want someone to carry

---

10 Phillip Friss, Private 128th Regiment Co. G: Age 18, mustered in as private on August 19, 1862; deserted May 13, 1864 at Alexandria, La. Lived with his parents and five siblings in Churchtown, NY.

my knapsack and such. I and Col. Waltermire are on pretty good terms and Dr. Briggs, and he has told me what he means to do, if he could keep the Regt. together and sober enough to march.

We can't tell which way we come home by land or sea, if by sea then we will have to march from here to Savannah a distance of 130 miles and that will take at least one week and a hard task at that. So, I will close – you must write once a week until the middle of Oct.

Most truly yours,
John W. Mambert

Direct:
Co. B. 159 Reg't. N.Y. Vols.
Augusta, GA (that's all)

My love to you all.[11]

THE AUGUSTA ARSENAL

The Federal arsenal where the regiment was encamped, was seized by the Confederates on January 24, 1861—five days after Georgia seceded from the Union. It remained under their command until it was surrendered back to the Union in May 1865. The arsenal was first constructed along the Savannah River in 1816 but was abandoned after the Black Fever epidemic killed all who resided within its walls, except the commanding officer and one lieutenant. A decision was made to move the arsenal to a new seventy-acre site. The building was deconstructed, the pieces were numbered, and then reconstructed at the chosen location with construction completed in 1828. The new arsenal had two sets of officer's quarters, a barracks, a storehouse, and was totally enclosed within a loop-holed wall. The men of the 159th NY were essentially imprisoned thanks to the disorderly conduct of the regiment's officers.

---

11 *Letter #51, August 24, 1865, Augusta Arsenal, Ga.* John W. Mambert letters from the Civil War.

'MADISON, TOO PRETTY TO BURN'

Quartermaster John H. Charlotte resigned August 22, 1865, and was discharged. He was replaced by 1st Lieutenant Spencer E. Elmer Company K. 1st Lieutenant Edward B. Duffy with Company A, was dishonorably dismissed by sentence of a court-martial August 30, 1865. After the war, Duffy wrote a book on the regiment: *History of the 159th Regiment, N.Y.S.V.*

On August 31, 1865, the regiment received orders to move out. They were detailed to garrison the "Sub-District of Madison, Georgia." They began by rail, traveling 104 miles to Madison, the capital of Morgan County. The 159th NY established headquarters with Lieutenant-Colonel Waltermire commanding the sub-district. All the regimental companies were posted in different towns within the district and assigned provost duty. Lieutenant John Day with companies B (John's Company) and F were sent to Covington in Newton County, but John remained in Madison at the request of Colonel Waltermire.[12]

Madison, Georgia was known as the town General Sherman thought "too pretty to burn," so the rumor goes. The truth is the left wing of Sherman's well-known "march to the sea" was commanded by Major General Henry Slocum. As the Union army neared Madison, Joshua Hill, a prominent citizen and past representative to the U. S. House of Representatives was a strong Unionist and outspoken opponent of secession. When Georgia withdrew from the Union in January 1861, Hill resigned from the House. Taking advantage of his reputation, as the Union army approached, Hill met Slocum outside the town and urged the military to restrict their destruction to sites of military significance only. General Slocum was able to grant Hill's request by following Sherman's Special Field Orders No. 120:

> To army corps commanders alone is entrusted the power to destroy mills, houses, cotton-gins, &c., and for them this general principle is laid down: In districts and neighborhoods where the army is unmolested no destruction of such property should be permitted; but should guerrillas or bushwhackers molest our march, or should the inhabitants burn bridges, obstruct roads, or otherwise

---

12 Tiemann, William F. *The 159th Regiment Infantry, New York State Volunteers*, In the War of the Rebellion 1862-1865, (Brooklyn New York 1891), p. 124.

manifest local hostility, then army commanders should order and enforce a devastation more or less relentless according to the measure of such hostility.

Union soldiers destroyed the depot, slave pens, cotton warehouses, and related buildings; but residences and most businesses were left undisturbed. By this action Hill acquired the reputation as the man who "saved" Madison from destruction. It is indeed likely that Hill's intervention induced the Federal forces to be somewhat more circumspect in Madison. In 1868, Joshua Hill was selected for office by the Georgia Legislature to the United States Senate for Georgia and would forever be known as Georgia's first Republican Senator.[13]

John's life as a soldier has become peaceful; he gained freedoms lost in a life of constant turmoil and can now relax as he plans his return home. He is satisfied with the Union's success in crushing the institution he detests and blames for the rebellion: the aristocracy has been conquered.

Letter #52 - September 5, 1865

Madison, Ga.

To the family

I thought I would write you a few lines to let you know that I am well and enjoying good health for which I am thankful to God. I received a letter today from you dated August the 27 and I was glad to hear that you were all well. You did not mention whether you received the money I sent to you, $175 on the 11th of August by the Southern Express. I hope you will let me know. I have written one letter since. Since writing my last letter we have moved to this place 104 miles farther north on the Atlanta railroad. Our Regt. is divided up into Companies and spread over the Country at different villages in the different Counties. My Company is at Covington 25 miles farther north on the railroad. The Colonel wanted me to stay here with him, so I am doing well and enjoying myself upon luxuries such

---

[13] Madison Georgia, Morgan County. *The Legend of Joshua Hill: The man who Saved Madison from Burning.* https://visitmadisonga.com/

as peaches, watermelons etc. Watermelons are as plenty as punkies in the north, I bought 3 for 30 cents, as large as I could carry. Today I bought 2 ½ doz. eggs for 50 cts butter is from 20 to 30 cts I don't use any. Things are reasonable here. Madison is about as large as Claverack or a little larger and the Capital of Morgan County, we are quartered in the Courthouse and around it. We have 6 companies here yet. It is healthy, well watered – the weather begins to feel like fall. The people are in a state of destitution and have very little to live on. Our journey here was a very pleasant one, we left Augusta on the 31 of August at 6 A.M., and arrived at 3 o'clock P.M. You stated in your letter that you did not receive that letter from Ed Hollenbeck, he saw your letter and his mother sent him one stating that you had been there. He told me that he had let Robert Gardner's son have the letter to leave for Ed Rorabacks and he supposed he had done so. There was over $130, and I don't remember just how much money was in it of Confederate money, but I think $230 or $240, of course the money is good for nothing only to look at. They talk of paying off the Regt. again about the middle of this month, if so, I will send home $50 more. You will see by our move that we won't be home before our time is out. There is to be an election in this state on the 1st Wednesday in October and I suppose we are here for the purpose of keeping peace, etc. and I think by the middle of Oct. we will leave here for home. I also think we will come home on the overland route. I think we will come by way of Nashville, Tenn. I think we can reach New York in 4 days from the time of mustering out. Our time is getting short now. I suppose we would have been home before now, but our officers sent a petition signed by themselves asking to be left in the service until our time was out and it appears that they have succeeded thus far much to the perplexity and chagrin of the Regt. For my part I can stay but it is rather imposing 57 days and our time is up. I hope you will all enjoy yourselves as comfortably as possible, I shall be with you, the Lord being my helper. Tell Martha I thank her for her letter, I am very sorry she works in a public house and great care should be taken of her welfare. I am very glad that John writes and I want him to write until the middle of Oct. and direct your letters as you did this to Augusta for Jones Coons is in the Post Office at that place and I will get the letters very soon, as

I got this. I shall stop at Johns first if I come that way and to Rams Decker's if I come the other way. I suppose Martha has got the wine and I expect I shall get drunk before I leave. I shall come home in my full uniform, and I hope there won't be any ashamed of me, and I have thus far discharged my duty to our national government. We have raised our national banner over all the forts, arsenals, capitals of every state where it was torn down by our enemies, and long may that Star-Spangled Banner in triumph wave over the land of the free and the home of the brave. We have subdued the rebellion; we have wiped out an institution which has degraded our nation. We have conquered the southern aristocracy, the lords of the foul and the brute if not only of the fowl of the brutes, but they also wanted to be lords of 32,000,000 men (so much for want to be). Now about our Regt. the boys are all pretty well and very unruly, we have got quite a number in jail for robbing poor negroes and getting drunk, etc. Our Bear is the same old sixpence most always got up with some excitement or fun. I, Thomas Miller, and daddy Wheeler are in one tent, and we have quite a comfortable place in the Court House yard under a tree. Tell Jobe and Walt they must be good boys and go to school and learn smart, I will soon be home. You must also see that we have plenty of buckwheat flour for you know I like pancakes and honey, I have not had any in three years. You must see that we have the necessities, etc. for it will be late before I get home. I don't think it will be before the 1st of November. You must all write so I can hear from you, this letter you sent came in 8 days, direct as you did this; to Augusta instead of Washington.

I must close for the mail is going away. Write immediately and let me know whether you got the 175 dollars and put all in the bank you can use economy, and be prudent. I have done the best I could. Give my love to all the folk.

Yours
John W. Mambert [14]

---

14 *Letter #52, September 5, 1865, Madison, Ga., To the family.* John W. Mambert letters from the Civil War.

RECONSTRUCTION: ALL EYES ON GEORGIA

General Sherman understood the severity of war. He knew he needed to implement the Union's ideas of restoring states' autonomy once their arms were laid down. Governor Brown of Georgia acted on Sherman's theory and surrendered the state troops, resigned as commander-in-chief, and called the legislature together to meet in Milledgeville, Georgia on May 22, 1865. However, General Wilson of the Union command issued an order forbidding the legislature to meet. Wilson's order stated:

"Neither the Legislature nor any other political body will be permitted to assemble under the call of the rebel state authorities."

The order further stated:

"The people of the state are earnestly counseled to resume their peaceful pursuits and are assured that the President of the United States will, without delay, exert all the lawful powers of his office to relieve them from the bondage of rebel tyranny and to restore them to the enjoyment of peace and order with security of life, liberty and property under the constitution and laws of the United States and of their own state."

Governor Brown resigned, and the Honorable James Johnson of Columbus was appointed Provisional Governor of Georgia by President Andrew Johnson. He was tasked with reorganizing the state. Johnson called for a convention of the people to be elected to take place on the first Wednesday in October 1865. The oath of amnesty had to be taken for a citizen to vote.

On July 15, 1865, James Johnson made an address in Macon at the City Hall. In that address, he declared the single purpose of his appointment as Governor: to enable the people of Georgia to form a government without slavery (which would have to be constitutionally recognized). After that speech, there was conflict within the state over slavery. There were men who hoped that by some miracle, slavery would either be revived or kept going in some small way. It was difficult for them to accept the fact: Slavery was dead.

The reconstruction of Georgia was watched with great interest. The state was instrumental in the establishment of the war and held great importance in the Confederacy's success, but it was also the most devastated as a result. Since all eyes were on Georgia, the reconstruction effort was heightened more so than other Confederate States.

The convention remained in session until November 8, 1865, when a new constitution was adopted repealing the ordinance of secession, canceling the war debt, and abolishing slavery.[15]

MURDER IN THE RANKS

A troubling incident occurred in Monticello, Georgia, on September 26, 1865. It was between two veteran recruits of company G from Hudson. One, a Private by the name of Frank Miller, became drunk and disorderly and attempted to take the life of the other, who happened to be the post's commanding officer, Captain James S. Reynolds. Miller was arrested and confined to the guardhouse. It was reported that during the night he was released by the corporal of the guard unit. Miller, along with several other men, went to a house where a disagreement turned into a fight with Miller being stabbed in the bowels by Sergeant Robert C. Bruce of his company. Miller died from the wounds early the next morning; Bruce escaped, never apprehended. Bruce is listed on the Muster out Rolls as a deserter, with no mention of charges.[16]

TIME TO GO HOME!

The day the men of the 159th NY were waiting for with great anticipation finally arrived. On September 28, 1865, the regiment received orders to prepare the muster-out rolls and report to Augusta by October 13, 1865.

In John's next letter, he updates his family on how some of the local men from Columbia County in the regiment are faring, and due to scuffles with soldiers or citizens, not all men will be returning home.

Letter #53 - September 28, 1865

Madison Court House Ga.
Headquarters 159th N.Y. Vols.

To John and the rest of the family.

---

[15] Avery, I. W. (Isaac Wheeler), 1837 – 1897, *The History of the State of Georgia from 1850 to 1881, embracing the three important epochs: the decade before the war of 1861-5; the period of reconstruction.* (New York 1881), pp. 347-349.

[16] Tiemann, William F. *The 159th Regiment Infantry, New York State Volunteers*, In the War of the Rebellion 1862 -1865, (Brooklyn New York 1891), pp. 124-25.

I thought I would write you a few lines to let you know that I
am well thanks be to God. I hope this will find you the same. I
received your letters and was glad to hear from you. I hope John
will soon recover from his diarrhea, don't eat fresh meats, they
are very bad. Toast, bread and tea are all the diet you should eat. I
have to write to you and Epraim Miller's[17] family the sad news that
Frank Miller[18] is dead. The circumstances are these, Frank and
some others went out one night I think on the eve of the 25, it was
to a dance and got into a ruckus among themselves, and at the
dance a Sargent Bruce[19] of the same Co. G and Frank got in a fight
with each other and Bruce got his knife and stabbed Frank in
the bowels and killed him. They then fetched Frank up to the Co.
dead, and then Bruce deserted. Bruce is from Kinderhook This
I was informed by Capt. Reynolds[20] of Co. G. this morning. Co. G.
lays about 25 miles from here in a place called Monticello. As I
told you before, 4 of our Comp. are scattered in different counties.
I must also tell you that Allen Van Deusen's[21] brother has been
badly hurt by a citizen, by striking him (Elias Van Densen[22]) on
the head above his eye with a stone. Van Deusen was a guard over
the citizen who had been arrested, charged with being implicated
with some negroes stealing bacon. Name is Robert Bruce, from
Hudson, not Kinderhook. Van Deusen was guarding him in prison
and the citizen taking advantage of VanDeusen struck him with
a stone and escaped. VanDeusen is in a critical position not
expected to live, in the midst of life we are in death. I am sorry to
hear of the death of Stickles and others, etc. I was glad to receive

---

17 Ephraim Miller: Age 48, father to Franklin (below).
18 Franklyn Miller, Private 159th Regiment Co. G: Age 20, killed by being stabbed with a knife at Monticello G.A. by Robert Bruce. He was from Livingston, NY. Left behind his parents and nine siblings.
19 Robert Bruce, 159th Regiment Co G., Sergeant: Age 35, Hudson, NY. Lived with Noah Spaulding family.
20 James S. Reynolds, Captain, 159th Regiment Co. G: Age 38, promoted to 2nd Lieutenant Feb. 8, 1864. From Albany, NY.
21 Allen Van Deusen: Age 36, wagon maker in Churchtown. From Livingston, NY. Married with four children.
22 Elias Van Deusen, Private 159th Regiment Co. G: Age 25, a farm laborer and carpenter from Livingston, NY. Lived with his brother and his wife.

your letter and thank you all for writing. It does me good to read your letters. I expect to be in New York on the 1st of November if we have good luck from the 24th of Oct. to the 31st. we will be on the Broad Blue Ocean. I want you to keep a memorandum of the weather during that time and I will. I think we will rendezvous at Harts Island although our Colonel is going to try to be mustered out at Hudson or Albany, if so you will see us at Hudson when we pass, but I don't think he will succeed, I should be happy if he should. Some don't think we will leave here until the 20th of October, but I think before. I think I will write you a letter a few days before we start.[23]

On October 5, 1865, the remaining companies of the 159th NY regiment were rapidly marched from their provost's locations to headquarters in Madison, Georgia. After all remaining companies were gathered, the men loaded into railcars and steamed off for Augusta. When the regiment arrived at their destination, the men were quartered in the old railroad shops next to the depot. Three buildings were used by the rebels as a machine shop during the war. Due to the work, the wood floors were saturated with oil and grease. The buildings formed a large square with a vacant yard in the center.[24]

John sends his last letter to the family about his life as a soldier. He writes with anticipation and enthusiasm, expecting to be back home with them by November 1st.

Letter #54 - October 8, 1865

Augusta, GA.

To the Family,

I thought I would write you a few lines to inform you that I am well and enjoying good health. We left Madison on the morning of the 5th at 11 o'clock A.M. and arrived here at 7 o'clock of the

---

23 *Letter #53, September 28, 1865, Madison Court House Ga., To John and the rest of the family.* John W. Mambert letters from the Civil War.

24 Tiemann, William F. *The 159th Regiment Infantry, New York State Volunteers,* In the War of the Rebellion 1862 -1865, (Brooklyn New York 1891), p. 126.

same day and tomorrow we will be mustered out of the United States Service and then we leave here on Tuesday morning for Savannah. There we are to report on the 13th instead for transportation for New York, if transportation is ready, it will take 4 days to sail from Savannah to New York, but if we are delayed in Savannah, we will reach New York accordingly and there we will be mustered out of the State Service, we have to be mustered out twice.

When we arrived here, I received 2 letters from you all, and 1 from John, 1 from Eb, 1 from Permelia, 2 from Job, 1 from Walt. I was very glad to hear from you to think that you are all well and enjoying yourselves. I am also very sorry that I can't have pancakes for I have had none in 3 years and it seems just as if I could taste them now. About the cow, I think that butter and milk would be very good this winter, however you know best, do as you like about selling. I am sorry to hear from Allen R. VanDeusen. His brother Elias is doing very well, I think he will get along, he and I walked out the day before yesterday. I carried his knapsack to the City Hotel from there he goes to Savannah by the Steamboat, we go by Rail Road to Waynesboro, from there we march 55 miles and then the Rail Road again to Savannah.

Walt wrote he or mother had 2 qts of woodchuck oil that is just what I want for my hair is very thin and will soon be bald and that will perhaps thicken it up. Jobe says that he will go west and buy a farm for me to go on, well that is a good idea for I think we will get there if nothing happens. Permelia said Jobe had such a big mouth, well that is better than no mouth at all, that is so he can suck candy. I suppose he is almost a man.

Eb says he made $50 in 18 days well done Eb, that is the way to do it. Keep on doing so and you will do well. Permelia sees me in one place and then another, well that is really so for we are but a short time in one place but catching the horse that was a mule.

To John I am glad he has been very faithful in writing and also in having plenty of work and that he is doing well. I suppose I will soon be where we can see each other face to face. John's Martha, I suppose Martha you have got the wine, if so, I suppose you will have to make a bed for me, for I suppose I will get tipsy. I have not been tipsy since I have been in the service, the rest get drunk every pay day and then they stay drunk about a week and some 2 weeks. So, I will get tipsy, once in 3 years. Martha works in Johnstown I suppose, and I hope is doing well. As I trust she must be very fleshy to weigh 140 lbs, I weigh 154 lbs when I eat 2 big watermelons. Well a few lines to mother, I suppose we will have to have a wedding when I come home, that is if you will except me, but we shall have to sleep on the floor for I can't sleep in a bed, I am to use to lay on the floor and out of doors on the ground that I can't sleep good unless I can see the moon and stars and have the fresh aid and strange to say I never take cold. I will here say that I variably believe that those tight rooms in which you sleep are the cause of sickness and consumption and shortens the days of thousands. I think living in a tent is much healthier. There is quite a number in our Regt. got the fever again, all the bummies at least. I ate the last watermelon the day before yesterday. Sweet potatoes I don't like very well, they are not as good as they are up there. We have got a great Regt. we have got a bear a dog without hair, they call him a Mexican dog, he has a strip of hair from his nose between his eyes up as far as his ears about as wide as your finger, he is a little larger than Mig. The mornings are a little cool and he shivers and crawls in the sun. I forgot to say his tail is drove in or cut off. We have got a prairie wolf and a possum and a prairie squirrel and 2 gray squirrels, the squirrels when we march set on the boy's shoulders and on their heads and all over, so you see we are well stocked if we get them all through. The weather here is very nice, just warm enough to be comfortable. The nights are cool but the sun in the daytime is quite hot. I think the cold weather will have a bad effect on me, mother will have to keep me in a bin box this winter under the stove or on the mantelpiece under the stovepipe. The weather here agrees with me very well; my blood is very thin and cold has a bad effect.

Well, have I got any clothing at home or not, I suppose not, I expect they are all worn out and if so alright for then I will get new clothes. Dr. Briggs says I must write a few words for him. Well, I will say that he is a darn clever feller for he gives me whiskey when I get a belly ache, and that is pretty often.

I must close, I suppose you are all doing well, and I hope soon to be at Hudson. I hope you won't, any of you come to New York, but I hope to meet some of you at Hudson, if so, you must fetch me with a carriage. I suppose you have saved my military overcoat and vest, at least I hope so. I sent you a box of clothing with J.B. Miller[25] to his sister Stephen Austin's wife in Hudson, each pays half of the expressage. The books belong to me and all except what is in Tom's rubber blanket. This will be my last letter I shall write for I soon expect to be with you. My love to you all.

John W. Mambert [26]

ANOTHER TREMENDOUS LOSS

On the morning of October 9, 1865, the regiment experienced another tragic incident. A fire broke out in one of the guardhouses at the end of their quarters in the depot building. It was assumed prisoners started the fire to liberate themselves. Like a flash of lightning, the fire fueled by the oil-soaked floors darted from one end of the building to the other; almost immediately the whole structure was a raging mass of flames. The men scrambled, barely having time to remove their arms and equipment, most of them losing all their clothing to the fire except for what they were wearing. One man, Alexander Keist[27] from Company C, was missing and assumed to have perished in the fire. The regiment's poor bear, who had been with them so long and to which the men had become greatly attached, was tethered to a post in the vacant

---

25 Jesse B. Miller, 159th Regiment Co E: Age 18 (actual 16), discharged from wounds in action Irish Bend, La. Sept.27, 1863. A farm laborer From Claverack, NY. Lived with his parents and three siblings.

26 *Letter #54, October 8, 1865, Augusta, GA., To the Family.* John W. Mambert letters from the Civil War.

27 Alexander Keist, Private 159th Regiment Co C: Age 24, died by burning to death in barracks fire in Augusta, Ga, October 9, 1865. From Livingston NY. Left behind his parents. His father, also Alexander Keist, age 47 served as a private in the 139th NYS Volunteers.

square. The heat from the fire was so intense it was impossible to get to him and release his chain. He was roasted to death. This was the second time the men of the 159th NY had all their belongings destroyed by fire. Due to this heart-wrenching tragedy, the men gave up hopes for a public parade back home, which they were already planning.[28]

An article in the Hudson Gazette documented the loss:

> The 159th serious loss: Some three weeks ago, while the 159th Regiment was awaiting transportation at Augusta, a fire broke out in the barracks and before it could be extinguished several men were seriously burned and much property destroyed. A private named Keist is said to have been suffocated and burned to death, and Mr. John Reed of Hudson, was so much injured that he had to be left in the hospital. The latter has a family here who have suffered much during his absence and are in need of assistance. The Regimental Colors and battle flags, so highly prized by the men, were also destroyed, together with much of their personal effects. They have also to mourn the loss of their "pet bear," which had afforded so much amusement to the men in their wearisome round of duty. A good story is told of this pet, which is said to have been presented to Colonel Molineux, on the Red River expedition. On the occasion of a visit from General Emory, the Colonel ordered a subordinate to send up the band. The orderly misunderstood him, and brought out the pet bear. General Emory was pleased with the gentle manners of Bruin, and permitted him to have a "whisky ration." This so elated the animal that he became familiar, and as the General approached raised his forepaws and, in spite of military etiquette gave him a very cordial hug. It is hinted that the response of the General to this situation would not appear well in print! [29]

---

28 Tiemann, William F. *The 159th Regiment Infantry, New York State Volunteers*, In the War of the Rebellion 1862 -1865, (Brooklyn New York 1891), p. 126.
29 *Hudson Gazette Newspaper*, NYS Military Museum & Veterans Research Center, Unit History Project, 159th NYS Regt. Newspaper Clippings.

MUSTERED OUT

On October 12, 1865, the 159th NY was called to assemble at the headquarters of the Department of Georgia where they were mustered out of service for the United States by Lieutenant S. S. Culbertson, United States mustering officer. The next morning, the soldiers of the 159th NY turned their faces homeward and began their march out of Augusta. Two days later in the early morning, they reached a small station on the Central Railroad and boarded rail cars transporting them to Savannah. Upon their arrival, the men went into camp by the train depot, anxiously awaiting their transport home.

On October 18, 1865, the men of the 159th NY were marched to the wharf and embarked on the transport *Varuna*, a screw steamer. The following day, the regiment began sailing for New York City, arriving four days later. They disembarked and marched a short distance boarding a ferry to Hart's Island. The regiment made up the final accounts for the quartermaster, ordnance stores, and received final pay.[30]

October 25, 1865, the 159th NYSV Regiment Infantry was mustered out of service for New York State and discharged. The regiment's active number at the time of mustering out was 346, a reflection of the loss of fighting men, a sixty-six percent reduction since its inception three years earlier.

During 1865, six men died from sickness (one from alcoholism). Fifty-five soldiers and officers were discharged with a disability—sickness related. One man was murdered by the hands of a comrade. Four men suffered and died in Confederate prison camps. There was also a large number of discharges, twenty-one. Lastly, those who scampered north without a furlough were minimal; only five deserted.

The following map depicts John's home and also those of the men who served alongside him. It represents one small area in, and around the town of Taghkanic which provided a large number of men serving in the Civil War. The stars identify John's neighbors who he mentions in the letters. Two men serving with the 128th NYSV from Taghkanic also gave their lives to the war.

---

30 Tiemann, William F. *The 159th Regiment Infantry, New York State Volunteers*, In the War of the Rebellion 1862-1865, (Brooklyn New York 1891), p. 127.

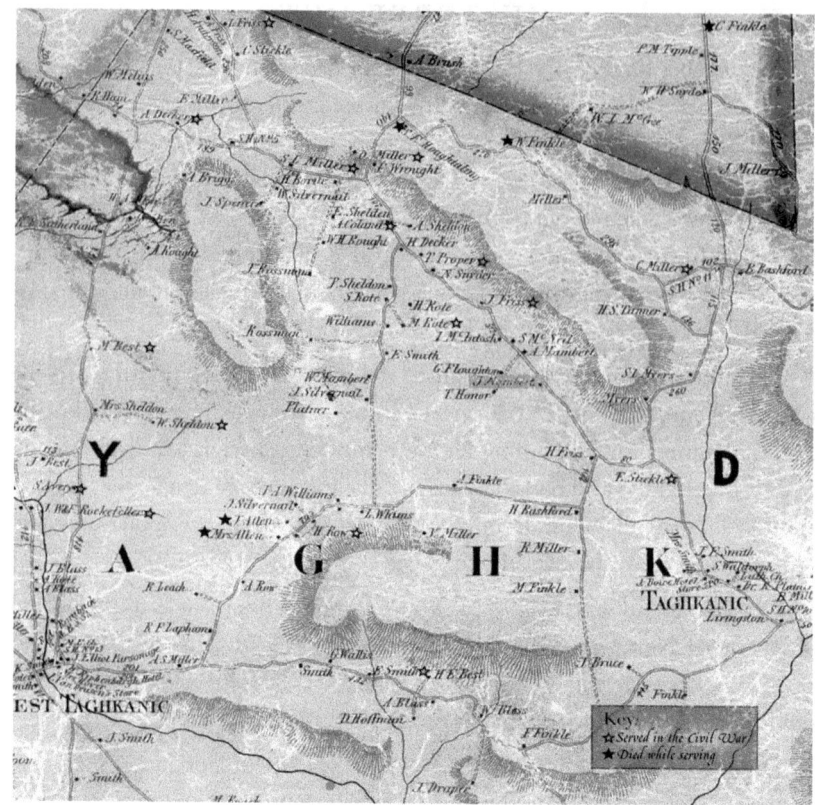

Resident map of Taghkanic in Columbia County, NY. [31]

After three years of service to the Union army, the 159th NY disbanded. The men from Brooklyn, having a rather short trip, soon returned to their homes and families. On October 27, 1865, the Columbia and Dutchess County companies anxiously loaded a train and departed New York City. They arrived in Hudson later that day and were given a hearty public reception, after which they separated. The regiment, as a body, became a thing of the past.[32]

The Columbia Republican Newspaper printed on Tuesday October 31, 1865:

The 159th Regiment, Its Return Home-Reception at City Hall: About one

---

[31] *1858 map of Columbia County, NY.*, Library of Congress. Editorial illustration by Garrett Kipp.

[32] Tiemann, William F. *The 159th Regiment Infantry, New York State Volunteers*, In the War of the Rebellion 1862 -1865, (Brooklyn New York 1891), pp. 127-128.

hundred and fifty of the brave soldiers composing the 159th Regiment, mostly from this county, returned home on Thursday and Friday last, having been paid off and discharged at Hart's Island. Colonel Waltermire, with several of the remaining officers and men, reached here at half past 11 on Friday, and were met at the depot by a large gathering of citizens, who gave them a most cordial greeting. The Mayor, Firemen and reception committee were also present with the Hudson and Claverack bands. A large proportion of the soldiers who had previously returned were also present by invitation, and all formed in line and were escorted through the streets in a drizzling rain (a long and muddy march for a peace footing,) to the City Hall. Residences and places of business all along the route were thrown open, and the returning heroes were welcomed by cheers, salutes and the waving of handkerchiefs and flags at almost every step. It was a very interesting and impressive scene- one perhaps never to be re-enacted in the life-time of those who participated in it. The "boys" seemed highly pleased with the demonstration and expressed their delight in a variety of select phases.

On reaching the Hall the soldiers were met by a large company of ladies and invited to take seats at tables which were spread with an ample feast. The officers of the day and bands passed upon the stage, and Mary Rogers welcomed the soldier's home in a few appropriate words. After music by the band and prayer by Rev. J. Mc Holmes, the soldiers were addressed by Honorable Theodore Miller.

Colonel Waltermire responded on behalf of the Regiment. Mr. Miller spoke in a very eloquent and feeling manner, and was listened to with good attention.

The soldiers were then waited upon by the ladies, and enjoyed a hearty repast, with tea and coffee, and desert of choice fruit. Cigars were also freely distributed, and for an hour or two the boys had a "smoking good time." They were also entertained by a musical performance and patriotic song from Wagner's minstrels then in the Hall.

The exercises were conducted by Colonel E. L. Gaul, formerly the 159th. Colonel Waltermire was introduced by the Marshal, and on stepping forward was loudly applauded by soldiers and citizens, but declined, Grant-like, to make a speech. He was overwhelmed with "attention" during the day and evening. The "boys" of the Regiment regretted very much the loss of their colors and uniforms at Augusta, which compelled many of them to appear in citizens clothing. The flag upon which the battle-fields of the 159th are inscribed was displayed at the Hall, together with the returned colors of the 128th Regiment. In common with all citizens, we welcome home the brave soldiers who have "fought a good fight" and helped to achieve the great victory which saves our nation from anarchy and ruin.[33]

John W. Mambert Discharge Papers [34]

RETURNING TO CIVILIAN LIFE

The men of the 159th NY endured much while serving, the heat and rain of the extreme South, the frost and snow of the North and the dust and mud of both. The soldiers proved their loyalty, always ready at the call of duty. The losses on the field of battle and the commendation of superior officer's attest to their faithful service and duty well done. The brave men of the command could be proud of having been members of the 159th New York

---

33 *The Columbia Republican Newspaper, Tuesday October 31, 1865*, NYS Military Museum & Veterans Research Center, Unit History Project, 159th NYS Regt. Newspaper Clippings.
34 John W. Mambert Discharge Papers. Property Michael L. Kipp.

State Volunteer Infantry Regiment.[35]

After the celebration, the veteran soldiers began the most important march they made in three years, one which would take them over old familiar ground as they made their way back to their families. How different life would be for the men who spent the last three years living in harsh elements and eating substandard food. Finally, they were on their way back to a life of proper shelter, a comfortable bed, warmth, and food that doesn't have to be soaked in vinegar to be edible. But more important than all those comforts was family. Finally, the men would be reunited with their wives and children, loved ones they dearly missed being so far away. Families grew during the war years. Children were born, new grandchildren arrived, some fathers returned to meet a new son or daughter-in-law for the first time. Without a doubt, the men themselves had changed, returning to a home and family somewhat altered from the one they left a few years ago, a lifetime ago.

As for John's journey home, one has to wonder what direction he took. Did his family come to the city and await his arrival, or did he take one of the two routes he discussed in his letters? Whatever direction he chose, John was on the final leg of a long tiresome adventure and was returning to his family.

John's return was just in time for the birth of his second grandchild, born to John Jr. and his wife Martha, now living in their new house. John's plans for his oldest daughter Permelia to marry the fisherman from Long Island never happened. A local boy, Ebenezer Duntz, courted her in the fall of 1863; they married soon after John arrived home. John's middle child, Martha, did not make the trip quite as far west as she had hoped. She did however, venture out on her own, moving to the little hamlet of Johnstown six miles west of her parents' home, taking employment at the local public house.

John's two youngest, Walt and Jobe, now fourteen and twelve respectfully, were home helping their mother. The two adventurous young men kept busy while John was away, starting a fur business, fishing for eels, teaching new tricks to the family dog Mig, and being hired out for day jobs. Walt appeared to have steady employment throughout 1865, which John encouraged. Ironically, in many letters John wrote home, he included a plea for the two boys to go to school, continuously pressing the issue.

John Jr. and Maria appeared to have handled the family business

---

35 Tiemann, William F. *The 159th Regiment Infantry, New York State Volunteers*, In the War of the Rebellion 1862-1865, (Brooklyn New York 1891), p. 134.

satisfactorily while John was serving. They were in charge of collecting rent on the different properties, some being residences and others farm property, additionally paying notes they were obliged to. John was in partnership with Andrew Decker and Sylvester Bortle on property owned by J. W. Elsaffer, a timberman from bordering Dutchess County. They were all farmers, so the assumption would be they planted crops on the land; John was concerned throughout the war about losing the property.

LATER IN LIFE

John Jr. and Martha sold the property purchased in 1863, along with the adjoining lot, to the City of Hudson, NY. A stream ran through the property which was dammed creating a reservoir to supply water to the city of Hudson, NY. Today, the abandoned house still stands next to the dam on Reservoir Road, just south of the hamlet of Churchtown. A carpenter by trade, Junior moved his wife and two daughters to Hartford, NY. Permelia, and her husband Ebenezer, with their daughter also made the move; Ebenezer, like Junior and John, was a carpenter.

Seven years after the war, John and Maria sold their home in Taghkanic making the move to Hartford, NY, in 1872. The family members residing in Hartford may have been working together in the same trade. The Southern Central Railroad was newly completed through Hartford, and members of the town raised funds to build an attractive three-story hotel. As carpenters, Ebenezer, Junior and John may have been employed on this project. The region was rich in timber and the town had a sawmill, gristmill, and distillery–a farming community with all the necessities.

While many in the family were heading to Hartford, the two youngest decided to venture west to Iowa where their sister Martha and her husband Peter B. Davis (another carpenter from Taghkanic) had already moved. She and Peter apparently took the written advice of her father and made the decision to move west to Iowa. They settled in the hamlet of Beaver, Butler County, where they raised three children.

On April 11, 1875, Jobe married a girl from Galesburg, Illinois, by the name of Sarah Jackson. Jobe, with his new wife, made the decision to move back to Taghkanic in 1876 after his grandfather Job Sheldon passed away. He then took charge of the Sheldon family farm while raising four boys. The second youngest of John's sons, Walter, married Annie Church from Illinois on October 31, 1876. Soon after, they established a farm in Albion, Iowa, and raised two boys.

THE DEATH OF JOHN W. MAMBERT, VETERAN OF THE CIVIL WAR

On April 2, 1879, at the age of fifty-nine, John W. Mambert, a veteran of the Civil War, died from stomach cancer. Some believe a leading cause of stomach cancer was salt and salt-rich foods. With all the rancid salted meat John consumed for three years serving the Union, it makes one wonder if that, along with all his stomach ailments throughout the war, contributed to his disease.

After John's death, Maria moved back to Taghkanic to live with her youngest son Jobe and his family in the farm where she grew up. In 1890, Maria sold the farm to Jobe for one dollar; she died two years later at seventy-one years of age.

John Jr. and his family returned to Columbia County the following year, establishing a new residence in Claverack, NY, where he remained until his death at the age of ninety-three. Permilia and her family remained in Hartford for ten additional years, then moved 200 miles southwest to the town of Richford, Tioga County, NY. After Ebenezer's death in 1907, Permelia also returned to Columbia County, residing in Hudson, NY. She passed away in 1920 from breast cancer.

Martha and Peter also returned to Columbia County during the 1890s, settling in Hudson. Martha died at the age of seventy-two, and Peter passed away three years later at the age of eighty-three.

Unlike the rest of the Mamberts, Walter and his family did not return to New York, but remained in Iowa where he was postmaster in the town of Swanton, Butler County. Walter died from diabetes at the young age of forty-five. His wife moved to Oklahoma City, Oklahoma, and died at the age of sixty.

Jobe, the youngest of John and Maria's children, died May 23, 1923; his wife Sarah passed away seven years later on March 5, 1930. Jobe and Sarah's house was located on Mambert Road, an extension of Reservoir Road. The homestead remained in the Mambert family for two more generations until it was sold after the widow of Jobe's grandson departed from the property.

In the 1780s, John W. Mambert's grandfather, John Manbrut, moved from Pennsylvania to settle in Taghkanic. After six generations, the Mambert name in Taghkanic, NY, came to an end.

Epilogue

Receiving my great-great-great-grandfather's letters and having the opportunity to transcribe them I realized there was more to his writings than just the regiment's military missions. The additional information on politics, health, friends, and family gave me a "broad brush" to paint the life of the common foot soldier. But the one subject John wrote about in many of his letters was one I never gave much thought to, that of the small contingency of Southerners known as aristocrats, the people ultimately responsible for the largest tragedy (to date) this country has experienced since its inception—the great rebellion.

I soon realized the extensive influence this group of greedy, wealthy individuals had over the Southern population. They were the authority holding all the power, the institute manipulating control, the ones creating the deception leading poor Southerners off to a war that was never intended to benefit them, only strengthen the Southern aristocrats.

In the end, after sustaining great loss of life and economic ruin to the South, our nation survived, a United States. I feel the words written by John W. Mambert in Letter #52 sum up the situation:

"We have subdued the rebellion; we have wiped out an institution which has degraded our nation. We have conquered the southern aristocracy, the lords of the foul and the brute if not only of the fowl of the brutes, but they also wanted to be lords of 32,000,000 men (so much for want to be)."

I can't help but realize similarities in society today where small minorities of individuals are capable of persuading our government to agree with their desires over those of the majority. Witnessing this, I can't help but think power and deception are once again tearing our country apart. I wonder if the level of animosity between people of this nation has currently become so intense that common ground is no longer an option. Could our differences create the same level of opposition? Am I being unrealistic in my reasoning? I don't know, but across this nation clashes are happening between opposing groups, and just like in the past, the media is doing a fine job stoking the fire.

In the epigraph of this historical narrative I quoted Mark Twain, "History never repeats itself, but it often rhymes." After further research, I discovered he likely never stated that. However, I did find the words of Psychoanalyst Theodor Reik, "It has been said that history repeats itself. This is perhaps not quite correct; it merely rhymes."

I can only pray that the people of our great nation find common ground in the midst of our current political instability soon. I think often about the legacy we leave our children, grandchildren, and all future generations, wondering if they will have the same rights and freedoms we so enjoy.

WILLIAM F. TIEMANN

William F. Tiemann (Frank) provided a great deal of historical data on the 159th NY regiment in 1891, compiling and publishing: The 159th Regiment Infantry, New York State Volunteers, In the War of The Rebellion 1862–1865. William was a Captain serving with the 159th NY Co. A. He enrolled as a Sergeant on September 5, 1862, at the age of eighteen. He was promoted to Sergeant Major November 2, 1862, promoted Second-Lieutenant April 30, 1863, and to Captain February 23, 1864, and then commissioned to the rank of Major.

Frank was wounded at the Battle of Irish Bend on April 14, 1863. He became a prisoner of war at the 3rd battle of Winchester on September 19,1864 and was held for five months until a prisoner exchange returned him to the regiment. Frank's older brother Julius, also enlisted in the 159th NYS Volunteers at the age of twenty-two. He enlisted as a First Lieutenant on September 24, 1862, for Company B and was discharged on December 24, 1863.

Historian of the regiment William "Frank" Tiemann died May 2, 1926 in East Orange, New Jersey. He was one of the last surviving members of the 159th New York State Volunteers. In the preface of his compiled history of the 159th NY, Frank thanks the Adjutant-General U.S.A. for full information provided and the permission to copy and verify the muster-out rolls. He thanks General Molineux, Sergeant Berridge, and musician Dunham for the use of their war diaries. He also thanks Colonel Gaul, Lieutenant-Colonel Burt and Lieutenant-Colonel Waltermire for the information they furnished.[1]

---

1 Tiemann, William F. *The 159th Regiment Infantry, New York State Volunteers*, In the War of the Rebellion 1862 -1865, (Brooklyn New York 1891), p. 3.

William "Frank" Tiemann [2]

The 159th New York State Volunteers Recapitulation:

STRENGTH AT MUSTER IN	OFFICERS	MEN	AGGREGATE
Original strength	38	802	840
Promoted from ranks	13	-	13
Promoted by appointment	12	-	12
Recruits	-	175	175
Total	**63**	**977**	**1040**
Less promoted from ranks	-	13	13
Total in Regiment	**63**	**964**	**1027**

The total losses to the 159th during service:

---

[2] E. L. Molineux Photo Collection, *William "Frank" Tiemann*, photo. NYS Military Museum and Veteran Research Center

	OFFICERS	MEN	AGGREGATE
Died	10	176	186
Discharged	35	201	236
Transferred	-	38	38
Dishonorably dismissed	5	-	5
Deserted	-	178	178
Unaccounted for	-	38	38
Total	**50**	**631**	**681**

Strength at muster out:

	OFFICERS	MEN	AGGREGATE
Present	13	290	303
Absent: Prisoners of War	-	2	2
Sick	-	24	24
Detached	-	8	8
Missing	-	1	1
In confinement	-	6	6
Without leave	-	2	2
Total mustered out	13	333	346[3]

THE 159TH NEW YORK STATE VOLUNTEERS' JOURNEY	
MODE	MILES
Transport	7,491
Rail	508
Foot	1,333
Total	**9,332**

---

3 Tiemann, William F. The 159th Regiment Infantry, New York State Volunteers, In the War of the Rebellion 1862 -1865, (Brooklyn New York 1891), pp. 132-134.

EPILOGUE

LOCATION OF 159TH NYS VOLUNTEERS	TEMPORARY LOCATION OF JOHN W. MAMBERT
Baton Rouge, LA December 17, 1863–March 28, 1863, US Arsenal grounds: location where regiment trained for 3 months.	
Port Hudson, LA March 15, 1863–March 20, 1863, Clinton Plank Road engagement used as a decoy for Navy to move past Port Hudson.	
Thibodaux, LA April 1, 1863–April 3, 1863, Camp.	
Bayou Boeuf, LA April 4, 1863–April 6, 1863, Camp; stored supplies.	
Brashear City, LA April 8, 1863–April 10, 1863.	
Irish Bend, LA April 13, 1863, McWilliams Plantation on west bank of Grand Lake Regiment landed and the battle began.	
Franklin, LA April 14, 1863: Fort Bisland Battle.	**Berwick City Hospital** May 1, 1863–June 1, 1863, admitted to hospital for exhaustion and severe diarrhea. **Brashear City Convalescents Hospital** June 1, 1863, camp within the fort where the sick get healthy. **Battle at Brashear City** June 23, 1863, held there as a prisoner until paroled on July 4, sent to Union line. **Ship Island, MI** July 11, 1863, John was transported to this location as part of a prisoner exchange. July 20, 1863, he was then transported to NOLA where he would eventually find his way back to his regiment at Fort Kearney.
Siege at Port Hudson May 23, 1863–July 8, 1863.	
Bayou La Fourche, LA July 13, 1863: Battle.	
Fort Kearney, LA Late July–August 29, 1863, Training camp (Camp Kearney) located 4-5 miles above New Orleans.	
Thibodaux, LA September 1, 1863–March 15, 1864, winter quarters; camped on Madam Guion's Plantation 65 miles southwest of New Orleans.	
Alexandria, LA March 30, 1864–May 11, 1864, camped at this location the regiment had two engagements: Red River campaign, March 10–May 22, 1864, Alexandria, May 2, 1864.	

MICHAEL LEE KIPP

LOCATION OF 159TH NYS VOLUNTEERS	TEMPORARY LOCATION OF JOHN W. MAMBERT
Mansura Plains, LA May 16, 1864, participated in major battle.	
Morganza, LA May 22, 1864–July 2, 1864, camped at this location between the river levee in a grove of cottonwood trees at the mouth of the Red River. One engagement on May 24, 1863.	
Algiers, LA July 4, 1863–July 17, 1864, regiment was quartered in the Belvidere Iron Foundry.	
Bermuda Hundreds, VA July 25, 1863–July 31, 1863, camped in various locations along the defenses.	
Washington, DC August 1, 1864–August 12, 1864, soldiers rest located at RR depot, August 2 marched to camp 6 miles northwest of the Capitol on a hill next to Fort Gaines.	
Halltown, WV August 22, 1864–August 28, 1864, (6 miles from Harpers Ferry) Battle.	
Charles Town, VA August 28, 1864–September 3, 1864, regiment posted behind breastworks, drilled daily.	
Berrytown, VA September 3, 1864–September 18, 1864, Battle then went into camp and did picket duty and drills.	
Winchester, VA – September 19, 1864:Battle (Ouan).	
Strasburg, VA September 21,1864–September Battle at Fishers Hill, regiment detailed to take charge of prisoners, captured arms and transport to Winchester.	**Winchester, VA Field Hospital** September 20, 1864 – November 15, 1864 John has been assigned to work in the hospital taking care of the wounded, eventually becoming the brigade carpenter whose duty is to make coffins
Mt Crawford, VA September 28, 1864–October 7, 1864, camp.	
Cedar Creek, VA October 10, 1864–October 19, 1864, camp: detailed to construct breastworks then went into a major battle on October19, Major-General Sheridan turning the engagement around and into a Union victory.	
Strasburg, VA October 20, 1864–November 9, 1864, camp.	
Newton, VA November 9, 1864–November 28, 1864, Camp; constructed breastworks and named it Camp Waltermire or Bear's Retreat or "A Rough Hug for Rebs." John rejoins regiment at this location.	
Strasburg, VA November 29, 1864–December 30, 1864, construction of log huts for winter quarters named Camp Royal (Front Royal Road).	

LOCATION OF 159TH NYS VOLUNTEERS	TEMPORARY LOCATION OF JOHN W. MAMBERT
Stephens City, VA December 30, 1864–January 5, 1865, camped at the RR depot terminus to Harpers Ferry, WV.	
Baltimore, MD January 7, 1865–January 13, 1865, camped at Camp Carroll waiting for transport to Savannah, GA.	
Savannah, GA January 20, 1865–January 26, 1865, camped at the Georgia Central RR depot.	
Savannah, GA January 27, 1865–March 9, 1865, camped at fortifications on west side of the city (Ogeechee Road).	
Morehead City, NC March 21, 1865-May 3, 1865, camp set up next to railroad, while at this location the regiment guarded the commissary stores.	
Savannah, GA May 7, 1865–May 11, 1865, camped at old Quarters.	
Augusta, GA May 16,1865–August 29,1865, regiment quartered in the Old City Hotel. Provost duty.	
Madison, GA August 31, 1865–October 5, 1865, regiment assigned to provost duty.	
Augusta, GA October 5, 1865–October 13, 1865, regiment quartered in the old railroad shops; Mustered out of US service.	
Harts Island, NY October 22, 1865–October 25, 1865, regiment mustered out of state service.	

MUSTER OUT ROLLS

Copied from those on file in the Adjutant-General's Office, Albany, New-York, by permission of the Adjutant-General, State of New-York. Revised and corrected from Bi-monthly Muster and Pay Rolls in possession of the company commanders as far as obtainable.[4]

---

4 Tiemann, William F. The 159th Regiment Infantry, New York State Volunteers, In the War of the Rebellion 1862 -1865, (Brooklyn New York 1891), pp. 137.

The 159th Regiment Infantry, NYS Volunteers - Field and Staff

NAME	AGE	RANK	DATE JOINED	PLACE JOINED	NOTES
William Waltermire	30	Lieutenant Colonel	10/15/1862	Hudson	Promoted from Captain 1/10/1864
Caleb C. Briggs	36	Assistant Surgeon	10/30/1862	Albany	Commissioned Surgeon
Homer A. Nelson		Colonel	10/31/1862	Hudson	Resigned 11/25/1862
Edward L. Molineux	29	Colonel	08/28/1862	Brooklyn	Promoted to Brevet Brigadier General 01/07/1864
Chas. A. Burt		Lieutenant Colonel	11/25/1862	New York	Promoted 4/14/1863
Edward L. Gaul	25	Lieutenant Colonel	09/18/1862	Hudson	Promoted 1/10/1864
Charles A. Robertson		Surgeon	08/30/1862	Albany	Resigned
William Y. Provost	25	Surgeon	09/29/1862	Brooklyn	Promoted 1st Assistant Surgeon 07/28/1865
Isaac L. Kipp		Chaplain		New York	Resigned 09/09/1863
Mark D. Wilber	33	Regiment Quarter Master	09/18/1862	Hudson	Resigned 12/05/1863
John H. Charlotte	20	Regiment Quarter Master	09/18/1862	Hudson	Promoted Quarter Master Sergeant Major 08/22/1865
Gilbert A. Draper	27	Lieutenant Colonel	08/28/1862	Brooklyn	Died in action Irish Bend 4/14/1863
Robert McD. Hart	22	Major	09/06/1862	Brooklyn	Died in action Cedar Creek, VA 10/19/1864
Robert D. Lathrop	22	Adjutant	09/17/1862	Hudson	Died in action Irish Bend LA 04/14/1863
Non-Commissioned Staff					
Gilbert S. Gullen	21	Sergeant Major	09/23/1862	Brooklyn	Promoted 08/16/1865
William A. Jaquins	26	Quarter Master Sergeant	09/26/1862	Hudson	Promoted 01/25/1864
Thomas Bergen	29	Commissary Sergeant	08/28/1862	Brooklyn	Promoted 07/25/1863
Edward E. Baker	26	Hospital Stewart	09/25/1862	Brooklyn	Promoted 11/22/1863
Thomas B. Miller	15.5	1st Principal Musician	09/13/1862	Livingston	Promoted 01/28/1864
George D. Dayton	21	2nd Principal Musician	09/04/1862	Brooklyn	Promoted 11/02/1862
William F. Tiemann	18	Sergeant Major	09/05/1862	Brooklyn	Promoted 11/02/1862
Marshall A. Dunham	22	Sergeant Major	09/17/1862	New Lebanon	Promoted 03/19/1864
William E. Palmer	18	Sergeant Major	09/05/1862	Brooklyn	Promoted 04/09/1864, Discharge for disability on 09/07/1865
Alfred H. S. Moore	24	Hospital Steward	09/05/1862	Brooklyn	Promoted 11/02/1862, Discharge for disability on 11/10/1863
John W. Mambert	43	Musician	09/20/1862	Taghkanic	Reduced from Principal Musician 01/28/1864
William F. French	19	Commissary Sergeant	09/18/1862	Hudson	Not accounted for at Muster out
Alfred H. Bruce	23	Sergeant Major	09/15/1862	Kinderhook	Discharge 03/06/1863 Enrolled 1st Sergeant Company G
Herman Smith	18	Sergeant Major	09/03/1862	Brooklyn	Discharge 06/10/1863 Promoted 2nd Lieutenant Company G

The 159th Regiment Infantry, NYS Volunteers – Company A

NAME	AGE	RANK	DATE JOINED	PLACE JOINED	NOTES
William F. Tiemann	18	Captain	09/05/1862	Brooklyn	Promoted 02/23/1864, Commissioned Major
Edwin F. Atwood	20	1st Lieutenant	08/18/1862	Hudson	Discharged 01/19/1863
Edward Duffy	32	1st Lieutenant	10/06/1862	Hudson	Promoted 05/23/1864, Dishonorably discharged 09/30/1865
John W. Manley, Jr.	22	1st Lieutenant	09/24/1862	Brooklyn	Died in action Irish Bend, La. 01/14/1863
Joshua D. Harrington	24	1st Sergeant	09/12/1862	Hudson	Promoted 03/16/1864
Thomas Berridge	22	Sergeant	09/06/1862	Greenport	
Henry A. Osborn	21	Sergeant	09/19/1862	Claverack	Promoted 09/09/1865
Richard M. Mosier	22	Sergeant	09/12/1862	Hudson	Promoted 09/09/1865
Robert Gardner	23 1/2	Sergeant	09/09/1862	Hudson	
Rossman Huested	25	Corporal	09/05/1862	Hudson	
Samuel Parks	20 1/2	Corporal	09/11/1862	Greenport	Promoted 09/09/1865
William Brenzel	22	Corporal	09/20/1862	Livingston	Promoted 09/09/1865
William H. Scott	18 1/2	Musician	09/10/1862	Hudson	
Thomas Akins	24	Private	09/12/1862	Hudson	
Victor Contois	35	Private	09/11/1862	New Lebanon	
Joseph Ferris	39	Private	09/04/1862	Greenport	Absent at Muster out, sick in hospital
Morgan Funk	19	Private	09/27/1862	Livingston	
George Howes	18	Private	09/05/1862	Hudson	
William Hawver	18 1/2	Private	09/08/1862	Livingston	Absent, in confinement by sentence Court-martial
Justus Junes	32	Private	09/15/1862	Hudson	
Daniel Jennings	19	Private	09/29/1862	Greenport	
Henry H. Loucks	22 1/2	Private	09/08/1862	Hudson	
John Leonard	45	Private	09/05/1862	Hudson	
John J. Morgan	33	Private	09/04/1862	Hudson	
John Maguire	38	Private	09/11/1862	Hudson	
James Moore	33	Private	09/08/1862	Hudson	
John Reid	32 1/4	Private	09/13/1862	Hudson	Enrolled Corporal, reduced 11/4/1862
Amos Roraback	18	Private	09/26/1862	Hudson	

The 159th Regiment Infantry, NYS Volunteers – Company A

NAME	AGE	RANK	DATE JOINED	PLACE JOINED	NOTES
Christian M. Rupfl	28	Private	09/26/1862	Claverack	
Granville M. Shaver	34	Private	09/06/1862	Hudson	Enrolled Wagoneer. Reduced 6/21/1863
Slone Smith	18	Private	09/08/1862	Stuyvesant	
Myron Wheeler	23	Private	09/17/1862	Greenport	
Washington N. Birdlong	46	C'd Cook	01/01/1864	Thibodeaux, La	Recruit
Edward Tynan	20 2/3	1st Sergeant	09/05/1862	Hudson	Discharged 3/16/1864, disability from wounds
Laban A. White	36	Sergeant	09/05/1862	Hudson	Discharged 3/23/1863
Jacob Sagendorph	31 1/3	Corporal	09/08/1862	Claverack	Discharged 8/24/1863
Silas A. Tator	21	Corporal	09/02/1862	Claverack	Discharged 8/24/1863
Silas W. Peary	18	Corporal	09/22/1862	Germantown	Discharged G.O. 77, A.G.O., W.D.
Jeremiah Benneway	44 1/2	Private	09/06/1862	Greenport	Discharged 12/3/1862
Joseph Clearwater	25	Private	09/26/1862	Fishkill	Discharged G.O. 77, A.G.O., W.D.
Thomas Daley	31	Private	09/17/1862	Hudson	Discharged 3/5/1864, disability
John H. Ford	18	Private	09/25/1862	Livingston	Discharged 6/30/1863, disability
Patrick Keegan	32	Private	09/29/1862	Hudson	Discharged 2/1/1864, disability
Jacob Keller	45	Private	10/3/1862	Livingston	Discharged 3/7/1864, disability
Theodore Kinzler	40	Private	9/12/1862	Hudson	Discharged 1/8/1865, disability
Charles Lawton	32	Private	9/8/1862	Livingston	Discharged 6/11/1863, disability
John M. Lines	28	Private	9/27/1862	Livingston	Discharged 6/11/1863, disability
John S. Lown	40	Private	1/15/1864	New Lebanon	Discharged 5/22/1865, recruit
Rensellaer Loucks	22	Private	8/25/1862	Poughkeepsie	Discharged 11/25/1862, G.O. 94, A.G.O., W.D.
Charles Raukins	18	Private	9/12/1862	Greenport	Discharged Writ Habeas Corpus 11/25/1862
William H. Traver	44	Private	9/4/1862	Claverack	Discharged 12/3/1862, disability
Henry Thompson	44	Private	10/20/1862	Hudson	Discharged 10/1/1862 disability
Garrett S. Van Hoesen	26	Private	9/9/1862	Ghent	Discharged 3/23/1863, disability
Lewis Vaughn	18	Private	10/10/1862	Ghent	Discharged 11/4/1862, Writ Habeas Corpus

The 159th Regiment Infantry, NYS Volunteers – Company A

NAME	AGE	RANK	DATE JOINED	PLACE JOINED	NOTES
Charles I. Winans	18	Private	9/8/1862	Hudson	Discharged 3/17/1864; disability
Philip Wallace	44	Private	9/9/1862	Greenport	Discharged 6/1/1863
Thomas B. Miller	15	Musician	9/13/1862	Livingston	Promoted 1/28/1864 Principal Musician
James Cooney	35	Private	9/7/1862	Hudson	Transferred to Co. C
Marshall A. Dunham	22	Private	9/17/1862	New Lebanon	Promoted 6/19/1863 Sergeant-Major
George Finney	42	Private	9/10/1862	Hudson	Transferred Veterans Reserve Corps 4/6/1864
William H. Hollenbeck	30	Private	9/18/1862	Hudson	Transferred Veterans Reserve Corps 1/10/1865
Jacob Hollenbeck	40	Private	9/6/1862	Hudson	Transferred Veterans Reserve Corps 1/10/1865
Thomas Herbert		Private		Brooklyn	Transferred to Co. F 11/6/1863, Ret'd as deserter
Francis Mahon	28	Private	9/8/1862	Brooklyn	Transferred to Co. F 11/6/1863, Ret'd as deserter
Solomon Maurer	34	Private	9/18/1862	Stockport	Transferred Veterans Reserve Corps 4/6/1864
John Higgins	18	Corporal	9/6/1862	Hudson	Died of wounds 5/18/1863 from battle at Irish Bend 4/14/1863
Edward V. Winans	18 1/2	Private	09/12/1862	Hudson	
Jonathan J. Race	45	Corporal	9/13/1862	Greenport	Died in action Port Hudson, La 5/27/1863
Peter P. Niver	45	Corporal	9/13/1862	Greenport	Died of disease 5/13/1863
John D. Tator	33	Corporal	9/4/1862	Claverack	Died in action Opequan, Va 9/19/1864
Patrick Connery	40	Private	9/11/1862	Hudson	Died in action Irish Bend, La 4/14/1863
John Dennis	32	Private	9/6/1862	Hudson	Died of disease 6/5/1863
John Kelly	44	Private	9/11/1862	Hudson	Died in action Irish Bend 4/14/1863
Robert Kipp	18	Private	9/22/1862	Greenport	Died in action Irish Bend 4/14/1863
Andrew W. Lape	30	Private	9/6/1862	Ghent	Died 7/25/1864
Peter H. Miller	25	Private	9/8/1862	Livingston	Died in action at Cedar Creek 10/19/1864
Lewis Moran	40	Private	9/7/1862	New Lebanon	Died of disease 8/1/1863
Fredrick Roraback	23	Private	9/26/1862	Hudson	Died of disease 4/6/1865
James Reynolds	42	Private	9/5/1862	Livingston	Died in action Irish Bend 4/14/1863
Joseph Snyder	32	Private	9/3/1862	Greenport	Died in action Irish Bend 4/14/1863
Joseph Smith	25	Private	9/16/1862	New Lebanon	Died of disease 4/12/1863
Delbert Van Deusen	18	Private	9/6/1862	Greenport	Died in action Opequan, Va 9/19/1864
Warren Winslow	26	Private	9/15/1862	Hudson	Died in action Irish Bend 4/14/1863

The 159th Regiment Infantry, NYS Volunteers – Company A

NAME	AGE	RANK	DATE JOINED	PLACE JOINED	NOTES
Thomas Ward	44	Private	9/22/1862	Hudson	Died prisoner of war in Salisbury, NC
George Maurer	22	Corporal	9/5/1862	Hudson	Deserted 1863
Edmond H. Alger	20	Private	9/4/1862	Claverack	Deserted 11/4/1862
James Agen	24	Private	9/5/1862	Greenport	Deserted 11/22/1862
Avery Smith Bradley	21	Private	9/12/1862	Hudson	Deserted 8/7/1864
John H. Badgley	21	Private	9/15/1862	Hudson	Deserted 11/14/1862
Charles Doyle	21	Private	4/25/1864	Poughkeepsie	Deserted 8/7/1864
Charles H. Gardner	21	Private	9/7/1862	Hudson	Deserted May 1863
Andrew Hermance	22	Private	4/25/1864	9th District	Deserted 6/20/1864
Jonas Hawver	21	Private	9/8/1862	Hudson	Deserted 11/14/1862
Thomas Junes	23	Private	9/15/1862	Hudson	Deserted 11/12/1862
Joseph Knowles	33	Private	10/17/1862	Ghent	Deserted 4/1/1863
James Stewart	40	Private	9/6/1862	Greenport	Deserted 11/22/1862
George Stevens	35	Private	9/15/1862	Hudson	Deserted from a parole camp 9/13/1863
John Tynan	36	Private	9/17/1862	Hudson	Deserted 11/2/1862
Richard Walters	37	Private	9/17/1862	Hudson	Deserted twice – 11/11/1862 & 11/30/1864

The 159th Regiment Infantry, NYS Volunteers – Company B

NAME	AGE	RANK	DATE	PLACE JOINED	NOTES
John Day	23	1st Lieutenant	8/23/1862	Brooklyn	Enrolled Corporal Co K, Promoted 9/2/1864
Augusta J. Dayton	33	Captain	8/28/1862	Brooklyn	Discharged 2/28/1863
Julius H. Tiemann	22	1st Lieutenant	9/24/1862	Brooklyn	Discharged 12/24/1863
Alfred Greenleaf Jr	24	2nd Lieutenant	9/24/1862	Brooklyn	Discharged 5/2/1864
George W. Hussey	24	1st Lieutenant	10/4/1862	Brooklyn	Transferred to Co. G
Frank P. Gavin	22	1st Sergeant	9/5/1862	Brooklyn	Absent
Amos Hark	19	Sergeant	10/20/1862	Brooklyn	Enrolled private, promoted 8/27/1863
Lewis H Smith	21	Sergeant	9/17/1862	Brooklyn	Enrolled private, promoted 4/1/1864
James Smyth	30	Sergeant	9/12/1862	Brooklyn	Enrolled Corporal, promoted 9/17/1863
George W. Hatfield	18	Corporal	9/29/1862	Brooklyn	Enrolled private, promoted 4/1/1864
Martin Smallix	35	Corporal	9/10/1862	Brooklyn	Enrolled Private, promoted 5/14/1864
Henry Dunham	20	Musician	9/10/1862	Brooklyn	
John W. Mambert	43	Musician	9/20/1862	Taghkanic	Transferred from Co. G 3/12/1864
Thomas Cavanagh	34	Private	9/6/1862	Brooklyn	Absent, sick in hospital
Thomas Corson	18	Private	9/17/1862	Brooklyn	
Ephraim P. Decker	18	Private	9/9/1862	Brooklyn	
James Drewett	44	Private	9/17/1862	Stockport	Transferred from Co. G 3/12/1864
John H. Fox	30	Private	9/6/1862	Brooklyn	
Robert Gass	18	Private	9/15/1862	Brooklyn	
John Gault	28	Private	9/26/1862	Claverack	Transferred from Co. G 3/12/1864
John Joy	21	Private	9/16/1862	Brooklyn	
John Keron	40	Private	9/8/1862	Brooklyn	Absent, sick in hospital
Frederick J. Lawrence	18	Private	9/11/1862	Brooklyn	Absent, sick in hospital
Stephen Loughlin	36	Private	9/3/1862	Brooklyn	
Patrick Mack	40	Private	9/18/1862	Brooklyn	
Hugh McElravey	38	Private	9/11/1862	Brooklyn	
Michael Murray	18	Private	9/15/1862	Brooklyn	Deserted, Baton Rouge 2/27/1863, returned 10/31/1863
James O'Neill	30	Private	9/18/1862	Brooklyn	Absent, detached

The 159th Regiment Infantry, NYS Volunteers – Company B

NAME	AGE	RANK	DATE	PLACE JOINED	NOTES
William Pitts	23	Private	9/5/1862	Brooklyn	
Silas W. Richmond	18	Private	9/28/1862	Kinderhook	Transferred from Co. G 3/12/1864
Fredrick Siegler	41	Private	9/26/1862	Brooklyn	
John Taylor	18	Private	9/10/1862	Brooklyn	Absent, in confinement by sentence G. C. M.
Balthazar Wurtz	28	Private	9/18/1862	Brooklyn	
George Washington	30	Cook	1/1/1864	Thibodeaux	Recruit
Patrick Larkin	34	Sergeant	9/12/1862	Brooklyn	Discharged disability 9/24/1863
Barzillai Ransom	32	Sergeant	9/8/1862	Brooklyn	Discharge 2/23/1864
Anthony J. Tiemann	31	Sergeant	9/18/1862	Brooklyn	Discharged 8/24/1864
Horace J. Brown	18	Private	9/8/1862	Brooklyn	Discharged 7/18/1865, enrolled musician
James Costello	43	Private	9/10/1862	Brooklyn	Disability discharge 3/14/1864
Joh Coyle	43	Private	9/17/1862	Brooklyn	
Alexander F. Irving	27	Private	9/6/1862	Brooklyn	Disability discharge 1/17/1864
John McCauley	21	Private	9/5/1862	Brooklyn	Disability discharge 7/29/1863
Jacob Sahler	44	Private	9/15/1862	Brooklyn	Disability discharge 6/29/1863
William Sherman	19	Private	1/1/1864	Hudson	Recruit, discharge 6/28/1865
Charles Snyder	18	Private	9/22/1862	Ghent	Transfer from Co. G 3/12/1864, discharge 7/21/1865
John J. Tappan	27	Private	9/17/1862	Brooklyn	Disability discharge 6/7/1863
Nicholas Wolfert	44	Private	9/29/1862	Brooklyn	Disability discharge 10/19/1863
William F. Tiemann	18	Sergeant	9/5/1862	Brooklyn	Promoted 11/2/1862 Sergeant Major (Author of 159th Regiment NYS Volunteers)
William H. Roberts	26	Corporal	10/10/1862	Brooklyn	Transferred 4/6/1864 to V.R.C.
Edward E. Baker	26	Private	9/25/1862	Brooklyn	Promoted to Hospital Stewart 11/22/1863
George D. Dayton	21	1st Sergeant	9/4/1862	Brooklyn	11/2/1862 Promoted Principal Musician
William Doolan	44	Private	9/11/1862	Brooklyn	Transferred to V.R.C. 1/10/1865
Stephen French	19	Private	9/13/1862	Brooklyn	Transferred to V.R.C. 4/6/1864
Sherman Hart	40	Private	9/27/1862	Brooklyn	Transferred to Navy 6/1/1864
John Hennessy	43	Private	9/12/1862	Kinderhook	Transferred from Co. G 3/12/1864

The 159th Regiment Infantry, NYS Volunteers – Company B

NAME	AGE	RANK	DATE	PLACE JOINED	NOTES
John Dailey	30	Private	9/22/1862	Kinderhook	Died Prisoner of War 11/12/1864 Salisbury NC
Richard Gaffney	38	Private	10/11/1862	Brooklyn	Died of disease 7/9/1865
James Hanlin	30	Private	9/5/1862	Brooklyn	Died of disease 9/26/1864
James J. Lenfesty	18	Private	9/24/1862	Brooklyn	Died of wounds in action Halltown WV 9/24/1864
Thomas McCartney	18	Private	9/22/1862	Brooklyn	Died of wounds in action Port Hudson LA 6/27/1863
Thomas Rierdon	38	Private	9/23/1862	Brooklyn	Died of Delirium tremens 9/18/1863
Gottlieb Schnepf	36	Private	9/11/1862	Brooklyn	Died Prisoner of war 12/5/1864 Salisbury NC
James Price	21	Corporal	9/22/1862	Brooklyn	Deserted 11/1/1862
John H. Steele	40	Corporal	9/30/1862	Brooklyn	Deserted 11/1/1862
Thomas Carney	18	Private	9/6/1862	Brooklyn	Deserted 11/1/1862
Owen Cassidy	44	Private	9/8/1862	Brooklyn	Deserted 11/1/1862
Clarence Chapin	24	Private	9/22/1862	Kinderhook	Transfer from Co. G 3/12/1864, deserted at that time
Thomas McCauley	18	Private	9/10/1862	Brooklyn	Deserted 9/30/1863
Thomas Mullen	19	Private	9/9/1862	Brooklyn	Deserted 11/6/1862
Adolph Kirchner	27	Private	10/6/1862	Brooklyn	Unaccounted for at Muster Out
Thomas Shaw	23	Private	10/6/1862	Brooklyn	Unaccounted for at Muster Out
Hezekiah Shaw	34	Private	10/13/1862	Brooklyn	Unaccounted for at Muster Out
Henry Van Loan	23	Private	9/22/1862	Brooklyn	Unaccounted for at Muster Out

The 159th Regiment Infantry, NYS Volunteers – Company C

NAME	AGE	RANK	DATE	PLACE JOINED	NOTES
Ranson Brazillai	34	1st Lieutenant	10/1/1864	Cedar Creek, VA	Appointed from Civil life
Gamwell M. Ariel	39	Captain	9/28/1862	Hudson	Discharged 2/11/1863
William Crawford	27	1st Lieutenant	10/4/1862	Hudson	Discharged 2/12/1864
Edgar G. Hubbell	20	2nd Lieutenant	10/4/1862	Hudson	Discharged 7/30/1863
Charles Lewis	31	Captain	10/14/1862	Hudson	Transferred 1/15/1864 Promoted Major 176th NYSV
Herman Smith	18	2nd Lieutenant	9/3/1862	Brooklyn	Died of wounds at Opequan, Va, 9/19/1864
Edgar Hollenbeck	17	1st Sergeant	9/9/1862	Livingston	Enrolled Private, promoted three times, last on 9/19/1864
William Coons	19	Sergeant	9/27/1862	Taghkanic	Enrolled Private, promoted twice, last on 11/7/1865
Hiram D. Pierce	19	Sergeant	9/30/1862	Copake	Enrolled Private, promoted twice, last on 9/9/1863
John Wheeler	20	Corporal	9/15/1862	Taghkanic	Enrolled Private, promoted 8/7/1865
Jacob Beatty	20	Corporal	9/27/1865	Taghkanic	Enrolled Private, promoted 1/1/1865
Louis S. Hart	16	Musician	9/13/1862	Hudson	
Andrew H. Williams	27	Private	9/29/1862	Hudson	Absent from Muster out, sick in hospital
Edgar H. Bingham	19	Private	9/22/1862	Hudson	Reduced from Corporal
Lewis Coppins	40	Private	10/3/1862	Hudson	
James Cooney	35	Private	9/7/1862	Hudson	Transferred from Co. A 3/3/1864
Patrick Connelly	21	Private	10/10/1864	Kingston	Recruit
Adam Duntz	18	Private	10/13/1862	Taghkanic	
Alb. Eugene Edwards	19	Private	9/26/1862	Copake	
Louis Fox	19	Private	9/25/1862	Huson	
Martin M. Garner	20	Private	9/20/1862	Hudson	Reduced from Corporal
Michael Hogan	19	Private	9/13/1862	Stuyvesant	
John Hellar	22	Private	10/2/1862	Taghkanic	
Edward H. Harrows	35	Private	12/29/1863	Hudson	Recruit
Alexander Keist	24	Private	9/24/1862	Livingston	Reported burned to death in Augusta Barracks 10/9/1865
George C. Miller	35	Private	9/25/1862	Hudson	
James Melius	22	Private	9/30/1862	Hudson	
Jared M. Nash	19	Private	10/13/1864	Jamaica	Recruit

The 159th Regiment Infantry, NYS Volunteers – Company C

NAME	AGE	RANK	DATE	PLACE JOINED	NOTES
William Nichols	17	Private	9/16/1862	Hudson	
Albert Ostrander	35	Private	10/26/1862	Copake	
M'Timer Rockefeller	18	Private	9/29/1862	Taghkanic	
Henry C. Smith	19	Private	9/9/1862	Stockport	
Leonard Smith	30	Private	9/11/1862	Taghkanic	Absent, detached
Grovvenor Smith	23	Private	9/10/1862	Taghkanic	
Christian Schnack	20	Private	9/27/1862	Hudson	
John Schermerhorn	18	Private	10/27/1862	Hudson	
Cornelius Van Valkenbergh	30	Private	9/26/1862	Stockport	
Anthony Brown	35	Cook	11/6/1863	Thibodaux, LA	Recruit
James Harrison	22	Cook	11/6/1863	Thibodaux, LA	Recruit
Robert V.L. Cameron	19	Sergeant	9/9/1862	Taghkanic	Discharged with disability 9/23/1863
William H. Spanburgh	22	Sergeant	9/12/1862	Hudson	Discharged 2/28/1864
James Fitzgerald	23	Sergeant	9/19/1862	Hudson	
Walter R. Conron	20	Corporal	9/9/1862	Hudson	
John H. Allen	35	Corporal	9/4/1862	Taghkanic	Discharged with disability
Robert DeSasia	15	Musician	9/4/1862	Hudson	Discharged 11/16/1862 Writ Habeas Corpus
James C. Armstrong	19	Private	9/3/1862	Hudson	Discharged 11/3/1862 Writ Habeas Corpus
Lucas Artil	30	Private	9/20/1862	Hudson	Discharged with disability
John Bradley	35	Private	9/25/1864	Brooklyn	Discharged 5/9/1865 G.O.77, A.G.O., W.D. Recruit
Fayette W. Barlow	18	Private	9/18/1864	Poughkeepsie	Discharged 5/9/1865 G.O.77, A.G.O., W.D. Recruit
Howard Coons	23	Private	9/25/1862	Taghkanic	Discharged with disability
Samuel Coons	35	Private	9/25/1862	Taghkanic	Discharged with disability
Austin Gailor	18	Private	3/11/1863	Kingston	Recruit, discharged with disability
John Hart	21	Private	9/19/1864	Jamaica	Discharged 5/9/1865 G.O.77, A.G.O., W.D. Recruit
Stephen Morey	25	Private	9/5/1862	Stuyvesant	Discharged disability 9/2/1865
Thomas Otis	30	Private	9/24/1864	Kingston	Discharged 5/9/1865 G.O.77, A.G.O., W.D. Recruit
Joseph Patterson	28	Private	9/13/1862	Hudson	Discharged with disability 4/1864

The 159th Regiment Infantry, NYS Volunteers – Company C

NAME	AGE	RANK	DATE	PLACE JOINED	NOTES
Ephraim Stickles	35	Private	9/19/1862	Taghkanic	Discharged with disability 8/25/1865
Myron Staats	25	Private	9/15/1862	Stockport	Discharged with disability
Hiram Scriver	25	Private	9/3/1662	Hudson	Discharged with disability
John Smith	21	Private	8/16/1864	Poughkeepsie	Discharged 5/9/1865 G.O.77, A.G.O., W.D. Recruit
William Tator	51	Private	10/13/1862	Taghkanic	Discharged with disability 8/25/1865
Freeland Wheeler	18	Private	9/9/1862	Hudson	Discharged due to deafness
Charles I. Winans	19	Private	9/12/1864	Albany	Discharged 5/9/1865 G.O.77, A.G.O., W.D. Recruit
Samuel A. Norman	40	Sergeant	9/12/1862	Stockport	Transferred 6/7/1864 to V.R.C.
Jonas A. Kellerhouse	25	Sergeant	10/26/1862	Copake	Transferred 5/3/1862 to V.R.C.
Tunis Hollenbeck	35	Wagoneer	9/13/1862	Hudson	Transferred 5/3/1862 to V.R.C.
John H. Charlotte	20	Private	9/18/1862	Hudson	Transferred 11/2/1862 to N.C.S & promoted to Q.M. Sergeant
Alexander L. Mayot	18	Private	9/10/1862	Hudson	Transferred 11/4/1862 to Cavalry Corps
William A. Jaquins	26	Private	9/26/1862	Hudson	Transferred 1/25/1864 to N.C.S. & promoted to Q.M. Sergeant
Henry A. Wilkinson	21	Private	9/8/1862	Hudson	Transferred 1/25/1864 to Signal Corps
Augustus W. Wendt	26	1st Sergeant	9/16/1862	Hudson	Died 9/19/1864 in action Opequan, Va
John Whitbeck	19	Sergeant	9/30/1862	Copake	Died 6/1863 from Typhoid
Michael Leonard	24	Corporal	10/27/1862	Copake	Died 9/19/1865 from compound fracture of skull
Calvin W. Finkle	30	Corporal	9/17/1862	Taghkanic	Died May 1863 from disease
William P. Allen	27	Private	9/17/1862	Taghkanic	Died 3/10/1862 from disease
Hiram Crumbie	34	Private	9/6/1862	Hudson	Died 9/16/1863
Ambrose Coon	19	Private	9/27/1862	Taghkanic	Died of wounds August/1863 from battle at Irish Bend
William Calkins	28	Private	9/26/1862	Taghkanic	Died May 1863
Dorr DeWitt	19	Private	10/5/1862	Stuyvesant	Died 10/5/1864 prisoner of war Richmond, Va
William H. Finkle	23	Private	9/17/1862	Taghkanic	Died from disease May 1863
John Gabriz	45	Private	10/4/1862	Stockport	Died 6/2/1863
William Houghtaling	20	Private	9/14/1862	Hudson	Died June 1863
James Houghtaling	17	Private	9/25/1862	Hudson	Died 4/14/1863 in action at Irish Bend
James Morrison	20	Private	9/13/1862	Hudson	Died 4/14/1863 in action at Irish Bend

The 159th Regiment Infantry, NYS Volunteers – Company C

NAME	AGE	RANK	DATE	PLACE JOINED	NOTES
Newell H. Olds	48	Private	9/14/1862	Chatham	Died 6/16/1863
John W. Pulver	19	Private	9/13/1862	Copake	Died February 1863
John Rilsing	35	Private	10/21/1862	Hudson	Died of wounds Oct. 1863 from battle at Irish Bend
Daniel Riley	26	Private	9/16/1862	Hudson	Died 4/14/1863 in action at Irish Bend
William Shufeldt	34	Private	9/17/1862	Germantown	Died 10/14/1864 of wounds in action Opequan, Va.
Stephen Wheeler	19	Private	9/21/1862	Taghkanic	Died 6/12/1863 of disease
Newton R. Benedict	19	Private	9/10/1862	Hudson	Deserted 11/12/1863
Harvey Clevend	21	Private	9/21/1862	Kingston	Deserted 1/27/1865
Jerimiah Duntz	30	Private	9/20/1862	Taghkanic	Deserted November 1862
George Heller	34	Private	10/4/1862	Hudson	Deserted November 1862
Obediah Miller	38	Private	9/8/1862	Taghkanic	Deserted November 1862
John Nichols	32	Private	9/24/1862	Stockport	Deserted at Baton Rouge, La
Ezra Stickles	30	Private	9/10/1862	Taghkanic	Deserted at Baton Rouge, La
Robert Simpson	21	Private	9/25/1862	Hudson	Deserted at New Orleans, La
Charles Raught	24	Private	10/30/1862	Hudson	Unaccounted for at Muster Out
Dedrick Smith	19	Private	9/16/1862	Taghkanic	Unaccounted for at Muster Out
Jacob Ten Eyck	43	Private	10/29/1862	Germantown	Unaccounted for at Muster Out

The 159th Regiment Infantry, NYS Volunteers – Company D

NAME	AGE	RANK	DATE	PLACE JOINED	NOTES
E. Parmley Brown	19	1st Lieutenant	11/20/1863	Brooklyn	Recruit, enrolled Private Co. I, promoted 6/22/1865
Joseph A. Hatry	32	Captain	9/24/1862	Brooklyn	Discharged Dishonorably 11/16/1863 by sentence G.C.M.
Charles A. Loretz	22	1st Lieutenant	9/24/1862	Brooklyn	Discharged with disability 4/3/1863
Henry M. Howard	20	1st Lieutenant	9/6/1862	Brooklyn	Discharged with disability 12/20/1864
John W. Manley, Jr.	22	2nd Lieutenant	9/24/1862	Brooklyn	Transferred 1/27/1863. Promoted to 1st Lieutenant Co. A
John T. Jennings	32	1st Sergeant	9/17/1862	Brooklyn	Enrolled Corporal, promoted twice, last on 9/1/1864
Americus Marzzi	27	Sergeant	3/11/1863	New Orleans, La	Recruit, promoted twice, last on 5/1/1864
Alanson Pearsall	36	Corporal	9/8/1862	Brooklyn	Enrolled Private, promoted 1/27/1863
Reuben Mayo	31	Corporal	9/6/1862	Brooklyn	
Joseph Riell	16	Musician	9/4/1862	Brooklyn	
Henry C. Velsor	16	Musician	9/4/1862	Brooklyn	
Phillip B. Becker	22	Private	9/4/1862	Brooklyn	
James Clapp	18	Private	9/6/1864	Brooklyn	
James Connors	18	Private	9/6/1862	Brooklyn	
Frank Doremus II	18	Private	9/12/1862	Brooklyn	Enrolled Corporal, promoted then reduced 9/1/1864
Lawrence Grissier	21	Private	9/11/1862	Brooklyn	
James Howe	44	Private	9/24/1862	Brooklyn	Absent at Muster Out, sick in hospital
Patrick Hore	19	Private	9/15/1862	Kinderhook	Transferred from Co. G 3/12/1864
Frank W. Kisters	42	Private	3/11/1863	New Orleans, La	Recruit
Lewis Messensole	18	Private	9/13/1862	Brooklyn	Deserted, apprehended and returned.
John McCanty	25	Private	9/22/1862	Brooklyn	
George W. Martin	18	Private	10/7/1862	Chatham	Transferred from Co. G 3/12/1864
John Martin	19	Private	10/7/1862	Chatham	Transferred from Co. G 3/12/1864
George H. Miller	20	Private	10/1/1862	Brooklyn	Deserted, apprehended and returned
Isaac Morris	42	Private	8/22/1864	Schenectady	Recruit, Absent sick in hospital
Thomas Nevins	42	Private	9/18/1862	Brooklyn	
William C. Powell	43	Private	10/1/1862	Brooklyn	Absent sick in hospital
Louis Pierre	24	Private	11/24/1863	New York	Recruit

The 159th Regiment Infantry, NYS Volunteers – Company D

NAME	AGE	RANK	DATE	PLACE JOINED	NOTES
John Pendergast	32	Private	9/19/1862	Brooklyn	
John C. Roberts	18	Private	9/5/1862	Brooklyn	
Isaac L. Rose	19	Private	9/8/1862	Brooklyn	Enrolled Corporal, promoted and reduced 8/16/1865
Michael Vorness	21	Private	9/11/1862	Brooklyn	
James B. Williams	20	Private	10/1/1862	Fishkill	Transferred from Co. G 3/12/1864
William Titus	19	Private	9/29/1862	Claverack	Transferred from Co. G 3/12/1864
William H. White	19	Private	9/15/1862	Brooklyn	
Henry M. Howard	20	Sergeant	9/6/1862	Brooklyn	Discharged 1/26/1863 for promotion to 2nd Lieutenant
Luke Bowdish	43	Corporal	9/4/1862	Brooklyn	Discharged with disability 8/27/1863
John W. Sherman Jr.	21	Corporal	9/12/1862	Brooklyn	Discharged with disability 3/22/1863
Joseph Acken	40	Private	9/15/1862	Brooklyn	Discharged with disability 3/22/1863
Alfred Doty	35	Private	10/22/1862	Brooklyn	Discharged with disability 6/27/1863
Michael Feisler	21	Private	9/1/1862	Brooklyn	Discharged with disability 8/30/1863
Thomas Farrell	44	Private	9/19/1862	Brooklyn	Discharged with disability 6/22/1864
Thomas B. Harrison	18	Private	4/5/1863	New Orleans	Discharged 4/13/1864 commissioned 2nd Lt. U.S.C.T.
David McKiney	25	Private	9/9/1862	Brooklyn	Discharged with disability 3/7/1864
Isaiah Porter (African American)	27	Private	11/11/1863	Thibodaux, La	Recruit, discharged term of service 8/11/1864
Mark Slaven	44	Private	9/8/1862	Brooklyn	Discharged with disability 8/24/1863
Conrad Smith	22	Private	9/18/1862	Brooklyn	Discharged with disability 6/24/1865
Samuel Youngs (African American)	37	Private	11/11/1863	Thibodaux, La	Recruit, discharged term of service 8/11/1864
William E. Palmer Jr.	18	Sergeant	9/5/1862	Brooklyn	Transferred to N.C.S 4/9/1864 as Sergeant-Major
Philip Kreuscher Jr.	18	Private	9/11/1862	Brooklyn	Transferred to V.R.C. 8/10/1864
John A. Schreiber	44	Private	9/2/1862	Brooklyn	Transferred to V.R.C. 6/22/1864
Adam Schuck	44	Private	9/2/1862	Brooklyn	Transferred to V.R.C. 4/10/1864
Peter Volkinur	21	Private	9/8/1862	Brooklyn	Transferred to V.R.C 8/10/1864. Deserted and returned
Charles H. Caldicott	32	1st Sergeant	9/5/1862	Brooklyn	Died 4/9/1863 from Dropsy (Edema)

The 159th Regiment Infantry, NYS Volunteers – Company D

NAME	AGE	RANK	DATE	PLACE JOINED	NOTES
Appleton W. Rackett	26	Corporal	9/8/1862	Brooklyn	Died, shot by rebels while drawing water 4/17/1863
James Bennett	39	Corporal	9/9/1862	Brooklyn	Died of disease 12/5/1862
Henry Newman	41	Wagoner	9/4/1862	Brooklyn	Died of disease 7/3/1864
John J. Brown	21	Private	9/22/1862	Brooklyn	Died of wounds in action at Cedar Creek, Va 10/27/1864
James Dorgan	37	Private	9/10/1862	Brooklyn	Died 12/24/1864 of Typhoid fever.
Patrick Fitzgerald	32	Private	10/21/1862	Ghent	Died 9/19/1864 in action at Opequan, Va.
Charles Halfus	21	Private	9/22/1862	Brooklyn	Died 4/14/1863 in action at Irish Bend, La.
John Hannoffy	40	Private	9/19/1862	Brooklyn	Died 6/9/1863 of wounds from Port Hudson 5/27/1863
George Hawks	44	Private	9/5/1862	Brooklyn	Died of disease 11/6/1863
John A. Klinsing	39	Private	9/6/1862	Claverack	Died of disease 7/22/1864
Thomas McEvoy	37	Private	9/6/1862	Brooklyn	Died of disease 10/6/1863
Abner Staunton	34	Private	10/10/1862	Poughkeepsie	Died 10/28/1864 prisoner of war, Camp Tyler, Texas
Joseph Treilein	19	Private	9/4/1862	Brooklyn	Died 1/1/1865 prisoner of war, Salisbury, N.C.
Michael Webber	29	Private	9/12/1862	Brooklyn	Died of disease 9/30/1863
Jacob Gabriel	21	Sergeant	9/4/1862	Brooklyn	Deserted 11/10/1862
Samuel Sanford	21	Corporal	9/26/1862	Brooklyn	Deserted 11/10/1862
Thomas Donovan	19	Private	9/9/1862	Brooklyn	Deserted 10/1/1865 (days before mustering out)
Frank Gabriel	18	Private	9/18/1862	Brooklyn	Deserted 11/10/1862
John Kripe	21	Private	9/9/1862	Brooklyn	Deserted 11/10/1862
Lewis Messensole	18	Private	9/13/1862	Brooklyn	Deserted July 1863, apprehended and returned.
John Murphy	18	Private	9/9/1862	Brooklyn	Deserted 11/10/1862
John Most	33	Private	9/4/1862	Brooklyn	Deserted 11/10/1862
Patrick Plunkett	44	Private	9/13/1862	Brooklyn	Deserted 11/10/1862
Michael Schmitt	30	Private	9/4/1862	Brooklyn	Deserted 11/10/1862
William Timmes	21	Private	9/11/1862	Brooklyn	Deserted 11/10/1862
Lewis Vughiner	20	Private	11/24/1863	New York	Recruit, deserted 7/1/1864
John S. Van Nosdall	17	Private	10/1/1862	Fishkill	Deserted 7/24/1865
Martin Ward	27	Private	9/8/1862	Brooklyn	Deserted 11/10/1862

The 159th Regiment Infantry, NYS Volunteers – Company D

NAME	AGE	RANK	DATE	PLACE JOINED	NOTES
Edward Carey	20	Cook	10/28/1863	Thibodeaux, La	Recruit, deserted 7/31/1864
Charles Archibald	32	Private	9/18/1862	Brooklyn	Unaccounted for at Muster out
James Anderson	21	Private	9/17/1862	Brooklyn	Unaccounted for at Muster out
John Sweeny	23	Private	9/9/1862	Brooklyn	Unaccounted for at Muster out
John Tobur	39	Private	10/3/1862	Fishkill	Unaccounted for at Muster out
Archibald DeLancy	32	Private	11/1/1862	New York	Transferred to Co. G.
Thomas Fitzgerald	21	Private	10/9/1862	Brooklyn	Unaccounted for at Muster out
John Gabriel	21	Private	9/4/1862	Brooklyn	Unaccounted for at Muster out
Charles Howard	22	Private	10/13/1862	Brooklyn	Unaccounted for at Muster out
Victor Thrust	36	Private	9/23/1862	Brooklyn	Unaccounted for at Muster out
William I. Irish	20	Private	9/25/1962	Brooklyn	Unaccounted for at Muster out
Michael Konig	21	Private	9/29/1862	Brooklyn	Unaccounted for at Muster out
John W. Keesler	22	Private	9/26/1862	Hudson	Unaccounted for at Muster out
James Morrison	20	Private	9/13/1862	Hudson	Transferred to Co. C.
Fredrick Nurber	31	Private	9/10/1862	Brooklyn	Unaccounted for at Muster out

The 159th Regiment Infantry, NYS Volunteers – Company E

NAME	AGE	RANK	DATE	PLACE JOINED	NOTES
Andrew Rifenburgh	32	1st Lieutenant	8/20/1862	Hudson	Enrolled Corporal, promoted three times, last commissioned Captain 6/10/1864
Nathan S. Post	34	1st Lieutenant	10/7/1862	Hudson	Discharged with disability 1/14/1863
Robert H. Traver	27	2nd Lieutenant	10/7/1862	Hudson	Discharged with disability 8/25/1863
William Waltermire	30	Captain	10/15/1862	Hudson	Promoted Major 2/25/1864
Wesley Bradley	32	1st Lieutenant	9/18/1862	Hudson	Died 5/10/1863 Typhoid fever.
John W. Phillips	19	1st Sergeant	9/10/1862	Ghent	Enrolled Corporal, promoted 5/3/1865
Nicholas Freehan	20	Sergeant	9/10/1862	Ghent	Enrolled Private, promoted twice, last on 7/11/1865
Jonas Coon	25	Sergeant	8/14/1862	Livingston	Enrolled Private, transferred from Co. D. promoted 9/13/1863
Rutson Blunt	28	Sergeant	10/1/1862	Claverack	Absent, detached.
David E. Waltermire	19	Corporal	9/10/1862	Ghent	
Fredrick Hamlin	17	Musician	9/22/1862	Hudson	
John E. Bristol	18	Private	9/11/1862	Ghent	Absent, detached.
James Bowdy, No. 1	18	Private	9/18/1862	Claverack	
James Bowdy, No. 2	26	Private	10/4/1862	Gallatin	Absent, in confinement by sentence G.C.M.
George A. Benzie	18	Private	9/20/1862	Hudson	
Edward W. Clapp	18	Private	10/1/1962	Stuyvesant	
Albert H. Clark	44	Private	10/4/1862	Copake	Absent, sick in hospital.
James Decker	18	Private	9/25/1862	Claverack	Absent, sick in hospital.
Sebastian Eppner	35	Private	9/3/1862	Claverack	
Fredrick Guyon	18	Private	9/22/1862	Hillsdale	Enrolled Corporal, reduced 3/26/1864
Charles N. Hawver	18	Private	9/18/1862	Claverack	Transferred from Co. D.
William H. Hart	18	Private	8/9/1862	Greenport	
Edward H. Hawver	23	Private	9/11/1862	Claverack	Enrolled Sergeant, reduced 11/16/1863
Charles Hamilton	19	Private	9/17/1862	Ghent	
John Krause	21	Private	9/22/1862	Ghent	Absent. Prisoner of War.
Stephen Lapeous	27	Private	9/20/1862	Livingston	
Anthony Maxwell	18	Private	9/20/1862	Claverack	

The 159th Regiment Infantry, NYS Volunteers – Company E

NAME	AGE	RANK	DATE	PLACE JOINED	NOTES
John W. Myers	24	Private	9/29/1862	Ghent	
Alonzo Mills	33	Private	9/19/1862	Fishkill	
Solon Macy	43	Private	9/19/1862	Taghkanic	
Orson A. Miller	18	Private	9/18/1862	Hillsdale	
George Proper	19	Private	9/7/1862	Claverack	
Robert Proper	19	Private	9/16/1862	Claverack	
John J. Poucher	38	Private	9/17/1862	Ghent	
Edward Rote	19	Private	10/11/1862	Hudson	
Francis R. Syre	28	Private	9/20/1862	Ghent	
John Thorpe	18	Private	10/14/1862	Hillsdale	
Egbert Webster	36	Private	10/4/1862	Hillsdale	
John Balthes	19	Cook	1/1/1864	Thibodeaux, La	Recruit
Frank Elie	20	Cook	1/1/1864	Thibodeaux, La	Recruit
Samuel B. Macy	29	1st Sergeant	10/13/1862	Ghent	Discharged 5/3/1865
James M. Ostrander	24	Sergeant	9/10/1862	Ghent	Discharged 7/11/1865
Dewitt C. McNeill	18	Sergeant	9/23/1862	Copake	Discharge 8/4/1865. Enrolled Private, promoted.
Martin Platner	18	Corporal	9/27/1862	Claverack	Discharged 11/16/1862. Writ Habeas Corpus.
Hiram P. Sagendorf	24	Corporal	9/20/1862	Ghent	Discharged 12/17/1863. Commissioned 1st Lieutenant U.S.C.T.
Peter H. Allen	44	Private	9/30/1862	Claverack	Discharged 11/24/1863 with disability
John W. Almstead	17	Private	9/11/1862	Ghent	Discharged 6/7/1865
Andrew Brush	44	Private	9/20/1862	Claverack	Discharged with disability 6/9/1864
Frank Coventry	26	Private	9/25/1862	Kinderhook	Discharged with disability 7/20/1864
Henry Decker	33	Private	9/20/1862	Livingston	Discharged with disability 12/22/1862
Henry Martin	44	Private	9/20/1862	Ghent	Discharged with disability 12/2/1862
Jesse Miller	18	Private	9/10/1862	Claverack	Discharged with disability 9/27/1863 from wounds at Irish Bend, La. 4/14/1863
Robert McCracken	28	Private	9/29/1862	Hillsdale	Discharged with disability 9/31/1863
Patrick O'Brien	34	Private	9/24/1862	Ghent	Discharged with disability 10/19/1863

The 159th Regiment Infantry, NYS Volunteers – Company E

NAME	AGE	RANK	DATE	PLACE JOINED	NOTES
Aaron Rouse	18	Private	10/6/1862	Ghent	Discharged 11/16/1862 Writ Habeas Corpus.
Thomas Reckett	39	Private	10/3/1862	Ghent	Discharged with disability 8/16/1863
Francis Smith	23	Private	9/19/1862	Claverack	Discharged with disability 12/2/1862
Racliff Shuragar	44	Private	9/18/1862	Livingston	Discharged with disability 5/21/1863
Albert Wheeler	28	Private	9/20/1862	Claverack	Discharged with disability 5/29/1863
Philip D. Shufelt	34	Corporal	9/25/1862	Ghent	Transferred 4/6/1864 to V.R.C.
James Hamlin	19	Musician	10/6/1862	Hudson	Transferred 7/17/1864 to V.R.C.
William H. Coyles	26	Private	9/12/1862	Copake	Transferred 5/12/1865 to V.R.C.
Charles H. C. Peterson	26	Private	9/1/1862	Ghent	Transferred 3/1/1864 to Navy.
John Smith	19	Corporal	9/4/1862	Claverack	Died of disease 9/18/1864
George W. Benzie	45	Wagoneer	10/20/1862	Hudson	Died 8/14/1863
James Burnes	18	Private	9/20/1862	Copake	Died 5/15/1863 of wounds at Irish Bend, La. 4/14/1863
Leonard (Laben) Boice	21	Private	9/29/1862	Claverack	Died of disease 2/11/1863. Brother to Richard Boice
Richard Boice	23	Private	9/27/1862	Claverack	Died in action, Irish Bend, La. 4/14/1863
Edward Citty	28	Private	9/27/1862	Ghent	Died of disease 2/16/1865
Jacob H. Christman	20	Private	9/12/1862	Ghent	Died in action Port Hudson, La. 5/27/1863
Edmond Decker	19	Private	10/16/1862	Claverack	Died of disease 2/21/1863
James Doran	45	Private	10/22/1862	Ghent	Died of wounds in action Irish Bend, La. 4/14/1863
William A. Doxey	18	Private	9/19/1862	Fishkill	Died of disease 2/20/1863
Henry F. Hawver	45	Private	9/20/1862	Ghent	Died of disease 7/27/1863
Alexander Kells	28	Private	9/6/1862	Claverack	Died of disease 2/2/1863
Owen Ludlow	42	Private	9/18/1862	Ghent	Died of disease 9/2/1863
John Maxwell	43	Private	10/7/1862	Claverack	Died in action Port Hudson, La. 5/27/1863
Edward McLean	24	Private	9/14/1862	Claverack	Died of disease 10/7/1863
William H. Proper	22	Private	10/22/1862	Taghkanic	Died in action Port Hudson, La. 5/27/1863
William Pugh	17	Private	9/17/1862	Ghent	Died in action Port Hudson, La. 5/27/1863
James T. Perkins	18	Private	10/20/1862	Hudson	Died in action Cedar Creek, Va. 10/19/1864
Robert Race	18	Private	9/24/1862	Greenport	Died of disease 6/18/1864

The 159th Regiment Infantry, NYS Volunteers – Company E

NAME	AGE	RANK	DATE	PLACE JOINED	NOTES
Obadiah Rockefeller	40	Private	9/22/1862	Hudson	Died of disease 5/20/1863
Hiram Spades	44	Private	10/10/1862	Copake	Died of disease 7/2/1863
William H. Smith	18	Private	9/16/1862	Fishkill	Died of disease 3/25/1863
Peter Silvernail	18	Private	10/21/1862	Ghent	Died in action Irish Bend, La. 4/14/1863
John T. Sherwood	19	Private	10/8/1862	Hillsdale	Died of disease 2/25/1863
John F. Warner	43	Private	10/2/1862	Claverack	Died of disease 8/10/1863
Henry D. Wolf	45	Private	9/11/1862	Ghent	Died in action Irish Bend, La. 4/14/1863
Benton S. Winchell	20	Corporal	9/29/1862	Claverack	Deserted 11/14/1862
Lester J. Chapman	19	Private	10/29/1862	Ghent	Deserted 11/3/1862
Peter Hagerdorn	35	Private	9/16/1862	Claverack	Deserted 1/25/1863
Casper Matt	22	Private	9/27/1862	Claverack	Deserted 9/27/1864
Reuben Rockefeller	22	Private	9/27/1862	Ghent	Deserted 11/14/1862
Reuben Roat	23	Private	9/26/1862	Taghkanic	Deserted 11/14/1862
Richard Raught	40	Private	9/13/1862	Claverack	Deserted 1/25/1863

The 159th Regiment Infantry, NYS Volunteers – Company F

NAME	AGE	RANK	DATE JOINED	PLACE JOINED	NOTES
George W. Hussey	24	Captain	10/4/1862	Brooklyn	Enrolled 2nd Lieutenant, promoted twice last on 9/3/1864
William Burtis		1st Lieutenant	10/4/1862	Brooklyn	Discharged November 1862
Christopher Branch	22	1st Lieutenant	9/27/1862	Oyster bay	Discharged 5/7/1864
Anthony J. Tiermann	31	1st Lieutenant	9/18/1862	Brooklyn	Discharged 1/4/1865
Robert McD Hart	22	Captain	9/6/1862	Brooklyn	Transferred 6/2/1864. Promoted Major
Henry M. Howard	20	1st Lieutenant	9/6/1862	Brooklyn	Transferred 8/25/1863 to Co. D.
Alfred H. Bruce	23	2nd Lieutenant	9/15/1862	Kinderhook	Transferred 4/30/1863 promoted 1st Lieutenant Co. K.
Philip McCormick	30	1st Sergeant	10/28/1862	Brooklyn	Enrolled private promoted twice, last on 9/1/1864
Daniel Corboy	22	Corporal	9/15/1862	Brooklyn	Enrolled Private promoted 7/15/1864
Augustine Ellis	17	Musician	10/4/1862	Brooklyn	Enrolled Corporal
William J. Brown	26	Private	9/24/1862	Brooklyn	Absent, sick in hospital. Wounded Vermilion Bay, La. 4/17/1863
Oliver Blackledge	30	Private	9/15/1862	Brooklyn	
John C. Bollinger	18	Private	9/6/1862	Poughkeepsie	Transferred from Co. G 3/12/1864
Benjamin Carroll	28	Private	10/23/1862	Brooklyn	
Edward Dillon	43	Private	9/17/1862	Brooklyn	Absent, sick in hospital.
John B. Hueston	29	Private	11/7/1862	Brooklyn	Substitute for Josiah Gittens. Deserted, apprehended and returned.
Isidore J. Harrison	19	Private	9/10/1862	Brooklyn	Enrolled Corporal promoted Sergt. Reduced 8/16/1865
Benjamin F. Mattoon	18	Private	9/22/1862	Canaan	Transferred from Co. G 3/12/1864
Richard McGurn	18	Private	9/22/1862	Canaan	Transferred from Co. G 3/12/1864
Terrance Mackay	18	Private	9/17/1862	Brooklyn	
Edward Morris	18	Private	9/20/1862	Brooklyn	
William McKenny	32	Private	9/6/1862	Brooklyn	
Edward J. Mackay	20	Private	9/13/1862	Brooklyn	Promoted Sergeant 1/1/1864. Reduced 4/2/1864
Smith J. Noe	21	Private	10/11/1862	Brookhaven	
John O'Mara	18	Private	9/8/1862	Brooklyn	
Leonard Smith	18	Private	9/20/1862	Claverack	Transferred from Co. G 3/12/1864
Frederick L. Schell	29	Private	9/6/1862	Islip	
Isaac D. Stilwell	22	Private	9/12/1862	Brooklyn	

The 159th Regiment Infantry, NYS Volunteers – Company F

NAME	AGE	RANK	DATE JOINED	PLACE JOINED	NOTES
Terry Brewster	31	Private	9/26/1862	Brookhaven	
Oliver Taylor	20	Private	2/23/1865	Brooklyn	Recruit
Christopher Branch	22	1st Sergeant	9/27/1862	Oyster Bay	Discharged 4/18/1864
Samuel C. Tompkins	24	Sergeant	9/4/1862	Brooklyn	Discharged with disability 8/5/1863 from wounds in action at Port Hudson, La. 5/27/1863
William Callaghan	32	Corporal	9/10/1862	Brooklyn	Discharged 7/19/1865
Henry A Barnes	19	Private	9/17/1862	Brooklyn	Discharged with disability 3/5/1864
John Doyle	44	Private	9/17/1862	Brooklyn	Discharged with disability March 1864
Josiah Gittens	35	Private	9/24/1862	Brooklyn	Discharged 11/7/1862, furnished substitute.
James Hill	44	Private	9/5/1862	Brooklyn	Discharged with disability 7/27/1863
Thomas Herbert		Private	1862	Brooklyn	Discharged 3/27/1865
Joseph V.R. Huntington	42	Private	10/1/1862	Brooklyn	Discharged with disability 5/24/1863
John Kelly	30	Private	9/8/1862	Brooklyn	Discharged with disability 6/30/1863
David Mott	30	Private	9/8/1862	Brookhaven	Discharged with disability 5/24/1863
Charles W. Mott	33	Private	9/8/1862	Brookhaven	Discharged 3/15/1865
Aaron Miller	44	Private	9/6/1862	Brooklyn	Discharged with disability 10/17/1864
David Parmenter	42	Private	9/13/1862	Brooklyn	Discharged with disability 6/1/1863
Francis Reilly	40	Private	9/19/1862	Brooklyn	Discharged with disability 6/6/1863
William H. Smith	44	Private	9/22/1862	Brookhaven	Discharged with disability 10/10/1863
Samuel C. Wicks	27	Private	9/15/1862	Brookhaven	Discharged with disability 8/11/1863
Gilbert S. Gullen	21	Sergeant	9/23/1862	Brooklyn	Transferred 8/16/1865 to N.C.S., Promoted Sergeant Major
John H. Ferguson	18	Corporal	9/19/1862	Brooklyn	Transferred 6/24/1864 to V.R.C. Enrolled Private
Amos B. Laws	29	Corporal	10/9/1862	Brookhaven	Transferred 5/1/1864 to Navy
Manley S. Rowland	18	Musician	10/9/1862	Brooklyn	Transferred 6/30/1865 to V.R.C.
Thomas Barrett	35	Private	10/3/1862	Brooklyn	Transferred 3/10/1863 to Co. H. Enrolled Corporal
Robert A. Smith	21	Private	9/6/1862	Brookhaven	Transferred 5/1/1864 to Navy
William T. Wells	21	Private	9/6/1862	Brookhaven	Transferred 5/1/1864 to Navy
Thomas White	19	Private	9/20/1862	Brooklyn	Transferred to V.R.C.

The 159th Regiment Infantry, NYS Volunteers – Company F

NAME	AGE	RANK	DATE JOINED	PLACE JOINED	NOTES
Bartholomew Doser	32	Corporal	9/16/1862	Brooklyn	Died 11/5/1864 as Prisoner of War, Salisbury N.C. enrolled Private, promoted 7/15/1864
John G. Laws	18	Corporal	10/9/1862	Brookhaven	Died in action Irish Bend, La. 4/14/1863. Younger brother to Amos B. Laws
Byron L. Smock	18	Musician	9/8/1862	Brooklyn	Died 1/19/1863 of Typhoid Fever.
John R. Brokee	19	Private	9/6/1862	Brooklyn	Died 11/1863 of wounds in action, Port Hudson, La
William Colgan	20	Private	10/1/1862	Brooklyn	Died 9/30/1864 of wounds in action, Opequan Creek Va. 9/19/1864
Henry Eaton	28	Private	9/8/1862	Brooklyn	Died in action Irish Bend, La. 4/14/1863
Zebulon V. Flowers	16	Private	9/9/1862	Brooklyn	Died in action Irish Bend, La. 4/14/1863
Azariah F. Hawkins	33	Private	9/26/1862	Brookhaven	Died 12/5/1864 from disease.
Richard Keron	44	Private	9/6/1862	Brooklyn	Died 4/2/1865
Thomas Kenny	19	Private	9/13/1862	Brooklyn	Died 8/27/1863
John McCauley	42	Private	9/6/1862	Brooklyn	Died 5/27/1863 in action Port Hudson, La.
Michael McMahon	35	Private	9/16/1862	Brooklyn	Died in action Irish Bend, La. 4/14/1863
Jerome McDonald	35	Private	9/17/1862	Claverack	Died 5/23/1865, transferred from Co. G, 3/12/1864
Lawrence Martin	25	Private	9/13/1862	Brooklyn	Died 6/23/1863 in action, Brashear City, La.
Floyd C. Nichols	18	Private	9/15/1862	Brookhaven	Died 10/5/1864 of wounds in action Opequan Creek, Va. 9/19/1864
Philip Stumpf	44	Private	9/26/1862	New Lebanon	Died 6/18/1863 of disease, transferred from Co.G. 3/12/1864 due to notice of death not received.
Robert T. Smith	42	Sergeant	10/22/1862	Brookhaven	Deserted 11/6/1862
John Dillon		Musician	1862	Brooklyn	Deserted 5/5/1864 in route to Regiment as an apprehended deserter.
Ephraim L.S. Bond	28	Private	9/18/1862	Brookhaven	Deserted 11/8/1862. Enrolled Teamster.
Fredrick Barrett	21	Private	10/1/1862	Brookhaven	Deserted 11/9/1862
William H. Bowers	27	Private	9/6/1862	Brookhaven	Deserted 11/10/1862. Enrolled Corporal.
James Coyne	21	Private	9/12/1862	Brooklyn	Deserted 11/2/1862
John Durie	21	Private	9/29/1862	Brooklyn	Deserted 9/13/1863. Missing in action 4/14/1863. Deserted from Camp Parole, St. Louis, Mo.
Isaac S. Downs	22	Private	9/6/1862	Brookhaven	Deserted 11/9/1862
William Griffith	37	Private	9/13/1862	Brooklyn	Deserted 11/2/1862

The 159th Regiment Infantry, NYS Volunteers – Company F

NAME	AGE	RANK	DATE JOINED	PLACE JOINED	NOTES
Thomas W. Gammage	40	Private	9/10/1862	Brooklyn	Deserted 12/3/1862
Henry Hall	30	Private	10/30/1862	Brooklyn	Deserted 11/2/1862
Edward F. Mott	21	Private	9/30/1862	Brooklyn	Deserted 11/10/1862.
Samuel H. Marsh	21	Private	9/23/1862	Brooklyn	Deserted 11/18/1862
Francis Mahon	28	Private	9/12/1862	Brooklyn	Deserted 6/27/1865 while being returned to regiment.
Henry Overlander	39	Private	9/22/1862	Brooklyn	Deserted twice 11/17/1862 and again on 9/30/1864
James S. Russell	32	Private	9/6/1862	Brookhaven	Deserted 11/2/1862
George N. Shaw	21	Private	9/29/1862	Brooklyn	Deserted 9/20/1863
Daniel Sheehan	39	Private	9/6/1862	Brooklyn	Deserted 11/2/1862
James A. Terry	32	Private	9/15/1862	Brookhaven	Deserted 11/10/1862
Charles Van Rospatch	40	Private	9/29/1862	Brookhaven	Deserted 11/2/1862
James Whelan	42	Private	9/8/1862	Brooklyn	Deserted 11/3/1862

The 159th Regiment Infantry, NYS Volunteers – Old Company G

NAME	AGE	RANK	DATE	PLACE JOINED	NOTES
William H. Slier	26	Captain	10/04/1862	Hudson	Discharged 07/30/1863
Charles Lewis	31	1st Lieutenant	10/14/1862	Hudson	Promoted Captain Company C 03/06/1863
George W. Hussey	24	1st Lieutenant	10/14/1862	Brooklyn	Promoted 2nd Lieutenant Company F 03/06/1863
Herman Smith	18	2nd Lieutenant	09/03/1862	Brooklyn	Promoted Sergeant Major Company C 06/10/1863
Andrew Rifenburgh	32	2nd Lieutenant	08/20/1862	Hudson	Promoted Sergeant Company E 02/25/1864
Byron Lockwood	25	2nd Lieutenant	10/14/1862	Hudson	Died in action Irish Bend LA 04/14/1863
Andrew Rifenburg	32	Sergeant	08/20/1862	Hudson	Promoted 2nd Lieutenant 02/25/1864
Lambert Dingman	23	Sergeant	09/04/1862	Kinderhook	Promoted 2nd Lieutenant 01/27/1863
Frank Horton	20	Corporal	09/05/1862	Claverack	Disability
Cyrus Mesick	24	Corporal	09/02/1862	Claverack	Disability
Oliver B. Whitney	20	Corporal	09/06/1862	Fishkill	Disability
Eli Miller	45	Musician	10/03/1862	Germantown	Disability
Charles Prior	28	Musician	10/29/1862	Hillsdale	Disability
Martin Best	41	Private	09/18/1862	Taghkanic	Writ Habeas Corpus 11/19/1862
Lewis Bauer	43	Private	09/15/1862	Kinderhook	Disability 03/22/1863
Henry Darling	39	Private	09/18/1862	Ghent	Disability 08/25/1863
William Fitzgerald	43	Private	10/03/1862	Chatham	Disability
Charles Houghtailing	14	Private	09/17/0862	Hudson	Writ Habeas Corpus 11/05/1862
Charles G. Romanzoff	28		09/25/1862	Hudson	Discharged 01/16/1864
George Whitlock	18	Private	10/14/1862	Stockport	Disability
William J. Warner	43	Private	09/16/1862	Hudson	Disability 11/24/1863
Alfred H. Bruce	23	1st Sergeant	09/15/1862	Kinderhook	Promoted Sergeant Major
William F. French	19	Sergeant	09/18/1862	Hudson	Promoted Commander Sergeant
Edward Duffy	32	Corporal	10/06/1862	Hudson	
Martin Traver	22	Corporal	09/23/1862	Kinderhook	
George H. Van Alstyne	25	Corporal	09/19/1862	Kinderhook	
Clark B. Suydam	18	Corporal	09/30/1862	Fishkill	Reduced

The 159th Regiment Infantry, NYS Volunteers – Old Company G

NAME	AGE	RANK	DATE	PLACE JOINED	NOTES
John A. Klinsing	39	Corporal	09/06/1862	Claverack	
William H. Wortman	45	Wagoneer	10/01/1862	Hudson	
John A. Abbott	39	Private	10/16/1862	Kinderhook	
Edward, Brophy	18	Private	09/14/1862	Kinderhook	
John C. Bollinger	18	Private	09/06/1862	Poughkeepsie	
Clarence Chapin	24	Private	09/22/1862	Kinderhook	
James Drewett	44	Private	09/17/1862	Stockport	
John Dailey	30	Private	09/22/1862	Kinderhook	
John Devlin	18	Private	09/17/1862	Kinderhook	
Bernard Doolin	28	Private	10/20/1862	Ghent	
William Farrington	32	Private	10/30/1862	Fishkill	
Patrick Fitzgerald	32	Private	10/21/1862	Ghent	
Andrew Goshia	29	Private	09/30/1862	Chatham	
John Gault	28	Private	09/26/18621	Claverack	
John Hennessay	43	Private	09/12/1862	Kinderhook	
Patrick Hore	19	Private	09/15/1862	Kinderhook	
Patrick Haverty	33	Private	10/16/1862	Hudson	
George Hoffman	26	Private	09/25/1862	Claverack	
John Lynch	20	Private	09/20/1862	Ghent	Promoted Corporal
William Lynch	18	Private	09/20/1862	Ghent	Musician
Benjamin F. Mattoon	18	Private	09/22/1862	Canaan	
George Millott	25	Private	09/26/1862	New Lebanon	
George Martin	18	Private	09/07/1862	Chatham	
John W. Mambert	43	Private	09/20/1862	Taghkanic	Principal Musician, transferred to Co. B 3/12/1864
Richard McGurn	18	Private	09/22/1862	Canaan	
Thomas McCormick	28	Private	09/23/1862	Hudson	
Jerome McDonald	35	Private	09/17/1862	Claverack	

The 159th Regiment Infantry, NYS Volunteers – Old Company G

NAME	AGE	RANK	DATE	PLACE JOINED	NOTES
John Martin	19	Private	10/07/1862	Chatham	
Austin Nevels	22	Private	10/07/1862	Ghent	
Silas W. Richmond	18	Private	09/28/1982	Kinderhook	Promoted Corporal
Thomas Shea	24	Private	10/13/1862	Kinderhook	
Abner Staunton	34	Private	10/10/1862	Poughkeepsie	
Charles Snyder	18	Private	09/22/1862	Ghent	
Henry S. Stickles	36	Private	09/29/1862	Kinderhook	
Leonard Smith	18	Private	09/20/1862	Claverack	
Cornelius Stickles	24	Private	10/07/1862	Greenport	
Philip Stumpf	44	Private	09/26/1862	New Lebanon	
William B. Tanner	36	Private	10/24/1862	Kinderhook	
John S. Van Nosdall	17	Private	10/01/1862	Fishkill	
Benjamin Van Dewater	41	Private	10/01/1862	Fishkill	
Francis Van Hoesen	28	Private	09/22/1862	Ghent	
William Titus	19	Private	09/29/1862	Claverack	
James B. Williams	20	Private	10/01/1862	Fishkill	
David C. Bean	45	Private	09/26/1862	Livingston	Died 06/25/1863
Platt DeGroff	45	Private	09/27/1862	Hudson	Died 03/29/1863
John Gallagher	24	Private	09/23/1862	Chatham	Died 05/27/1863 Port Hudson LA
John Murphy	27	Private	10/12/1862	Chatham	Died 04/14/1863 Irish Bend LA
John Morgan	42	Private	09/06/1862	Claverack	Died 06/13/1863
Harvey G. Pultz	18	Private	09/22/1862	Ghent	Died 05/27/1863 Port Hudson LA
James W. Sharon	37	Private	09/09/1862	Chatham	Died 04/14/1863 Irish Bend LA
Wesley Tanner	28	Private	10/06/1862	Chatham	Died 06/3081863 Irish Bend LA
George Williams	18	Private	09/29/1862	Ghent	Died 06/09/1863
Edward Shannon	27	Sergeant	09/26/1862	Chatham	Deserted 11/23/1862
Abram F. Boyce	35	Corporal	09/30/1862	Stuyvesant	Deserted 11/05/1862

The 159th Regiment Infantry, NYS Volunteers – Old Company G

NAME	AGE	RANK	DATE	PLACE JOINED	NOTES
Thomas Bannon	18	Private	09/30/1862	Kinderhook	Deserted 11/10/1862
Ambrose Brezal	26	Private	09/22/1862	Stockport	Deserted 11/15/1862
Conrad L. Coon	22	Private	09/25/1862	Ghent	Deserted 11/23/1862
Archibald De Lancey	32	Private	11/01/1862	New York	Deserted 08/15/1863 Carrollton LA
Michael Fortin	24	Private	09/12/1862	Ghent	Deserted 10/16/1863
John L. Haight	39	Private	10/07/1862	Pleasant Valley	Deserted 11/02/1862
David P. Hacker	39	Private	09/29/1862	Stockport	Deserted 11/02/1862
Platt Knickerbocker		Private	09/17/0862	Hudson	Deserted 12/02/1862
Frank McEnnessay	27	Private	10/27/1862	Chatham	Deserted 11/02/1862
Isaac C. Schermerhorn	23	Private	10/13/1862	Claverack	Deserted 11/22/1862
James Wilson	20	Private	10/29/1862	Chatham	Deserted 11/03/1862
Alfred Reynolds	18	Private	09/12/1862	Kinderhook	Unaccounted for at Muster Out

The 159th Regiment Infantry, NYS Volunteers – New Company G - Recruits

NAME	AGE	RANK	DATE JOINED	PLACE JOINED	NOTES
James S. Reynolds	38	Captain	12/22/1863	Albany	Promoted from 2nd Lieutenant 2/8/1864
E. Spencer Elmer	25	1st Lieutenant	2/8/1864	Poughkeepsie	Commissioned Captain
Peter Van Deusen	44	2nd Lieutenant	2/8/1864	Poughkeepsie	Discharged 7/17/1864
Egbert E. Covey	28	1st Sergeant	12/30/1863	Hudson	Veteran
Jennings J. Covey	23	Sergeant	1/4/1864	Hudson	Veteran
William H. Mesick	21	Sergeant	1/25/1864	Hudson	Veteran
Martin Day	25	Sergeant	1/26/1864	Hudson	Enrolled Private, promoted twice, last on 6/30/1865
David Post	23	Corporal	1/25/1864	Hudson	Veteran
Nicholas R. Shultis	24	Corporal	1/25/1864	Hudson	Veteran. Absent sick in hospital.
Charles A. Michael	21	Corporal	2/3/1864	Poughkeepsie	Veteran
George Rettig	40	Corporal	1/4/1864	Hudson	Veteran
Oscar Lewis	18	Corporal	1/25/1864	Hudson	Enrolled Private. Promoted 9/1/1865
Edward Tryon	18	Musician	1/23/1864	12th District	Recruit
George W. Bristol	23	Private	1/1/1864	Hudson	Veteran. Enrolled Sergeant, promoted and reduced 9/25/1865
Abram Bunt	18	Private	1/4/1864	Hudson	
Robert R. Butts	18	Private	12/30/1863	Hudson	
Rowland Brooks	18	Private	1/28/1864	Hudson	
Dennis Calligan	19	Private	1/25/1864	Hudson	
Thomas Doyson	29	Private	1/21/1864	Hudson	
John W. Darkins	18	Private	1/25/1864	Hudson	Absent. Sick in hospital.
Alexander Day	18	Private	1/26/1864	Hudson	
Alexander Ellison	22	Private	1/6/1864	12th District	Recruit. Absent without leave.
Lotan Fuller	27	Private	1/4/1864	Hudson	Veteran. Enrolled Corporal, reduced 7/10/1864
Edward Fitzgerald	22	Private	2/1/1864	Hudson	
Patrick Guilfoil	35	Private	1/30/1864	Hudson	
Jacob H. Groat	26	Private	2/22/1865	Ancram	Veteran. Recruit.
Jacob Hallenbeck	28	Private	1/28/1864	Hudson	Enrolled Private, promoted Corporal, reduced 12/31/1864

The 159th Regiment Infantry, NYS Volunteers – New Company G - Recruits

NAME	AGE	RANK	DATE JOINED	PLACE JOINED	NOTES
Lewis H. Hermance	27	Private	12/26/1863	Hudson	Deserted 2/8/1865, apprehended and returned 6/30/1865
Charles H. Howes	21	Private	12/30/1863	Hudson	Absent. Sick in hospital.
Spencer Helmes	18	Private	12/30/1863	Hudson	
Lodi Kitchell	36	Private	1/6/1864	Copake	
Lane Daniels	24	Private	1/29/1864	Hudson	
Simeon Morris	21	Private	1/25/1864	11th District	Recruit
Jacob H. Melius	18	Private	2/3/1864	Hudson	Recruit
Anthony M. Michael	18	Private	2/16/1864	Hudson	
Warren H. Miller	20	Private	1/27/1864	Hudson	
Franklin Perry	18	Private	1/26/1864	Hudson	
Joshua Plumb	18	Private	3/21/1865	Poughkeepsie	Recruit
Charles Root	37	Private	1/27/1864	Hudson	Enrolled Corporal, reduced 10/1/1864
Elliot Rowlinson	22	Private	1/23/1864	Hudson	Absent. Sick in hospital.
Henry M. Simpson	21	Private	3/22/1864	12th District	Recruit.
William H. Shultis	23	Private	1/25/1864	Hudson	Veteran. Promoted Corp. 12/31/1864, reduced 8/16/1865
Charles E. Shultis	18	Private	1/25/1864	Hudson	Promoted Corp. 9/30/1864, reduced 12/31/1864
Jordan Shultis	25	Private	12/30/1863	Hudson	Enrolled Corporal, reduced 3/31/1864
John Shaughnessy 1st	28	Private	1/11/1864	Hudson	
Henry H. Steel	30	Private	1/20/1864	Hudson	
Henry C. Van Deusen	20	Private	1/20/1864	Hudson	Veteran. Enrolled Corporal, reduced 2/3/1865
Russel Van Deusen	17	Private	1/18/1864	Hudson	Promoted Corp. 10/1/1864, reduced 12/31/1864
Charles E. Van Valkenburgh	18	Private	1/25/1864	Hudson	
Elias Van Deusen	25	Private	12/30/1863	Hudson	
Fidelle Wise	18	Private	1/6/1864	Hudson	
Benjamin Moore	30	Cook	1/1/1864	Thibodeau, La.	Recruit
Levi Bird	18	Cook	1/1/1864	Thibodeau, La.	Recruit
Luzerne Stewart	18	Musician	1/26/1864	Hudson	Discharged with disability 7/22/1864

The 159th Regiment Infantry, NYS Volunteers – Company H

NAME	AGE	RANK	DATE JOINED	PLACE JOINED	NOTES
William O. Pettit	45	Captain	9/10/1862	Brooklyn	Commissioned Major and Lieutenant-Colonel
George B. Stayley	26	1st Lieutenant	8/8/1864	Charlestown, Va.	Promoted 1st Sergt. From 48th NYSV, commissioned Captain.
George R. Herbert	21	2nd Lieutenant	9/1/1862	Brooklyn	Discharged 5/7/1865. Detached to Signal Corps. 11/13/1862
Charles C. Baker	21	1st Lieutenant	9/1/1862	Brooklyn	Transferred 1/27/1863. Promoted Captain Co. I.
Duncan Richmond	21	1st Lieutenant	11/3/1862	Brooklyn	Transferred 2/20/1864. Promoted Captain Co. K.
William J. Kennedy	24	1st Sergeant	9/5/1862	Brooklyn	Enrolled Sergeant. Promoted 1st Sergeant 9/9/1865
William F. French	19	Sergeant	9/18/1862	Hudson	Absent, detached. Transferred from Co. G. 2/12/1864
Martin Traver	22	Sergeant	9/23/1862	Kinderhook	Enrolled Corporal promoted 6/13/1864. Transferred from Co. G. 3/12/1864
Andrew Canonier	19	Corporal	9/11/1862	Brooklyn	Enrolled Private, Promoted 9/9/1865
John P. Abbott	39	Corporal	10/16/1862	Kinderhook	Absent sick in hospital. Enrolled Private promoted 5/3/1864. Transferred from Co. G. 3/12/1864
William F. Churchward	20	Private	8/25/1862	Brooklyn	
Herman Esser	18	Private	9/11/1862	Brooklyn	
James Graham	25	Private	9/17/1862	Brooklyn	
Thomas Graham	40	Private	10/21/1862	Brooklyn	
William C. Handy	44	Private	9/5/1862	Brooklyn	
John S. Hovell	18	Private	10/16/1862	Brooklyn	
Bryan Hopkins	31	Private	9/9/1862	Brooklyn	
Benjamin F. Herrick	25	Private	1/5/1864	Stockport	Deserted apprehended and returned 5/23/1864. Veteran recruit.
Martin Kauffman	18	Private	9/8/1862	Brooklyn	
Henry Lane	21	Private	10/11/1862	Brooklyn	Absent. Deserted twice, last returned 7/5/1865
William Lynch	18	Private	9/20/1862	Ghent	Transferred from Co. G. #/12/1864. Detailed as Musician
Ditmas Ludlow	44	Private	9/10/1862	Brooklyn	Enrolled Corporal, reduced 12/26/1862
Henry William Lutz	18	Private	9/4/1862	Brooklyn	
Peter Miller	21	Private	9/26/1862	Brooklyn	
Joseph Martin	26	Private	9/20/1862	Brooklyn	
William Murphy	27	Private	9/23/1862	Brooklyn	

The 159th Regiment Infantry, NYS Volunteers – Company H

NAME	AGE	RANK	DATE JOINED	PLACE JOINED	NOTES
George Millott	25	Private	9/26/1862	New Lebanon	Transferred from Co. G. 3/12/1864
Neil Peterson	21	Private	9/26/1862	Brooklyn	Absent, in confinement, sentence G.C.M. Deserted 8/29/1863 apprehended 2/21/1864.
Charles Powell	19	Private	11/25/1863	2nd District	Recruit
Lucas R. Smith	18	Private	8/28/1862	Brooklyn	
Walter Simonson	23	Private	9/24/1862	Brooklyn	Deserted 11/18/1862. Apprehended and returned 7/15/1863.
John Wardrop	35	Private	9/18/1862	Brooklyn	
Charles R. Wenstrom	21	Private	10/31/1862	Brooklyn	Deserted 1/8/1865. Apprehended and returned 4/16/1865.
Perry P. Perry	18	Corporal	9/12/1862	Brooklyn	Discharged with disability 8/24/1863. Enrolled Private promoted 12/26/1862
William H. Wortman	45	Wagoneer	10/1/1862	Hudson	Discharged 6/7/1864. Transferred from Co. G. 3/12/1864
Henry Alderton	44	Private	10/1/1862	Brooklyn	Discharged 1/30/1863
Thomas Barrett	35	Private	10/3/1862	Brooklyn	Discharged with disability11/3/1863. Transferred from Co. F.
James Bennett	34	Private	9/27/1862	Brooklyn	Discharged with disability 6/1/1864
Alfred Brown		Private	11/11/1863	New Orleans, La	Discharged with disability 8/12/1864. Recruit
James Brown	44	Private	9/6/1862	Brooklyn	Discharged with disability 10/1/1863
Stephen M. Brown		Private	11/11/1863	New Orleans, La	Discharged 8/17/1864. Recruit. Expiration of service.
David French	26	Private	9/25/1862	Brooklyn	Discharged 6/30/1865. Enrolled wagoner, reduced 6/21/1863
Robert Hurley	23	Private	9/12/1862	Brooklyn	Discharged with disability 3/27/1865
William Herbert	32	Private	9/13/1862	Brooklyn	Discharged with disability 6/7/1865
Davis Jacobs	31	Private	10/21/1863	Brooklyn	Discharged with disability 3/8/1864
Augustus G. Jenkins	18	Private	8/27/1862	Brooklyn	Discharged with disability 2/17/1865
Benjamin Lemond	35	Private	9/4/1862	Brooklyn	Discharged with disability 6/10/1865
Edward McGreen	32	Private	10/21/1862	Brooklyn	Discharged 9/27/1863
William H. Onderdonk	25	Private	9/16/1862	Brooklyn	Discharged with disease 2/1/1863. Enrolled 1st Sergeant Reduced 12/4/1862
Wickham Powell	20	Private	11/23/1863	2nd District	Discharged 8/4/1865. Recruit
George Roden	43	Private	9/29/1862	Brooklyn	Discharged with disability 3/9/1864
Joseph W. Thomas	24	Private	10/31/1862	Brooklyn	Discharged 12/15/1863. Deserted 11/2/1862 apprehended and returned.
Herman Smith	18	1st Sergeant	9/3/1862	Brooklyn	Transferred 3/6/1863 to N.C.S. Promoted twice last on 12/4/1862.

The 159th Regiment Infantry, NYS Volunteers – Company H

NAME	AGE	RANK	DATE JOINED	PLACE JOINED	NOTES
Charles P. Price	35	Sergeant	9/6/1862	Brooklyn	Transferred 1/27/1863 to Co. K. Enrolled Corporal, promoted twice last on 1/21/1863
Henry Martin	21	Sergeant	9/11/1862	Brooklyn	Transferred 5/1/1864 to Navy. Enrolled Private, promoted twice last on 5/6/1864
Benjamin Vandewater	41	Corporal	10/1/1862	Fishkill	Transferred 6/18/1864 to V.R.C. Transferred from Co. G. 3/12/1864
William H. Frier	21	Private	8/27/1862	Brooklyn	Transferred 6/5/1865 to V.R.C. Wounded in Opequan, Va. 9/19/1864
Alfred H.S. Moore	24	Private	9/5/1862	Brooklyn	Transferred 11/2/1862 to N.C.S. Promoted Hospital Stewart
Michael Murtha	18	Private	9/15/1862	Brooklyn	Transferred 4/28/1864 to V.R.C.
Edward Pittinger	19	Private	9/10/1862	Brooklyn	Transferred 9/11/1863 to V.R.C. Enrolled Corporal, promoted Sergeant 11/4/1862, reduced 1/21/1863
Cornelius Stickles	24	Private	10/7/1862	Greenport	Transferred 4/10/1864 to V.R.C. Transferred from Co. G. 3/12/1864
Thomas E. Crowell	18	Sergeant	10/16/1862	Brooklyn	Died 5/27/1863 in action, Port Hudson, La. Enrolled Corporal promoted 1/21/1863
Edmund B. King	18	Sergeant	9/6/1862	Brooklyn	Died of disease 11/5/1863
David W. Darling	25	Corporal	8/28/1862	Brooklyn	Died of disease 6/14/1863
John Neefus	18	Corporal	9/6/1862	Brooklyn	Died 5/2/1863 of wounds in action at Irish Bend, La 4/14/1863
William Uggla	38	Corporal	9/16/1862	Brooklyn	Died 5/27/1863 in action at Port Hudson, La.
Washington Adams	44	Private	9/17/1862	Brooklyn	Died 9/29/1864 of wounds in action at Opequan, Va. 9/19/1864
Daniel Corson	41	Private	9/4/1862	Brooklyn	Died of disease 4/28/1863
John Daly	19	Private	9/15/1862	Brooklyn	Died 5/27/1863 in action at Port Hudson, La.
Mathew Donovan	23	Private	10/27/1862	Brooklyn	Died of disease 1/29/1864
Henry E. Lander	18	Private	10/18/1862	Brooklyn	Died 9/19/1864 in action at Opequan, Va.
John L. Lougea	40	Private	8/30/1862	Brooklyn	Died date unknown as Prisoner of War Salisbury, N.C.
James McCormick	38	Private	9/11/1862	Brooklyn	Died 5/27/1863 in action at Port Hudson, La.
Charles Rossiter	39	Private	9/16/1862	Brooklyn	Died 5/27/1863 in action at Port Hudson, La.
John Walker	21	Private	9/4/1862	Brooklyn	Died of disease 11/27/1862

The 159th Regiment Infantry, NYS Volunteers – Company H

NAME	AGE	RANK	DATE JOINED	PLACE JOINED	NOTES
Andrew Wilson	34	Cook	11/28/1863	Thibodaux, La.	Died of disease 11/4/1864. Recruit.
Charles Andrew	25	Sergeant	10/13/1862	Brooklyn	Deserted 11/2/1862
James Farrell	18	Musician	9/5/1862	Brooklyn	Deserted 11/3/1862
Samuel Argent	39	Private	9/26/1862	Brooklyn	Deserted 11/3/1862
Dennis Burns	39	Private	10/20/1862	Brooklyn	Deserted 11/2/1862
John Carroll	23	Private	10/1/1862	Brooklyn	Deserted 11/1/1862
Thomas Devine	23	Private	9/15/1862	Brooklyn	Deserted 11/2/1862
Charles Edward	32	Private	9/4/1862	Brooklyn	Deserted 11/10/1862
James Feeny	25	Private	9/29/1862	Brooklyn	Deserted 11/3/1862
Thomas Fitzsimmons	26	Private	9/11/1862	Brooklyn	Deserted 11/11/1862
Edward Flanagan	29	Private	9/22/1862	Brooklyn	Deserted 11/12/1862
William Gray	21	Private	10/9/1862	Brooklyn	Deserted 11/2/1862
John Greenfield	21	Private	9/16/1862	Brooklyn	Deserted 11/14/1862
Michael Haskins	40	Private	9/19/1862	Brooklyn	Deserted 11/15/1862
Bernard Keegan	28	Private	9/29/1862	Brooklyn	Deserted 11/14/1862
Oscar Koltshorn	22	Private	9/27/1862	Brooklyn	Deserted 11/17/1862
John Lynch	43	Private	9/17/1862	Brooklyn	Deserted 11/20/1862
Patrick McDonald	22	Private	9/5/1982	Brooklyn	Deserted 11/19/1862
Barney Olthoff	42	Private	9/21/1862	Brooklyn	Deserted March 1863
Samuel Riley	41	Private	10/15/1862	Brooklyn	Deserted 11/17/1862
Edward A. Sterling	26	Private	9/20/1862	Brooklyn	Deserted 11/10/1862
Philip F. Scoot	24	Private	9/4/1862	Brooklyn	Deserted 11/10/1862
Henry Seymore	38	Private	10/14/1862	Brooklyn	Deserted 11/21/1862
Gilbert Wanser	28	Private	10/3/1862	Brooklyn	Deserted 11/14/1862
Peter Welland	42	Private	10/4/1862	Brooklyn	Deserted 11/19/1862
William Wells	37	Private	9/11/1862	Brooklyn	Deserted 11/13/1862
Frederick R. Woller	30	Private	9/27/1862	Brooklyn	Deserted 11/11/1862

The 159th Regiment Infantry, NYS Volunteers – Company I

NAME	AGE	RANK	DATE JOINED	PLACE JOINED	NOTES
Edward Tynan	22	1st Lieutenant	9/6/1864	Albany	Appointed from civil life. 8/13/1864. Commissioned Captain
Edward Wardle	26	Captain	10/15/1862	Hudson	Discharged 1/26/1863
Joseph G. McNutt		Captain	5/1/1864	Morganza, La.	Discharged Dishonorably 8/20/1865. Sentenced G.C.M.
John W. Shields	28	1st Lieutenant	10/15/1862	Poughkeepsie	Discharged 12/7/1863. Cashiered, sentenced G.C.M.
William Prince		1st Lieutenant	12/26/1863	Albany	Discharged 9/13/1864. Appointed & never served with regiment.
Jacob Fingar	33	2nd Lieutenant	10/10/1862	Hudson	Discharged 1/26/1863
Lambert Dingman	23	2nd Lieutenant	9/4/1862	Kinderhook	Discharged 2/3/1864. Enrolled Sergeant Co. G. Prom. 1/27/1863
Charles C. Baker	21	Captain	9/1/1862	Brooklyn	Transferred 2/10/1864. Prom. Major 39th NYSV. Promoted from 1st Lieutenant Co. H. 1/26/1863
Cornelius V. Coventry	39	1st Sergeant	9/5/1862	Hudson	Enrolled Captain promoted twice, last on 9/28/1865
Scutt Grosvenor	24	Sergeant	10/3/1862	Gallatin	Enrolled Private, promoted.
James Brazier Sr.	41	Corporal	10/3/1862	Poughkeepsie	Enrolled Private, promoted.
James Brazier Jr.	18	Private	10/3/1862	Poughkeepsie	Absent, in confinement. Deserted and apprehended.
William Oscheitz	45	Corporal	10/30/1862	Poughkeepsie	Absent, sick in hospital. Enrolled Private, promoted.
Michael Butler	18	Private	10/2/1862	Poughkeepsie	
Fredrick Bogardus	21	Private	10/18/1862	Poughkeepsie	Wounded at Port Hudson 6/14/1863.
Philip H. Coon	21	Private	9/27/1862	Gallatin	
Alfred R. Coon	21	Private	9/22/1862	Gallatin	
Alvarus Coon	16	Private	9/27/1862	Gallatin	None of the three Coons are brothers.
Martin Cahalen	21	Private	10/4/1862	Livingston	
John E. Cleminshire	21	Private	10/15/1862	Poughkeepsie	
Eugene A. Corey	19	Private	12/9/1863	Brooklyn	Veteran recruit.
Henry Dennis	23	Private	9/16/1862	Gallatin	
John Devlin	18	Private	9/17/1862	Kinderhook	Wounded at Irish Bend, La. 4/14/1863. Transferred from Co. G. 4/12/1864
Almond Edson	29	Private	10/17/1862	Poughkeepsie	
Charles Feller	21	Private	10/30/1862	Clermont	
William Foster	24	Private	10/18/1862	Hudson	

The 159th Regiment Infantry, NYS Volunteers – Company I

NAME	AGE	RANK	DATE JOINED	PLACE JOINED	NOTES
Charles Frost	19	Private	11/20/1863	Brooklyn	Recruit
William Farrington	32	Private	10/30/1862	Fishkill	Transferred from Co. G. 3/12/1864
James Howell	28	Private	10/15/1862	Poughkeepsie	
Walter C. Houck	45	Private	10/8/1862	Gallatin	Absent in hospital. Wounded Opequan, Va. 9/19/1864
Stephen Kilmer	42	Private	10/4/1862	Hudson	Absent, sick in hospital.
David W. Lawrence	19	Private	10/1/1862	Gallatin	Enrolled Private, promoted 1st Sergeant, reduced 9/28/1865
James R. Lee, Jr.	19	Private	10/6/1862	Poughkeepsie	Absent, in confinement, sentence G.C.M.
William Roach	19	Private	10/9/1862	Hudson	Enrolled Corporal, reduced.
George W. Schofield	21	Private	10/8/1862	Poughkeepsie	
William H. Shaw	42	Private	10/21/1862	Hudson	Absent, sick in hospital.
John Tiffany	17	Private	9/18/1862	Hudson	
Gottfried E. Veigle	40	Private	10/22/1862	Poughkeepsie	Enrolled Sergeant, reduced.
Richard Whalen	18	Private	9/29/1862	Poughkeepsie	
Norman Wagoner	21	Private	10/3/1862	Gallatin	Deserted apprehended and returned.
William H. Wagoner	38	Private	10/2/1862	Livingston	Wounded, Opequan, Va. 9/19/1864
Washington L. White	20	Private	9/30/1862	Claverack	
William Hayes	32	Cook	1/1/1864	Thibodeaux, La.	Recruit
Edward Duffy	32	Sergeant	10/6/1862	Hudson	Discharged 5/23/1864. Promoted 1st Lieutenant Co. A. Transferred from Co. G. 3/12/1864.
Eleazor Parmly Brown	19	Sergeant	10/20/1863	Brooklyn	Discharged 6/22/1865. Recruit promoted two times.
James Dennis	21	Corporal	9/30/1862	Hudson	Discharged 8/22/1863
Joseph O. Reed	22	Corporal	10/4/1862	Hudson	Discharged 5/17/1865. Disability from wounds in action Halltown, Va. 8/24/1864.
Charles Clark	17	Musician	9/15/1862	Hudson	Discharged with disability 3/22/1863
Jordan N. Lee	18	Musician	10/6/1862	Poughkeepsie	Discharged 1/26/1865
John Race	45	Wagoner	10/18/1862	Hudson	Discharged with disability 1/21/1865
John S. Campbell	45	Private	10/7/1862	Hudson	Discharged 11/23/1863
Peter W. Coon	47	Private	10/25/1862	Livingston	Discharged

The 159th Regiment Infantry, NYS Volunteers – Company I

NAME	AGE	RANK	DATE JOINED	PLACE JOINED	NOTES
Joshua Decker	36	Private	9/26/1862	Livingston	Discharged with disability 6/18/1863
Peter S. Garrettson	24	Private	4/10/1865	New York	Discharged 5/9/1865. Recruit.
George Griffin	31	Private	3/14/1865	Fishkill	Discharged 5/9/1865. Recruit.
Charles H. Hine	21	Private	2/23/1865	Brooklyn	Discharged 5/9/1865. Recruit.
William H. Hyler	19	Private	4/10/1865	New York	Discharged 5/9/1865. Recruit.
George N. Kirchner	45	Private	10/18/1862	Poughkeepsie	Discharged 6/7/1863
John M. Lawrence	23	Private	10/4/1862	Gallatin	Discharged with disability 3/26/1864.
James R. Lee Sr.	42	Private	10/6/1862	Poughkeepsie	Discharged 6/27/1863
Archibald Osborn	44	Private	9/29/1862	Pleasant Valley	Discharged 12/3/1862. Writ Habeas Corpus
Gustave Roth	18	Private	2/16/1865	Hudson	Discharged 5/9/1865. Recruit
Carliston T. Smith	45	Private	10/16/1862	Hudson	Discharged 5/30/1865
Peter H. Snyder	37	Private	10/1/1862	Gallatin	Discharged with disability 3/31/1865
Thomas Scutt	21	Private	10/3/1862	Gallatin	Discharged 6/28/1865. Prisoner of war Opequan, Va. 9/19/1864
George Starr	36	Private	12/14/1863	Brooklyn	Discharged with disability 5/17/1865. Recruit
James L. Tiffany	44	Private	1/24/1864	Poughkeepsie	Discharged 5/9/1865. Recruit
David Troy	19	Private	4/11/1865	New York	Discharged 5/9/1865. Recruit
Edward Vail	44	Private	10/8/1862	Pleasant Valley	Discharged with disability.
Henry F. Vermilyea	21	Private	4/10/1865	New York	Discharged 5/9/1865. Recruit
Jacob Coon	36	Private	10/14/1862	Livingston	Transferred 3/23/1863 to V.R.C.
Edward Cosgrove	40	Private	9/29/1862	Livingston	Transferred 6/24/1864 to V.R.C.
David Simmons	44	Private	9/30/1862	Clinton	Transferred 6/24/1864 to V.R.C
Mark Baker	19	1st Sergeant	10/13/1862	Poughkeepsie	Died 4/14/1863 in action at Irish Bend, La.
Pleasant Kline	18	Corporal	9/9/1862	Taghkanic	Died of disease 9/28/1863
John Lynch	20	Corporal	9/20/1862	Ghent	Died. Transferred from Co. G. 3/12/1864
Charles Alger	36	Private	9/9/1862	Hudson	Died of alcoholism 9/7/1865
Theodore Bohrer	23	Private	9/4/1862	Taghkanic	Died 4/14/1863 in action at Irish Bend, La.
Henry Bugel	39	Private	9/24/1862	Livingston	Died 6/13/1863

The 159th Regiment Infantry, NYS Volunteers – Company I

NAME	AGE	RANK	DATE JOINED	PLACE JOINED	NOTES
Daniel Carey	44	Private	9/8/1862	Poughkeepsie	Died 9/25/1863
Henry A. Conroe	30	Private	9/30/1862	Gallatin	Died 7/15/1863
William Coon	17	Private	9/27/1862	Gallatin	Died 5/25/1863 in action at Port Hudson, La.
John W. Coon	27	Private	10/4/1862	Livingston	Died 8/3/1865 of accidental gunshot wound.
Charles Dutcher	44	Private	10/7/1862	Poughkeepsie	Died 3/8/1863
Philip Houghtaling	32	Private	10/11/1862	Gallatin	Died 6/8/1863
William H. Haws	42	Private	9/4/1862	Hudson	Died 4/14/1863 in action at Irish Bend, La.
Collins Jackson	35	Private	9/4/1862	Poughkeepsie	Died 8/3/1863
Jeremiah Kellerhouse	41	Private	9/26/1862	Livingston	Died 9/18/1863
William Kellerhouse	37	Private	9/4/1862	Gallatin	Died 5/31/1863
Philip B. Kipp	44	Private	10/9/1862	Gallatin	Died 5/29/1863 from Chronic Diarrhea.
William Knickerbocker	18	Private	9/23/1862	Taghkanic	Died 8/29/1863 from Chronic Rheumatism.
Milton Kilmer	21	Private	10/16/1862	Livingston	Died 8/7/1863
James McCanley	19	Private	10/29/1862	Stockport	Died from disease 3/6/1863
Peter McDailey	28	Private	10/1/1862	Gallatin	Died 11/26/1863
William Moon	45	Private	10/4/1862	Milan	Died from disease 8/21/1864
James Morton		Private	9/29/1864	Troy	Died 1/3/1865 while Prisoner of War Salisbury N.C. Recruit
Abram F. Palmer	18	Private	10/4/1862	Livingston	Died 3/22/1863 from Typhoid Fever.
Oliver Palmater	19	Private	10/8/1862	Poughkeepsie	Died 8/4/1863
Peter Proper	25	Private	9/23/1862	Taghkanic	Died
Richard Tesel	31	Private	10/17/1862	Poughkeepsie	Died July 1864
John Terry	18	Private	10/8/1862	Poughkeepsie	Died 3/10/1863 from Typhoid Fever.
Thomas Van Hosen	44	Private	9/23/1862	Hudson	Died of disease 4/27/1863
Martin Whalen	30	Private	10/13/1862	Poughkeepsie	Died 3/25/1864, drowned from transport.
John B. Worden	44	Private	10/20/1862	Poughkeepsie	Died of disease 5/15/1863
John Cook	33	Corporal	9/29/1862	Gallatin	Died 4/14/1863 in action at Irish Bend, La.
John K. Haney	25	1st Sergeant	10/7/1862	Poughkeepsie	Deserted April 1863

The 159th Regiment Infantry, NYS Volunteers – Company I

NAME	AGE	RANK	DATE JOINED	PLACE JOINED	NOTES
William Brown	33	Sergeant	10/4/1862	Poughkeepsie	Deserted 11/2/1862
John Keegan	30	Sergeant	10/12/1862	Poughkeepsie	Deserted 11/2/1862
Stephen E. Best	44	Private	9/4/1862	Hudson	Deserted 11/10/1862
Peter Clark	30	Private	9/12/1862	Pleasant Valley	Deserted 11/2/1862
William B. Dennis	44	Private	10/1/1862	Clinton	Deserted 11/10/1862
John Donnelly	28	Private	10/16/1862	Poughkeepsie	Deserted 11/10/1862
Henry Ingalls	22	Private	9/27/1862	Gallatin	Deserted 11/3/1862
Jacob Kilmer	36	Private	9/6/1862	Claverack	Deserted 9/18/1863
George Mayley	44	Private	9/29/1862	Gallatin	Deserted July 1863
James McAvoy	25	Private	10/21/1862	Poughkeepsie	Deserted 11/2/1862
Samuel J. Miller	21	Private	10/28/1862	Livingston	Deserted 11/2/1862
Henry Smith	23	Private	10/8/1862	Poughkeepsie	Deserted 11/15/1862
William Winters	28	Private	10/8/1862	Poughkeepsie	Deserted 11/20/1862

The 159th Regiment Infantry, NYS Volunteers – Company K

NAME	AGE	RANK	DATE JOINED	PLACE JOINED	NOTES
Joe B. Ramsden	26	Captain	9/18/1862	Brooklyn	Discharged 7/30/1863
Edward Sherer	19	1st Lieutenant	9/1/1862	Brooklyn	Discharged 11/3/1862
Alfred H. Bruce	23	1st Lieutenant	9/15/1862	Kinderhook	Discharged 5/17/1865 Enrolled 1st Sergeant Co. G. Promoted three times last on 4/30/1863
Charles H. Brundage		2nd Lieutenant	12/18/1863	New Orleans, La.	Discharged 4/9/1864 Cashiered by sentence G.C.M. Promoted from Private 7th NYHA
Duncan Richmond		Captain	11/3/1862	Brooklyn	Died 10/19/1864 in action at Cedar Creek, Va. Appointed 2nd Lieutenant, promoted twice, last on 2/20/1864
William R. Plunkett	37	1st Lieutenant	9/27/1862	Brooklyn	Died 4/17/1863 of wounds in action at Irish Bend, La. 4/14/1863 Enrolled 2nd Lieutenant promoted 11/3/1862
Charles P. Price	35	2nd Lieutenant	9/6/1862	Brooklyn	Died 4/18/1863 of wounds in action at Irish Bend, La. 4/14/1863 Enrolled Corporal Co. H. promoted twice, last on 1/27/1863
James M. Mills	19	1st Sergeant	8/21/1862	Brooklyn	Enrolled Private, promoted three times, last on 9/9/1865
John Pelletreau	25	Sergeant	8/23/1862	Brooklyn	Enrolled Private, promoted twice, last on 6/25/1864
George H. Van Alstyne	25	Sergeant	9/19/1862	Kinderhook	Enrolled Corporal promoted 9/19/1864. Transferred from Co. G. 3/12/1864
William D. Tanner	36	Corporal	10/24/1862	Kinderhook	Enrolled Private, promoted 9/19/1864. Transferred from Co. G. 3/12/1864
Austin Nevals	22	Corporal	10/7/1862	Ghent	Enrolled Private, promoted 9/19/1864. Transferred from Co. G. 3/12/1864
George Myers	15	Musician	10/8/1862	Brooklyn	
Joseph Horan	17	Musician	10/6/1862	Brooklyn	
John Brush	30	Private	8/30/1862	Brooklyn	
Sales Bender	37	Private	9/30/1862	Brooklyn	
Edward Brophy	18	Private	9/14/1862	Kinderhook	Transferred from Co. G. 3/12/1864
Joseph Corcoran	40	Private	9/30/1862	Brooklyn	
James Davis	25	Private	8/29/1862	Brooklyn	
Peter Dunn	18	Private	10/4/1862	Brooklyn	
Timothy Dolan	20	Private	1/25/1864	Poughkeepsie	Absent, Prisoner of War, Opequan, Va. Recruit.
John Feeny	18	Private	8/28/1862	Brooklyn	
Henry Hahn	33	Private	8/25/1862	Brooklyn	
George A. Hoffman	26	Private	9/25/1862	Claverack	Absent, sick in hospital. Transferred from Co. G. 3/12/1864

The 159th Regiment Infantry, NYS Volunteers – Company K

NAME	AGE	RANK	DATE JOINED	PLACE JOINED	NOTES
Edward A. Johnson	32	Private	9/13/1862	Brooklyn	
James Kelly	21	Private	8/23/1862	Brooklyn	
John M. Kewan	34	Private	9/27/1862	Brooklyn	
John Kane	37	Private	12/31/1863	12th District	Recruit.
Daniel Lafferty	25	Private	8/28/1862	Brooklyn	
James S. Leonard	21	Private	8/20/1862	Brooklyn	
John H. McDivitt	33	Private	8/30/1862	Brooklyn	
William Mills	22	Private	8/22/1862	Brooklyn	
Thomas McCormick	28	Private	9/23/1862	Hudson	Absent, sick in hospital. Transferred from Co. G. 3/12/1864
Cornelius Myers	35	Private	9/15/1862	Brooklyn	Enrolled 1st Sergeant, reduced twice, last on 8/16/1865
John Stout	42	Private	10/30/1862	Brooklyn	
Henry S. Stickles	36	Private	9/29/1862	Kinderhook	Transferred from Co. G. 3/12/1864
Thomas Shea	24	Private	10/13/1862	Kinderhook	Transferred from Co. G. 3/12/1864
Clark B. Suydam	18	Private	9/30/1862	Fishkill	Absent, detached. Transferred from Co. G. 3/12/1864
John Schrader	34	Private	2/7/1865	Kingston	Recruit.
Daniel Leonard		Cook	1/13/1864	Thibodeaux, La.	Recruit
William Kelly		Cook	1/13/1864	Thibodeaux, La.	Recruit
John Day	23	1st Sergeant	8/23/1862	Brooklyn	Discharged 9/1/1864 for promotion to 1st Lieutenant Co. B. Engineers Corp.
Richard Mathews	29	Sergeant	8/20/1862	Brooklyn	Discharged with disability 5/24/1863
Francis L. Vandergaw	22	Corporal	8/21/1862	Brooklyn	Discharged 5/21/1864. Commissioned 1st Lieutenant 56th NYSV
John Coughlin	30	Private	8/30/1862	Brooklyn	Discharged 5/3/1865
James Close	43	Private	8/26/1862	Brooklyn	Discharged with disability 5/22/1865
Eugene V. Cruger	23	Private	8/21/1862	Brooklyn	Discharged 1/22/1864. Deserted apprehended & returned 7/20/63
Michael Green	26	Private	10/28/1862	Brooklyn	Discharged 8/4/1865

The 159th Regiment Infantry, NYS Volunteers – Company K

NAME	AGE	RANK	DATE JOINED	PLACE JOINED	NOTES
Richard Gardner	45	Private	11/14/1863	Thibodeaux, La.	Discharged 8/11/1864, Recruit, expiration of service.
James M. Garrison	40	Private	11/14/1863	Thibodeaux, La.	Discharged 8/11/1864, Recruit, expiration of service.
Daniel Hyde	28	Private	8/26/1862	Brooklyn	Discharged with disability 3/22/1863
Patrick Kelly	28	Private	8/28/1862	Brooklyn	Discharged with disability 8/27/1863
William Mathews	39	Private	8/29/1862	Brooklyn	Discharged with disability 5/22/1865
Thomas Maguire	21	Private	8/28/1862	Brooklyn	Discharged with disability 1/11/1863
Alden Powers	44	Private	9/27/1862	Brooklyn	Discharged with disability 1/9/1863
James P. Rooney	23	Private	8/23/1862	Brooklyn	Discharged June 1863
George W. Simonton	30	Private	8/29/1862	Brooklyn	Discharged with disability 2/8/1863
Francis Van Hoesen	28	Private	9/22/1862	Ghent	Discharged 6/27/1865. Transferred from Co. G. 3/12/1864
Fredrick Young	36	Private	8/25/1862	Brooklyn	Discharged with disability 1/12/1863
Thomas Bergen	29	1st Sergeant	8/28/1862	Brooklyn	Transferred 7/22/1863, promoted Commissary Sergeant
Samuel Ashton	21	Private	8/28/1862	Brooklyn	Transferred 5/1/1864 to Navy
Francis Mahon	28	Private	9/12/1862	Brooklyn	Transferred. Deserted & apprehended & was reported to Co. A. as held for that Co. in Fort Columbus, New York harbor.
John Van Mater	23	Corporal	8/28/1862	Brooklyn	Died 3/29/1863
Andrew Asbell	22	Corporal	9/18/1862	Brooklyn	Died 4/14/1863 in action at Irish Bend, La. Deserted November 1862 apprehended 2/18/1863
Edward F. Bridges		Private	1862	Brooklyn	Died 5/27/1863 in action Port Hudson, La.
George Carr	21	Private	9/28/1862	Brooklyn	Died 4/14/1863 in action at Irish Bend, La.
Peter R. Conklin	35	Private	11/14/1863	Thibodeaux, La.	Died of disease 7/21/1864. Recruit
Bernard Dolan	28	Private	10/20/1862	Ghent	Died 1/15/1865 as Prisoner of War, Salisbury, N.C. Transferred from Co. G. 3/12/1864
Robert W. Drake	43	Private	8/25/1862	Brooklyn	Died from Pleurisy April 1863
Andrew Goshia		Private	9/30/1862	Chatham	Died 1/21/1865 as Prisoner of War, Salisbury, N.C. Transferred from Co. G. 3/12/1864
David Miller	41	Private	8/26/1862	Brooklyn	Died 4/14/1863 in action at Irish Bend, La.
Albert C. Trumbull	25	Private	8/25/1862	Brooklyn	Died 6/24/1863 of wounds in action at Port Hudson, La. 5/27/63

The 159th Regiment Infantry, NYS Volunteers – Company K

NAME	AGE	RANK	DATE JOINED	PLACE JOINED	NOTES
Augustus Garholt	43	Private	8/26/1862	Brooklyn	Died 6/14/1863 in action Port Hudson, La.
Walter Smith	26	Sergeant	10/31/1862	Brooklyn	Deserted 11/20/1862
George C. Paine	30	Sergeant	8/22/1862	Brooklyn	Deserted 12/2/1862
Louis Roberts	24	Corporal	8/23/1862	Brooklyn	Deserted 11/22/1862
Alex R. Patterson	37	Corporal	8/26/1862	Brooklyn	Deserted 11/22/1862
Joseph A. Bailey	28	Private	9/12/1862	Brooklyn	Deserted 12/3/1862
Edward D. Baker	26	Private	9/15/1862	Brooklyn	Deserted 11/5/1862
William Bevan	18	Private	10/4/1862	Brooklyn	Deserted 11/2/1862
Michael Condon	23	Private	8/25/1862	Brooklyn	Deserted 11/5/1862
Joseph Colyer	21	Private	9/22/1862	Brooklyn	Deserted 11/5/1862
Henry Carson	21	Private	9/20/1862	Brooklyn	Deserted 11/5/1862
Thomas Donnelly	21	Private	9/20/1862	Brooklyn	Deserted 11/2/1862
John Emmons	30	Private	8/25/1862	Brooklyn	Deserted 8/27/1863
John Ganley	39	Private	8/29/1862	Brooklyn	Deserted 11/22/1862
William Hawker	20	Private	8/30/1862	Brooklyn	Deserted 11/1/1862
James Harding	28	Private	8/25/1862	Brooklyn	Deserted 11/2/1862
Sherman Hart	40	Private	8/27/1862	Brooklyn	Deserted 11/2/1862
Patrick Haverty	33	Private	10/16/1862	Hudson	Deserted twice, first time apprehended and returned, last time 8/10/1865. Transferred from Co. G. 3/12/1864
Thomas Jones	21	Private	10/27/1862	Brooklyn	Deserted 11/2/1862
Patrick King	22	Private	8/26/1862	Brooklyn	Deserted 11/5/1862
John P. Kelly	19	Private	9/25/1862	Brooklyn	Deserted 11/10/1862
John Kelly Jr.	37	Private	9/30/1862	Brooklyn	Deserted 11/10/1862
Thomas B. Lynch	21	Private	8/27/1862	Brooklyn	Deserted 11/20/1862
Francis Mahon	28	Private	9/12/1862	Brooklyn	Deserted 11/5/1862, apprehended and reported to Co. A.
William Miller	21	Private	8/20/1862	Brooklyn	Deserted 11/2/1862
Patrick Mallon	31	Private	8/30/1862	Brooklyn	Deserted 11/22/1862
Thomas Miller	23	Private	8/26/1862	Brooklyn	Deserted 11/5/1862

The 159th Regiment Infantry, NYS Volunteers – Company K

NAME	AGE	RANK	DATE JOINED	PLACE JOINED	NOTES
Michael McMahon	35	Private	9/19/1862	Brooklyn	Deserted 11/23/1862
James McNulty	26	Private	9/23/1862	Brooklyn	Deserted 11/2/1862
Patrick McParland	34	Private	10/28/1862	Brooklyn	Deserted 11/23/1862
James Morrison	22	Private	9/30/1862	Brooklyn	Deserted 11/15/1862
James Mylet		Private	10/1/1862	Brooklyn	Deserted in 1863 in Baton Rouge, La.
Peter Quinn	27	Private	8/29/1862	Brooklyn	Deserted 11/22/1862
Samuel Townsend	21	Private	8/23/1862	Brooklyn	Deserted 11/2/1862
Richard Wallace	26	Private	9/15/1862	Brooklyn	Deserted 11/5/1862

BIBLIOGRAPHY

Remember Me, A Soldier's Life in the 159th N.Y.S.V. Infantry Regiment

INTRODUCTION

NYS Historic Newspapers-Kinderhook Herald, April 18, 1861.

Colonel Daniel S. Cowels – New York State Military Museum & Veterans Research Center. 128th NY Infantry Regiments Civil War Newspaper Clippings.

New York State Military Museum & Veterans Research Center, Columbia County NY in the Civil War (3rd Annual Report of the Bureau of Military Statistics of the State of NY, (C. Wendell), 1866.

Merritt, Keri Leigh, Masterless Men, Poor Whites and Slavery in the Antebellum South, (Cambridge University Press 2017). p. 287.

NYS Historic Newspapers-Kinderhook Herald, Jan 17, 1861, p. 2 image 2.

Taylor, Richard. Destruction & Reconstruction, Personal Experiences of the Late War, New York 1879, p. 217.

Article, A Short Overview of the Reconstruction Era & Ulysses S. Grant's Presidency. Ulysses S. Grant National Historic Site.

CHAPTER 1

Homer Nelson, Civil War Photo Collection United States Army Heritage and Education Center, Carlisle, PA.

Diary of an Enlisted man, Lawrence Van Alstyne (New Haven Conn. 1910), pp. 3-6.

Military Books advertisement, Adriance Memorial Library, Poughkeepsie, NY. Poughkeepsie Daily Eagle, Oct. 25,1862, p. 3.

Letter #1, October 28, 1862, to Maria Mambert. John W. Mambert letters from the Civil War.

Tiemann, William F. The 159th Regiment Infantry, New York State Volunteers, In the War of the Rebellion 1862 -1865, (Brooklyn New York 1891), pp. 9-22.

159th Regiment NY Volunteer Infantry | Regimental Color | Civil War, New York State Military Museum & Veterans Research Center.

Poulin, David, FIELD MUSIC OF THE CIVIL WAR Including extracts from, The 1861 Revised Regulations, Enactments of Congress, And Customs of Service. Organization and Duties, pp. 13-24.

Homer A. Nelson, Biographical Directory of the United States Congress.

On Board Steamer North Star, Gulf of Mexico, December 29, 1862, Page 1, New York Times.

Blodgett Memorial Library, Fishkill NY–Fishkill Journal, Rev. Isaac L. Kipp, Jan. 1, 1863, p. 2.

Description of ship *Northern Lights*, The Maritime Heritage Project, Ship Passengers: 1846–1899.

New York State Military Museum, 161st Regiment Newspaper Clippings, Army correspondence. From New Orleans. On board *Northern Light*, off New Orleans, Dec. 15, 1862. Editors Telegraph.

Irwin, Richard B. History of the 19th Army Corps. G. P. Putnam's Sons New York, (The Knickerbocker Press 1892), ch. 5, par. 6.

Nathaniel Banks- Anthony, E. & Brady, M. B. (1861) Major General Nathaniel P. Banks. Massachusetts United States of America, 1861. [New York: Edward Anthony, to 1876] [Photograph] Retrieved from the Library of Congress, https://www.loc.gov/item/2021669484/.

Cuvier Grover – Harper's Weekly November 12, 1864.

U.S.S. Monitor: The first ironclad warship in the United States Navy, commanded by Admiral John L. Worden. The vessel had a large, round gun turret on top of a flat raft-like bottom, which caused some spectators to describe it as a "cheese box on a raft."

Duffy, Edward B., History of the 159th Regiment, N.Y.S.V. (New York 1890), p. 5.

Letter #2, December 20, 1862, Baton Rouge, To Mrs. John W. Mambert. John W. Mambert letters from the Civil War.

Mud Turtles, The Descent of the Mississippi, published in Frank Leslie's Illustrated, December 1861.

Civil War naval chronology, 1861-1865. Sec. 1 & 2. pp. 17-114.

Brummer, Sidney David, A.M. Political History of New York State During the Period of the Civil War. (New York 1911), p. 255.

Sprague, Homer B. (Homer Baxter), b.1829., History of the 13th Infantry Regiment of Connecticut Volunteers, during the Great Rebellion, (Hartford Conn. 1867), p. 11.

Molineux Camp Sketches, James I Robertson Jr Civil War Sesquicentennial Legacy Collection at The Library of Virginia – Edward L. Molineux collection.

W.F. Tiemann Collection at The N.Y.S. Military Museum, Letter Jan 26, 1863, Baton Rouge.

James I Robertson Jr Civil War Sesquicentennial Legacy Collection at The Library of Virginia – Edward L. Molineux collection, Letter written March 25, 1863.

Letter #3, January 28, 1863, Louisiana, To John Mambert and Marie Mambert. John W. Mambert letters from the Civil War.

Butler, B. F. (Benjamin Franklin). (1892). Autobiography and personal reminiscences of Major-General Benj. F. Butler: Butler's book: a review of his legal, political, and military career. Boston [Mass.]: A.M. Thayer & Co. pp. 454-455.

O.R., Series I, Vol. 15. Butler to Stanton May 8, 1862.

Civil War Naval Chronology, 1861-1865. Part I – 1861. pp. 17-114 / Part II – pp. 30-34.

O.R., Series I, Vol. 15. Reports of Brig. Gen. Thomas Williams, Army, commanding Expeditionary Corps, operations May 26- Aug 2, instructions from Major-General Butler. Headquarters Second Brigade, Baton Rouge Arsenal, 29, 1862.

O.R., Series I, Vol. 15. Major-General Earl Van Dorn, Headquarters District of Mississippi, Jackson, Miss., September 9, 1862.

O.R., Series I, Vol. 15. Maj. Gen. Benjamin F. Butler, U. S. Army, commanding the Department of the Gulf of engagement at Baton Rouge, La., with orders and resulting correspondence. Headquarters Department of the Gulf, New Orleans, La., August 10, 1862.

Encampment at Baton Rouge, La 1862, E. L. Molineux Photo Collection, NYS Military Museum.

Letter #4, Baton Rouge, Louisiana, in care of Capt. Sliter. John W. Mambert letters from the Civil War.

Letter #5, Baton Rouge, Louisiana. John W. Mambert letters from the Civil War.

CHAPTER 2

Letter #6, January 29, 1863, Baton Rouge, Louisiana. John W. Mambert letters from the Civil War.

United States hotel guide and railway companion for 1867: being a directory to the best hotel in nearly all the principal cities and towns throughout the United States. With supplement containing useful and interesting information, p. 28.

Letter #7, February 6, 1863, read my letters in secret. John W. Mambert letters from the Civil War.

A Manual of Instruction for Enlisting and discharging Soldiers with Special Reference to the Medical Examination of Recruits and the Detection of Disqualifying and Feigned Diseases, by Roberts Bartholow A.M., M.D. pp. 9-102.

Sprague, Homer B. (Homer Baxter), b.1829., History of the 13th Infantry Regiment of Connecticut Volunteers, during the Great Rebellion, (Hartford Conn. 1867), p. 15.

James I Robertson Jr Civil War Sesquicentennial Legacy Collection at The Library of Virginia, Edward L. Molineux collection, Infusia Diabolus, War stories.

New York State Military Museum and Veterans Research Center, Nine Month Men, Military Affairs In New York 1861—1865.

Bacon, Edward [1830-1901], Among the Cotton Thieves. Detroit 1867. p. 152.

W.F. Tiemann Collection, Letter Feb 20, 1863, Camp Grover, Baton Rouge. The N.Y.S. Military Museum and Veteran Research Center.

M.D. Wilber, Quartermaster 159th NYSV. Daily Eagle on April 10, 1863, p. 3. Adriance Memorial Library, Poughkeepsie, NY.

E. L. Molineux Photo Collection, Quartermaster's Store House, NYS Military Museum & Veterans Research Ctr.

W.F. Tiemann Collection, Letter February 23, 1863, Baton Rouge. The N.Y.S. Military Museum and Veteran Research Center.

Letter #8, March 2, 1863, Headquarters, 3rd Brigade, Baton Rouge, To Mrs. Mambert. John W. Mambert letters from the Civil War.

G.O., No. 113, General Headquarters State of NY. Adjutant General Office, Albany Nov. 26, 1861. NYS Military Museum and Veteran Research Center.

Butler, B. F. (Benjamin Franklin). (1892). Autobiography and personal reminiscences of Major-General Benj. F. Butler: Butler's book: a review of his legal, political, and military career. Boston [Mass.]: A.M. Thayer & Co. pp. 396-402.

Irwin, Richard B. History of the 19th Army Corps. G. P. Putnam's Sons New York, (The Knickerbocker Press 1892), ch.8, par.1-13 / ch.16, par.2.

O.R. Banks to Hollock. Ser. I, Vol. XXVI, p.44.

Tiemann, William F. The 159th Regiment Infantry, New York State Volunteers, In the War of the Rebellion 1862-1865, (Brooklyn New York 1891), pp. 23-27.

Letter #9, March 11, 1863, Baton Rouge. John W. Mambert letters from the Civil War.

W.F. Tiemann Collection at The N.Y.S. Military Museum, Letter Mar 23, 1863, Baton Rouge.

From the collection of E. L. Molineux, unknown newspaper copy. NYS Military Museum & Veteran Research Center.

From the collection of E. L. Molineux, extracts from his diary, horse return. NYS Military Museum & Veteran Research Center.

Farragut's Union Fleet past Port Hudson, Harper's Weekly Newspaper. Drawing by Hamilton, April 18, 1863

Duffy, Edward B., History of the 159th Regiment, N.Y.S.V. (New York 1890), p. 7.

James I Robertson Jr Civil War Sesquicentennial Legacy Collection at The Library of Virginia – Edward L. Molineux collection - Letter written home Mar. 20, 1863. Letter home Mar. 27, 1863.

Letter #10, March 27, 1863, Baton Rouge. John W. Mambert letters from the Civil War.

Surgeon Daniel M. Holt as quoted in, Coryell, Janet L., James M. Greiner, James R. Smither, and Janet L. Coryell. A Surgeon's Civil War the Letters and Diary of Daniel M. Holt, M.D. Ashland, (The Kent State University Press, 2012), p. 33.

W.F. Tiemann Collection at The N.Y.S. Military Museum, letter describing Thibodaux to his father.

Letter #11, April 2, 1863, Thibodaux, LA, To Wm H. Hawver, Esq. John W. Mambert letters from the Civil War.

W.F. Tiemann Collection at The N.Y.S. Military Museum, Letter describing enemy actions, April 7, 1863.

Duganne, A. J. H. Camps and Prisons, Twenty Months in the Department of the Gulf. (New York J.P. Robens, Publisher 1865), p157-158.

Letter #12, April 6, 1863, Bayo Beauff, To Mrs. John W. Mambert Family. John W. Mambert letters from the Civil War.

Billings, John D., Hardtack and Coffee. Published 1888. pp. 113-131.

Partial Letter #13, Sent in April 1863. John W. Mambert letters from the Civil War.

Morgan City Archives, 501 Federal Ave, Morgan City, La. Letter from Sam white to brother Solo, May 15, 1863 Brashear City.

United States Postal Service, Mail service and the Civil War, USPS.com.

CHAPTER 3

Irwin, Richard B. History of the 19th Army Corps. G. P. Putnam's Sons New York, (The Knickerbocker Press 1892), ch. 9, par. 1-6 / ch. 10, par. 4-11 / ch. 11, par. 5-36.

Tiemann, William F. The 159th Regiment Infantry, New York State Volunteers, In the War of the Rebellion 1862-1865, (Brooklyn New York 1891), pp. 27-37.

Taylor, Richard. Destruction & Reconstruction, Personal Experiences of the Late War, (New York 1879), p. 115.

The Battle of Irish Bend Louisiana, sketch by William Hall of the 22nd Maine, Harper's Weekly, May 16,1863, https://commons.wikimedia.org/w/index.php?title=File:Battle_of_Irish_Bend.jpg&oldid=818663725.

Federal Troops Marching Through Franklin, April 15, 1863, After General Richard Taylor's Withdrawal Following the Battle of Irish Bend, Courtesy of Morgan City Archives, 501 Federal Ave, Morgan City, La.

Illustration of the Battle at Irish Bend, April 14, 1863. History of the 19th Army Corps, 1892.

Young Sanders Center, Franklin LA, Letter from William F. Tiemann dated April 18, 1863.

Genealogical and Biographical of the Molineux Families, by Nellie Zada Rice Molineux & Edward Leslie Molineux.

Colonel Molineux being carried off the field at the Battle of Irish Bend, Artist unknown was a soldier with the 159th N.Y.S.V., E. L. Molineux Photo Collection, NYS Military Museum, Property of Will Molineux.

James I Robertson Jr Civil War Sesquicentennial Legacy Collection at The Library of Virginia – Edward L. Molineux collection, War Stories 1863.

NYS Military Museum & Veterans Research Center, Unit History Project, 159th NYS Regt. Newspaper Clippings, paper unknown.

Young Sanders Center, Franklin LA, General Richard (Dick) Taylor, article from the Civil War Times Jan. 1985, v. 23, no. 9.

NYS Military Museum & Veterans Research Center, Unit History Project, 159th NYS Regt. Newspaper Clippings - New Orleans correspondent of the Boston Traveler the assassination of a federal soldier.

Young Sanders Center, Franklin LA: Official Report relative to the Conduct of Federal Troops in Western Louisiana during the Invasion of 1863 and 1864. Compiled from Sworn Testimony, Under Direction of Governor Henry W. Allen, Shreveport, April 1865, p. 9.

Letter #14: April 29, 1863, La. Brashear City. John W. Mambert letters from the Civil War.

Young Sanders Center, Franklin LA: Official Report relative to the Conduct of Federal Troops in Western Louisiana during the Invasion of 1863 and 1864. Compiled from Sworn Testimony, Under Direction of Governor Henry W. Allen, Shreveport, April 1865, p. 12.

Young Sanders Center, Franklin LA, Letter from William F. Tiemann dated May 9, 1863.

The Billy Wilson Zouaves, artist Frank Leslie.

6th NYS Infantry Regiment Civil War Historical Sketch, NYS Military Museum & Veterans Research Center.

Letter #15: May 3, 1863, Berwick City, La. John W. Mambert letters from the Civil War.

Lemuel Shattuck reported on a general plan for the promotion of general and public health devised, prepared and recommended by the Commissioners appointed under a resolution of the Legislature of Massachusetts, relating to a sanitary survey of the State (1850). This report is one of the fundamental documents in public health in the United States.

History of the United States Sanitary Commission, being The General Report of its Work During the War of the Rebellion, by Charles J. Stille. (New York 1868) pp. 258, 532.

United States Sanitary Commission. Western Department & Newberry, J. S. (1871) The U. S. Sanitary Commission in the valley of the Mississippi, during the war of the rebellion, -1866. Cleveland, Fairbanks, Benedict & Co., printers. [Pdf] Retrieved from the Library of Congress, https://www.loc.gov/item/01010332/. pp. 51-52.

Moss, L. (1868) Annals of the United States Christian commission. Philadelphia, J. B. Lippincott & co. [Pdf] Retrieved from the Library of Congress, https://www.loc.gov/item/02018786/. p. 364.

Young Sanders Center, Franklin LA, Letter from William F. Tiemann dated May 9, 1863.

NYS Military Museum & Veterans Research Center, Unit History Project, 159th NYS Regt. Newspaper Clippings, Return of R. D. Lathrop.

First-Lieutenant Byron L. Lockwood, photo, United States Army Heritage and Education Center, Carlisle, PA.

Lieutenant Colonel Gilbert A. Draper, photo, United States Army Heritage and Education Center, Carlisle, PA.

First-Lieutenant Bradley, photo, United States Army Heritage and Education Center, Carlisle, PA.

Letter #16: May 16, 1863, Louisiana - Berwick City, To Family. John W. Mambert letters from the Civil War.

The Third New Hampshire and All About It, by D. Eldredge, Captain 3rd NH Volunteer Infantry, Boston, Mass. Press of E.B. Stillings and Company, 1893.

The National Archives. The Emancipation Proclamation. Archives.gov/

Britannica, The Editors of Encyclopedia. "Emancipation Proclamation." Encyclopedia Britannica, https://www.britannica.com/event/Emancipation-Proclamation. Accessed 21 March 2023.

CHAPTER 4

Tiemann, William F. The 159th Regiment Infantry, New York State Volunteers, In the War of the Rebellion 1862 -1865, (Brooklyn New York 1891), pp. 38 - 63.

Duffy, Edward B., History of the 159th Regiment, NYSV (New York 1890), p. 15.

Irwin, Richard B. History of the 19th Army Corps. G. P. Putnam's Sons New York, (The Knickerbocker Press 1892), ch.17, par.12 / ch.18, par.10-48 / ch19, par.36.

NYS Military Museum & Veterans Research Center, Unit History Project, 159th NYS Regt. Newspaper Clippings published by Francis H. Webb.

Wendell, C., Columbia County NY in the Civil War (3rd Annual Report of the Bureau of Military Statistics of the State of NY, New York State Military Museum & Veterans Research Center.

O.R. Colonel Molineux, letter written to Committee of Correspondence of the War Fund Association, Series I – Vol. XXVI – Part 1.

The 159th Infantry Unit, nicknamed Second Dutchess and Columbia Regiment. NYS Military Museum and Veterans Research Center.

O.R. Department of the Gulf, General Banks, General Report, Series I – Vol. XXVI – Part 1, pg.14, par 4.

W.F. Tiemann Collection at The N.Y.S. Military Museum, Letter June 26, 1863, Port Hudson.

Port Hudson State Historic Site, Lieutenant Tiemann "flag of Truce," historic plaque located near Fort Desperate.

Sneden, Robert Knox. Siege of Port Hudson. [S.I., to 1865, 1863] Map. Retrieved from the Library of Congress, www.loc.gov/item/gvhs01.vhs00139/.

Sherman, Andrew M. (Andrew Magoun), 1844 – 1921; In the Lowlands of Louisiana in 1863, pp. 21-36.

Finch, Boyd L., Tucson Arizona. Surprise at Brashear City Sherod Hunter's Sugar Cooler Cavalry. Young Sanders Center, Franklin, La.

Illustration, Brashear City, La. Young Sanders Center, Franklin LA.

Duganne, A. J. H. Camps and Prisons, Twenty Months in the Department of the Gulf. (New York J.P. Robens,1865), p. 135-151.

Noel, Theo. A Campaign from Sante Fe to Mississippi, p54. Hunter to Mouton, June 26 and Green to Mouton, June 30, 1863, O.R., Series I Vol. 26, Part 1, pp. 224, 225.

O.R. Mouton to Surget, letter written, July 4, 1863, Series I, Vol .26, Part1, p. 215.

Taylor, Richard. Destruction & Reconstruction, Personal Experiences of the Late War, (New York 1879), p141.

O.R. Series I-Volume XXVI: Battle of Brashear City 23 June 1863, Report of Maj. Sherod Hunter, Baylor's (Texas) Cavalry, Commanding Mosquito Fleet, of the capture of Brashear City.

Map, Positions at Brashear & Berwick Cities. Editorial illustration by Garrett Kipp.

Sherman, Andrew M. (Andrew Magoun), 1844 – 1921; In the Lowlands of Louisiana in 1863, pp. 31-37.

O.R. Emory to Banks July 3, 1863, letter written, Series I–Vol. XXVI–Part 1, pp. 49-50.

O.R. Union & Confederate Navies. General Emory to Commodore Morris July 8, 1863, message written. p. 336.

Bearss, Edwin C., Historic Research Study Ship Island.

Dunham, Marshall Sgt. 159th NYSV Photograph Album, ca 1861-1865. Photo of soldiers. Louisiana digital library, LSU Libraries Special Collection.

Letter #17: August 6, 1863, A partial letter from Fort Kearney, LA. John W. Mambert letters from the Civil War.

G.O. Molineux to Regiment commandants on deserters. Regimental Company Books for the 159th NYSV, National Archives Washington DC.

Sprague, Homer B. (Homer Baxter), b.1829, History of the 13th Infantry Regiment of Connecticut Volunteers, during the Great Rebellion, (Hartford Conn. 1867), p. 180.

Beecher, Harris H., Record of the 114th Regiment N.Y.S.V. (Norwich, N.Y.,1866), p. 237.

Tiemann, William F. The 159th Regiment Infantry, New York State Volunteers, In the War of the Rebellion 1862-1865, (Brooklyn New York 1891), p. 63. Pena, Christopher G. Scared by War, Civil War in Southeast Louisiana, Author House September 2004, p. 387.

Beecher, Harris H., Record of the 114th Regiment N.Y.S.V. (Norwich, N.Y.,1866), p. 238.

From the collection of E. L. Molineux, photo encampment for the 159th NYS Volunteers and the 13th Connecticut Volunteers, Thibodaux, LA 1864, NYS Military Museum and Veteran Research Center.

From the collection of E. L. Molineux, photo Dress Parade, Thibodaux, LA 1864, NYS Military Museum and Veteran Research Center.

G.O. Molineux to Regiment hours of service. Regimental Company Books for the 159th N.Y.S.V., National Archives Washington DC.

Beecher, Harris H., Record of the 114th Regiment N.Y.S.V. (Norwich, N.Y.,1866), p. 238.

Pellet, Elias P., History of the 114th Regimental New York State Volunteers, (Norwich, N.Y. 1866), pp. 143-144.

Letter #18: September 11, 1863, To family, Thibodaux, LA. John W. Mambert letters from the Civil War.

Merritt, Keri Leigh, Masterless Men, Poor Whites and Slavery in the Antebellum South, (Cambridge University Press 2017).

New York Times, June 23, 1862.

G.O. Paroles back to service, September 5, 1863. Regimental Company Books for the 159th NYSV, National Archives Washington DC.

Butler, B. F. (Benjamin Franklin). (1892). Autobiography and personal reminiscences of Major-General Benj. F. Butler: Butler's book: a review of his legal, political, and military career. Boston [Mass.]: A.M. Thayer & Co. p. 324 / pp. 324-514.

The United States Sanitary Commission: A Sketch of Its Purposes and Its Works. Boston, Mass., 1863. p. 201.

CHAPTER 5

G.O. Molineux to Regiment on punishment. Regimental Company Books for the 159th NYSV, National Archives Washington, DC.

Letter #19, October 5, 1863, Thibodaux, To Miss John W. Mambert. John W. Mambert letters from the Civil War.

Van Pelt, Daniel, Leslie's History of the Greater New York, Volume I, pp. 413-419.

Hudson Gazette ran on September 14, 1863 / September 15, 1863. Unit History Project, 159th NYS Regt. NYS Military Museum & Veterans Research Center.

Letter #20, October 15, 1863, Thibodaux, A private letter to Mrs. John W. Mambert & John Mambert. John W. Mambert letters from the Civil War.

Tiemann, William F. The 159th Regiment Infantry, New York State Volunteers, In the War of the Rebellion 1862 -1865, (Brooklyn New York 1891), pp. 37-65.

Letter #21, November 5, 1863, Thibodaux, To Mrs. John W. Mambert. John W. Mambert letters from the Civil War.

New York Times, Aug. 25 and 27, 1864.

Kinderhook Herald. (Kinderhook, N.Y.) Dec 31, 1863, pg3, image 3. https://nyshistoricnewspapers.org/

Antebellum Cotton Plantation, painting, permission to use by Edwin L. Jackson.

Letter #22, November 22, 1863, La. Thibodaux, Dear Wife. John W. Mambert letters from the Civil War.

Letter #23, December 16, 1863, La. Thibodaux, To Mrs John W. Mambert and family. John W. Mambert letters from the Civil War.

"Anti-slavery lectures," New York Times, Jan. 17, 1855, 5, as quoted in: Merritt, Keri Leigh, Masterless Men, Poor Whites and Slavery in the Antebellum South, (Cambridge University Press 2017), ch. 5, p. 143.

Helper, The impending Crisis, quoted on 375, as quoted in: Merritt, Keri Leigh, Masterless Men, Poor Whites and Slavery in the Antebellum South, (Cambridge University Press 2017), ch. 5, pp. 144,145.

The Christian Recorder (PA), July 22,1865.

Gazette Newspaper, Unit History Project, 159th NYS Regt. Newspaper Clippings. NYS Military Museum & Veterans Research Center.

Letter #24, January 26, 1864, Thibodaux, La., Mrs. John W. Mambert and family. John W. Mambert letters from the Civil War.

Moss, L. (1868) Annals of the United States Christian commission. Philadelphia, J. B. Lippincott & co. [Pdf] Retrieved from the Library of Congress, https://www.loc.gov/item/02018786/. pp. 375, 439.

G.O. Captain Robert McD. Hart, Mambert demotion, January 28,1864. Regimental Company Books for the 159th NYSV, National Archives, Washington DC.

Letter #25, January 1864, 159th Regt. N.Y. Vols., Private letter to Mrs. J.W. and John. John W. Mambert letters from the Civil War.

Annual Message of Governor Henry Watkins Allen, To the State of Louisiana. Printed at the Office of the Caddo Gazette, January, 1865.

Resources of the southern fields and forests, by Francis P. Porcher (1825–1895) Publisher, Charleston: Steam-power Press of Evans & Cogswell, p. 552.

McClellan and his officers. At a finding of a court martial, not satisfactory to him, Gen. McClellan sent the following just and noble rebuke and admonition. New York. https://www.loc.gov/resource/rbpe.12303900/

Billings, John D., Hardtack and Coffee. Published 1888. pp. 217-219.

From the collection of E. L. Molineux, Thibodaux Feb. 19, 1864, Letter to sister. NYS Military Museum & Veterans Research Center.

Young Sanders Center, Franklin LA: Official Report relative to the Conduct of Federal Troops in Western Louisiana during the Invasion of 1863 and 1864. Compiled from Sworn Testimony, Under Direction of Governor Henry W. Allen, Shreveport, (April 1865), pp. 3-6.

G.O. Molineux to regiment, acts committed on peaceable citizens and harmless negroes, March 5, 1864, From the collection of E. L. Molineux NYS Military Museum & Veteran Research Center.

CHAPTER 6

Tiemann, William F. The 159th Regiment Infantry, New York State Volunteers, In the War of the Rebellion 1862-1865, (Brooklyn New York 1891), pp. 66-81.

Irwin, Richard B. History of the 19th Army Corps. G. P. Putnam's Sons New York, (The Knickerbocker Press 1892), ch. 23, par. 1-30 / ch. 24, par. 5-19 / ch. 25, par. 1-26 / ch. 28, par. 2-10 / ch. 29, par. 13-18.

Admiral Porter's Flotilla, Illustration from Harper's Weekly, Library of Congress # 2006691852.

Banks Army Crossing Cane River, Illustration, artist C.E.H. Bonwill, Library of Congress # 94507514.

O.R. Assistant Adjutant-General to General Grover, construct additional defenses, Series I – Vol. XXVI – Part 1, p. 19.

Letter #26, April 4, 1864, Alexandria, La. John W. Mambert letters from the Civil War.

Levee at Alexandria, La [Photographed between 1861 and 1865, printed between 1880 and 1889] Retrieved from the Library of Congress # 2013647481.

The History of Alexandria, The early years. https://www.alexandria-louisiana.com/alexandria-louisiana-history.htm

Bringhurst R. W, and Lith Braden & Burford. Map of the town of Alexandria, Louisiana. [Indianapolis, Ind.: Braden & Burford, 1872] Map. Retrieved from the Library of Congress # 2004629015.

O.R. Grover was pleading with headquarters, Series I – Vol. XXXIV – Part 3, p. 71.

O.R. Hunter to Grant May 2, 1864, Series I – Vol. XXXIV – Part 3, p. 390.

Letter #27, April 12th, 1864, Answer to letter No. 15, Alexandria La. John W. Mambert letters from the Civil War.

Merritt, Keri Leigh, Masterless Men, Poor Whites and Slavery in the Antebellum South, (Cambridge University Press 2017), pp. 5-6.

[HOUSTON] TRI-WEEKLY TELEGRAPH, August 29, 1863, p. 2, c. 3 The Louisiana Jayhawkers. Camp Stonewall Jackson, near Washington, La.} August 17th, 1863.} Editor Telegraph.

James I Robertson Jr Civil War Sesquicentennial Legacy Collection at The Library of Virginia, Letter written Apr. 10, 1864 – Edward L. Molineux collection.

O.R. Secret Special Order by General Mouton, June 12, 1863.

O.R., report, Rear Admiral Porter to General Sherman, Series I – Vol. XXXIV – Part 3, p. 171.

"Battle of Pleasant Hill, Louisiana, April 9, 1864 artist impression," House Divided: The Civil War Research Engine at Dickinson College, https://hd.housedivided.dickinson.edu/node/41848

The Opelousas Courier, May 14, 1864, Library of Congress, Natl. Endowment for the Humanities, Chronicling. America.

James I Robertson Jr Civil War Sesquicentennial Legacy Collection at The Library of Virginia – Edward L. Molineux collection, Letter written Apr. 17, 1864.

O.R., Halleck to Banks, April 27, 1864, Series I – Vol. XXXIV – Part 3, pp. 306-307.

O.R., Hunter to Grant April 28, 1864, Series I – Vol. XXXIV – Part 3, p. 308.

Transport John Warner, Frank Leslie's Illustrated Newspaper 1864, C.E.H Bonwell, Library of Congress # 9750156.

Sketch of the two breakwaters above Alexandria in the Red River, Gerdes, F.H. & Lewis G De Russy. [1864] Map. Library of Congress # 99446418

O.R., Molineux to Capt. Hibbert, May 8, 1864, Series I – Vol. XXXIV – Part 3, p. 508.

Bailey's Dam, Frank Leslie's Scenes and Portraits of the Civil War, Page 402.

O.R., Halleck to Banks, Series I – Vol. XXXIV – Part 2, p. 15.

O.R., Banks report to Halleck, Series I – Vol. XXXIV – Part 3, p. 192.

O.R., Halleck to Grant, Series I – Vol. XXXIV – Part 3, p. 409.

Cowardin, James A., and John D. Hammersley "Burning of Alexandria, La." The Daily Dispatch [Richmond]: August 11,1864. Richmond Dispatch. Microfilm. Ann Arbor, Mi: Proquest. 1 microfilm reel; 35mm.

Young Sanders Center, Franklin LA: Official Report relative to the Conduct of Federal Troops in Western Louisiana during the Invasion of 1863 and 1864. Compiled from Sworn Testimony, Under Direction of Governor Henry W. Allen, Shreveport, (April 1865), p. 4.

Sprague, Homer B. (Homer Baxter), b.1829., History of the 13th Infantry Regiment of Connecticut Volunteers, during the Great Rebellion, (Hartford Conn. 1867), pp. 207-217.

Illustration, Battle of Mansura Plains. Tiemann, William F. The 159th Regiment Infantry, New York State Volunteers, In the War of the Rebellion 1862 -1865 p. 77.

Letter #28, May 23, 1864, Camp along the bank of the Mississippi at the mouth of Red River. John W. Mambert letters from the Civil War.

James I Robertson Jr Civil War Sesquicentennial Legacy Collection at The Library of Virginia – Edward L. Molineux collection. Diary entry, no date.

W.F. Tiemann Collection, letter June 2,1864 Morganza. N.Y.S. Military Museum and Veterans Research Center Archives.

James I Robertson Jr Civil War Sesquicentennial Legacy Collection at The Library of Virginia – Edward L. Molineux collection, Letter written June 3, 1864.

W.F. Tiemann Collection, letter June 7, 1864, Morganza. N.Y.S. Military Museum and Veterans Research Center Archives.

Camp at Morganza Bend Louisiana. United States Louisiana, None [Photographed between 1861 and 1865, printed between 1880 and 1889] [Photograph] Library of Congress # 2013651623.

Letter #29, June 1, 1864, Answer to letters 21 & 22, Mississippi River. John W. Mambert letters from the Civil War.

Letter #30, June 22,1864, Camp Mud, Morganza. To Mrs. John Mambert & Family. John W. Mambert letters from the Civil War.

G.O. Waltermire to Regiment June 25, 1864. Regimental Company Books for the 159th N.Y.S.V., National Archives Washington DC.

CHAPTER 7

Irwin, Richard B. History of the 19th Army Corps. G. P. Putnam's Sons New York, (The Knickerbocker Press 1892), ch. 29, par. 21 / ch. 30, par. 26 / ch. 31, par. 10-13 / ch. 32, par. 3-24 / ch. 33, par. 1-5.

Tiemann, William F. The 159th Regiment Infantry, New York State Volunteers, In the War of the Rebellion 1862 -1865, (Brooklyn New York 1891), pp. 82 - 105.

159th Regiment NY Volunteer Infantry Regimental Color, Civil War. New York State Military Museum & Veterans Research Center.

Illustration, US Transport "Cahawba", artist Alfred Rudolph. Gift of J.P. Morgan, Library of Congress # 2004660624.

W.F. Tiemann Collection, Letter July 28, 1864, In the Field. The N.Y.S. Military Museum and Veterans Research Center.

Magnus, Charles. Soldiers Rest, Washington, D.C., Smithsonian, National Museum of American History (CC0).

W.F. Tiemann Collection, Letter Aug 4, 1864, Tenleytown D.C. The N.Y.S. Military Museum and Veterans Research Center.

James I Robertson Jr Civil War Sesquicentennial Legacy Collection at The Library of Virginia – Edward L. Molineux collection, Letter written June 6, 1864.

Telzrow, Michael, Russell Horton, Kevin Hampton. Wisconsin in the Civil War, Wisconsin Veterans Museum/Wisconsin Department of Veterans Affairs, p. 143.

Illustration, Soldiers fording the river, artist Edwin Forbes. Gift of J.P. Morgan. Library of Congress # 2004661504.

James I Robertson Jr Civil War Sesquicentennial Legacy Collection at The Library of Virginia – Edward L. Molineux collection, Letter written Aug. 4, 1864.

Illustration, Sheridan's Wagon Train in the Valley, artist Alfred Rudolph. Gift of J.P. Morgan. Library of Congress # 2004660442.

James I Robertson Jr Civil War Sesquicentennial Legacy Collection at The Library of Virginia – Edward L. Molineux collection, Letter written Aug. 30, 1864.

Letter #31, August 27, 1864, Camp near Halltown, West WV. John W. Mambert letters from the Civil War.

Sprague, Homer B. (Homer Baxter), b.1829, History of the 13th Infantry Regiment of Connecticut Volunteers, during the Great Rebellion, (Hartford Conn. 1867), p. 224.

Illustration, The 19th Army Corp, Grover's Division at 3rd Battle in Winchester, artist Alfred Rudolph

Illustration, 3rd Battle at Winchester, artist Alfred Rudolph.

O.R., Colonel Edward L. Molineux, Battle Reports on Winchester. Serial 090, p. 330.

Illustration, Charge upon the Rebels at the Stonewall, artist Alfred Rudolph. Gift of J.P. Morgan, Library of Congress # 2004660452.

Author, Bvt. Lt. Col. G. L. Gillespie, Major of Engineers, U.S.A. under the authority of the U.S. Secretary of War. Portion of an 1873 map titled "Battlefield of Winchester, Va (Opequon) [September 19, 1864]" prepared by an order of Lt. Gen. P.H. Sheridan under the authority of the Secretary of War.

Illustration, Wounded Being Carried Away, artist Alfred Rudolph. Gift of J.P. Morgan, Library of Congress, # 2004660079.

O.R., No. 8. Report of Surg. James T. Ghiselin, U. S. Army, Medical Director, Middle, Military Division, of operations August 27-December 31. Serial 090, p. 141.

Letter #32, September 21, 1864, Winchester, To Mrs. J.W. M. & Family. John W. Mambert letters from the Civil War.

O.R., No. 8. Report of Surg. James T. Ghiselin, U. S. Army, Medical Director, Middle, Military Division, of operations August 27-December 31. Serial 090, p. 141.

O.R., Colonel Edward L. Molineux reports on the Fisher Hill engagement. Serial 090, p. 331.

James I Robertson Jr Civil War Sesquicentennial Legacy Collection at The Library of Virginia – Edward L. Molineux collection, Letter Fisher Hill.

The Camden Confederate, apple brandy, Oct. 5, 1864. The Library of Congress, Natl. Endowment for the Humanities, Chronicling America.

Letter #33, September 25, 1864, Winchester, To Mrs. John W. & Family. John W. Mambert letters from the Civil War.

CHAPTER 8

Briton, John H., Personal Memoirs of John H. Brinton, Major and Surgeon U.S.V. 1861-1865, (New York 1914), pp. 298-303.

O.R., No. 8. Report of Surg. James T. Ghiselin, U. S. Army, Medical Director, Middle, Military Division, of operations August 27-December 31. Serial 090, p. 142.

Letter #34, Monday, September 26, 1864. John W. Mambert letters from the Civil War.

Tiemann, William F. The 159th Regiment Infantry, New York State Volunteers, In the War of the Rebellion 1862 -1865, (Brooklyn New York 1891), pp. 105-115.

Letter #35, October 7, 1864, Winchester, Va., To the Family. John W. Mambert letters from the Civil War.

Letter #36, (separate letter), To John. John W. Mambert letters from the Civil War.

Plum, William R., Military Telegraph during the Civil War in the United States. Chicago, IL: Jansen, McClurg, 1882. p. 267-439.

McKinney, Effie. The Cleveland Convention (1928).

The Argus – semi-weekly Democratic publication, Albany NY.

Brummer, Sidney David, A.M. Political History of New York State During the Period of the Civil War. (New York 1911), pp. 279-432.

Irwin, Richard B. History of the 19th Army Corps. G. P. Putnam's Sons New York, (The Knickerbocker Press 1892), ch.34, par. 7-40.

Illustration, Throwing Up Earthworks to prevent a Night Attack, artist Alfred R. Waud [Between 1860 and 1865] Photograph. Library of Congress, # 2004660724.

Illustration, Chasing the Rebs through Strasburg, artist Alfred Rudolph. Gift of J.P. Morgan, Library of Congress # 200466192.

From the collection of E. L. Molineux, extracts from his diary, NYS Military Museum and Veterans Research Center.

Illustration, Sheridan's army following Early up the Valley of the Shenandoah, artist Alfred R. Waud. [between August and 1865 March] Photograph. Library of Congress #2004660195.

Letter #37, October 24, 1864. John W. Mambert letters from the Civil War.

Putnam, George Haven, Abraham Lincoln: the Peoples Leader in the Struggle for National Existence, (The Knickerbocker Press 1909). p. 153.

Letter #38, November 8, 1864, Winchester, Va., To the Family. John W. Mambert letters from the Civil War.

Collected Works of Abraham Lincoln. Volume 8. Lincoln, Abraham, 1809-1865. AD, NDry. Concerning this speech and Lincoln's response to a serenade on November 8 (supra), John Hay's Diary records under date of November 11: "The speeches of the President at the two last serenades are very highly spoken of. The first I wrote after the fact to prevent the 'loyal Pennsylvanians' getting a swing at it themselves. The second one, last night, the President himself wrote late in the evening and read it from the window. 'Not very graceful,' he said, 'but I am growing old enough not to care much for the manner of doing things.'"

Moss, L. (1868) Annals of the United States Christian commission. Philadelphia, J. B. Lippincott & co. [Pdf] Retrieved from the Library of Congress, pp. 294-439. https://www.loc.gov/item/02018786/

G.O. Emory ordered an issue of whiskey, Regimental Company Books for the 159th NYSV, National Archives, Washington, DC.

Duffy, Edward B., History of the 159th Regiment, NYSV New York 1890, p. 38.

Rutherford B. Hayes Presidential Library & Museums. Rutherford B. Hayes Vol. 2 Chap. XXV In Garrison End of war, pp. 527-540.

McFeely, Williams S., Ulysses S. Grant, an Album. (New York 2003), p. 564.

Letter #39, December 1, 1864, Camp Russell, To my Family. John W. Mambert letters from the Civil War.

Sutherland, Daniel. Guerrilla Warfare in Virginia during the Civil War. (2020, December 07). In Encyclopedia Virginia. https://encyclopediavirginia.org/entries/guerrilla-warfare-in-virginia-during-the-civil-war.

Williamson, James J. [1834-1915], Mosby's Rangers, A Record of the Operations of the Forty-Third Battalion Virginia Cavalry. The Polhemus Press, New York 1895. pp. 21-107.

John Singleton Mosby, Confederate Guerrilla Leader, "Mosby's Confederacy." Encyclopedia.com

Letter #40, December 19, 1864, Camp Russell, Va. 159 Regt. NYS, To John Mambert, (jr). John W. Mambert letters from the Civil War.

Billings, John D., Hardtack and Coffee. Published 1888. p. 224.

Illustration, A Sutlers Tent Near HQ, artist Arthur Lumley. Gift of J.P. Morgan. Library of Congress #2004661305.

CHAPTER 9

Brummer, Sidney David, A.M. Political History of New York State During the Period of the Civil War. (New York 1911), pp. 445-446.

Tiemann, William F. The 159th Regiment Infantry, New York State Volunteers, In the War of the Rebellion 1862-1865, (Brooklyn New York 1891), pp. 116-122.

From the collection of E. L. Molineux, extracts from his diary, assigned garrison duty. NYS Military Museum & Veteran Research Center.

Letter #41, January 24, 1865, Savannah, To: Mrs. John W. & Family. John W. Mambert letters from the Civil War.

Sherman, Andrew M. (Andrew Magoun), 1844 – 1921; In the Lowlands of Louisiana in 1863, pp. 9-10.

Illustration, Steam Ship in Storm, Man Overboard, artist Alfred R. Waud [Between 1860 and 1865] Photograph. Library of Congress #2004660690.

Letter #42, February 6, 1865, G. A. Savannah. John W. Mambert letters from the Civil War.

Bumgardner, E., Disqualification for military service in the Civil War on account of loss of teeth. [Dental Cosmos 1894]. p. 429.

Letter #43, February 14, 1865, Tell Jobe to direct his letters in this way. John W. Mambert letters from the Civil War.

From the collection of E. L. Molineux, extracts from his diary, Molineux becomes a Jailor. NYS Military Museum and Veteran Research Center.

G.O., Waltermire establishes a policing force. National Archives Washington DC, Regimental Company Books for the 159th NYSV.

Letter #44, Savannah, Ga. John W. Mambert letters from the Civil War.

Van Alstyne, Lawrence, Diary of An Enlisted Man, (New Haven Connecticut 1910), p13-14.

54th New York Infantry Regiment Civil War Historical Sketch. NYS Military Museum and Veterans Research Center.

Little, Ruth M. Phd., The Historic Architecture of Morehead City, N.C. First Coastal Railroad Resort. p. 11.

Letter with no date # 45, Sent from Morehead City NC. John W. Mambert letters from the Civil War.

Letter #46, April 14, 1865, Camp 159th Regt. Morehead City, N.C., To Mrs. J.W. and Family. John W. Mambert letters from the Civil War.

From the collection of E. L. Molineux, extracts from his diary. NYS Military Museum & Veteran Research Center.

Herold, David E, United States Army. Military Commission, and Alfred Whital Stern Collection Of Lincolniana. Trial of the assassins and conspirators for the murder of Abraham Lincoln, and the attempted assassination of Vice-President Johnson and the whole cabinet: the most intensely interesting trial on record: containing the evidence in full, with arguments of counsel on both sides, and the verdict of the military commission: correct likenesses and graphic history of all the assassins, conspirators, and other persons connected with their arrest and trial. [Philadelphia: Barclay & Co., i.e. 1865, 1865] Image. https://www.loc.gov/item/12002579/.

Letter #47, April 24, 1865, Morehead City, N.C., To John. John W. Mambert letters from the Civil War.

Plum, William R. Military Telegraph during the Civil War in the United States. Chicago, IL: Jansen, McClurg, 1882. pp. 14-353.

Bates, D. H. & Alfred Whital Stern Collection Of Lincolnian. (1907) Lincoln in the telegraph office; recollections of the United States Military Telegraph Corps during the Civil War. New York, Century Co. [Pdf] Retrieved from the Library of Congress, https://www.loc.gov/item/07032385/

Letter #48, April 28, 1865, Morehead, N.C., 1st Part of this letter to all the family. John W. Mambert letters from the Civil War.

The Wilmington Herald, Assassination of President Lincoln. April 20, 1865. North Carolina Newspapers, p. 2.

New York Times, June 9, 1863; THE LIBERTY OF THE PRESS.; A Conference of New-York Editors A Platform Adopted.

Andrews, J. Cutler, The North Reports the Civil War (Pittsburgh: University of Pittsburgh Press, 1955), pp. 34-601.

Savannah Republican, June 16, 1862.

From the collection of E. L. Molineux, letter to Tiemann May 1, 1865. NYS Military Museum & Veterans Research Center.

From the collection of E. L. Molineux, extracts from his diary. NYS Military Museum & Veteran Research Center.

Illustration, Star of the South, artist Alfred R. Waud. Gift of J.P. Morgan, Library of Congress #2004661370.

From the collection of E. L. Molineux, Thomas Woodrow Wilson. NYS Military Museum and Veteran Research Center.

G.O., No. 5, May 23,1865 the daily schedule, National Archives Washington DC, Regimental Company Books for the 159th NYSV.

G.O., No. 3, Major General W. T. Sherman on May 30, 1865, go to your homes. From the collection of E. L. Molineux at NYS Military Museum and Veteran Research Center.

G.O., No. 11, Brevet General Molineux on June 6, 1865, mustered out of service. From the collection of E. L. Molineux at NYS Military Museum and Veteran Research Center.

128th NYSV Infantry Regiment, Civil War Historical Sketch, NYS Military Museum & Veterans Research Center.

O. R. from Q. A. Gillmore, Major-General commanding to General Molineux, June 5, 1865. Ser. I, Vol. XLVII, Part III. p. 629-630.

G.O. No. 8, Lieutenant-Colonel Waltermire, washroom filthy conditions, June 11, 1865. National Archives Washington DC, Regimental Company Books for the 159th NYSV.

G.O. No. 9, Lieutenant-Colonel Waltermire, stealing and pilfering, June 12, 1865. National Archives Washington DC, Regimental Company Books for the 159th NYSV.

Letter #49, June 19, 1865, City Hotel, Augusta, Ga., To Mrs. John W. Mambert. John W. Mambert letters from the Civil War.

G.O., No. 10 Lieutenant-Colonel Waltermire, abusing the colored race, June 21, 1865. National Archives Washington DC, Regimental Company Books for the 159th NYSV.

Butler, B. F. (Benjamin Franklin). (1892). Autobiography and personal reminiscences of Major-General Benj. F. Butler: Butler's book: a review of his legal, political, and military career. Boston [Mass.]: A.M. Thayer & Co. pp. 491-495.

G.O., No. 40 May 1863 The Major-General commanding the Department proposes the organization of a Corps d' Armee of colored troops.

Illustration, 2nd Louisiana Colored Troops, special artist. Library of Congress, # 2003668336.

The Augusta Chronicle, June 28, 1865, page 3: The Flag of the 159th New York.

CHAPTER 10

Tiemann, William F. The 159th Regiment Infantry, New York State Volunteers, In the War of the Rebellion 1862-1865, (Brooklyn New York 1891), pp. 122-137.

Edward Leslie Molineux, The Virtualology Project. http://famousamericans.net/edwardlesliemolineux/

E. L. Molineux Photo Collection. Edward Leslie Molineux. NYS Military Museum and Veterans Research Center.

Letter #50, August 6, 1865, Sand Hill Arsenal Augusta, Ga. To the family. John W. Mambert letters from the Civil War.

G.O. Colonel Waltermire, to regiment, proper attire. National Archives Washington DC, Regimental Company Books for the 159th NYSV.

Letter #51, August 24, 1865, Augusta Arsenal, Ga. John W. Mambert letters from the Civil War.

Madison Georgia, Morgan County. The Legend of Joshua Hill: The man who Saved Madison from Burning. https://visitmadisonga.com/

Letter #52, September 5, 1865, Madison, Ga., To the family. John W. Mambert letters from the Civil War.

Avery, I. W. (Isaac Wheeler), 1837 – 1897, The History of the State of Georgia from 1850 to 1881, embracing the three important epochs: the decade before the war of 1861-5; the period of reconstruction. (New York 1881), pp. 347-349.

Letter #53, September 28, 1865, Madison Court House Ga., To John and the rest of the family. John W. Mambert letters from the Civil War.

Letter #54, October 8, 1865, Augusta, GA., To the Family. John W. Mambert letters from the Civil War.

Hudson Gazette Newspaper, NYS Military Museum & Veterans Research Center, Unit History Project, 159th NYS Regt. Newspaper Clippings.

1858 map of Columbia County, NY., Library of Congress. Editorial illustration by Garrett Kipp.

The Columbia Republican Newspaper, Tuesday October 31, 1865, NYS Military Museum & Veterans Research Center, Unit History Project, 159th NYS Regt. Newspaper Clippings.

E. L. Molineux Photo Collection, William "Frank" Tiemann, photo. NYS Military Museum and Veteran Research Center.

Michael Lee Kipp is a newcomer to writing historical nonfiction. His family's history, dating back to the 1600s, is deep-rooted in the Hudson Valley of New York State. His desire to know his ancestors' history and stories has resulted in countless hours of research. Mike lives in Florida with his wife and enjoys time with family and friends.

www.ingramcontent.com/pod-product-compliance
Lightning Source LLC
Chambersburg PA
CBHW062005180426
43198CB00037B/2387